Dreaming Reality

Dreaming Reality

How Neuroscience and Mysticism
Can Unlock the Secrets of Consciousness

VLADIMIR MISKOVIC

& STEVEN JAY LYNN

THE BELKNAP PRESS OF
HARVARD UNIVERSITY PRESS

Cambridge, Massachusetts,
and London, England
2025

First printing

Library of Congress Cataloging-in-Publication Data
Names: Miskovic, Vladimir, author. | Lynn, Steven J., author.
Title: Dreaming reality : how neuroscience and mysticism can unlock
    the secrets of consciousness / Vladimir Miskovic and Steven Jay Lynn.
Description: Cambridge, Massachusetts ; London, England : The Belknap Press
    of Harvard University Press, 2025. | Includes bibliographical references and index.
Identifiers: LCCN 2024001364 | ISBN 9780674271869 (cloth)
Subjects: LCSH: Consciousness. | Psychology and religion. | Contemplation. |
    Dreams. | Brain. | Mind and reality.
Classification: LCC BF51 .M57 2024 | DDC 153—dc23/eng/20240321
LC record available at https://lccn.loc.gov/2024001364

To my parents, in gratitude for nourishing my beginnings and growth in being. To Life, precious loom, for teaching me so much every single day and drawing me ever deeper and closer toward the source of all beginnings and endings.

   —*VM*

To my grandson, Julian, the newest light of my life, and my hope for the future, with love.

   —*SJL*

Contents

Dreaming Reality

Preface

Imagine that you have inherited a precious family heirloom: a richly inlaid, mahogany, eighteenth-century English grandfather clock. This heirloom has been passed on from generation to generation, traveling across oceans. It has been transported thousands of miles, survived wars. It has been a witness to the lives of your ancestors, ticking away as the generations passed, marking the passage of time.

Every chip and stain, every facet of its carved surface, every burl accent seems somehow to contain a record inscribed within it, a living network of collective memories—the mundane, sure, but also the most intense joys, peaks of transcendence, heartbreaking losses, the melancholy of all those things that have slipped into an irrevocable past. It is as if the entire psychological life of your ancestors has seeped into the matter of this antique grandfather clock, and all of this now silently radiates from it as it stands in front of you.

Now, if someone were to say that this is merely a contraption of mahogany and metal and nothing more, you might protest. Your indignation might grow if the person were only interested in the clock's mechanics—or worse, suggested its market value was the sole index of its worth. And the reason for your protest is this: there is a deeper, more essential truth about that old grandfather clock that is not captured by its external appearance. The quantitative surface is complemented by qualitative features that are invisible but nonetheless vital to truly "knowing" about phenomena in their depths.

There is something operating here that we can call the principle of *coinherence*. This principle states that a thing, in addition to its outer, physical appearance, is at the same time enmeshed in an entire network of meaning and associations, permeated by an aspect that although unseen is every bit as "real" as what is observable to the naked eye; both this invisible dimension and the visible one inhere in the thing.

To have the clock scanned by the fanciest machines, and even to undertake a molecular analysis of it, would never turn up even one sliver of that invisible world, because these would be looking at it from the wrong vantage point. Looking at the clock in its wholeness—standing back and seeing both dimensions—would be the key to unlocking its full truth, which encompasses aspects of both its outer surface and inner depths, without ignoring either one.

We can think of the brain encased in your skull, and the nervous system that spreads its reach to every corner of your body, as a grander version, more mysterious and magical, of the grandfather clock: it is another ancient heirloom of which you are the temporary proprietor. Your body also obeys the principle of coinherence. Matter and mind, the outer and inner, visible and invisible—they are braided together into a single weave of fabric. This book is an attempt to stand back and see them that way, as an integrated whole, to see what lies before us in full "stereoscopic vision," looking at the outer and inner in a holistic gaze.

For primarily historical and cultural reasons, neuroscientists have most frequently approached their subject matter in the same way the imaginary person in our thought experiment did, believing that a detailed physical analysis would suffice to yield full knowledge of the grandfather clock. The subjective and qualitative was derided as being somehow "less real" or at least not a fit topic for science.

But there have also been attempts to correct for a one-sided concern with the external and quantitative, while neglecting the internal, experiential, and qualitative. Most notably, the field of neurophenomenology, initiated by Francisco Varela, Evan Thompson, and others, has reintroduced into contemporary neuroscience a sophisticated engagement with the nuances and complexities of subjective experience.[1] Stemming from those efforts, we now have valuable tools to uncover the rich subtleties of conscious experience with increasing precision and granularity.

The contention that the uniquely subjective point of view is valuable is also attested to by a recent wave of clinical trials examining the efficacy of psychedelic substances in treating chronic psychological maladies. There are now substantial data to bolster the claim that it is the subjective quality of these experiences that predicts treatment success. In other words, these substances do not derive all of their value by "merely" causing a transient alteration of brain chemicals. Rather, the efficacy of psychedelics seems heavily determined by the powerful cognitive and emotional experiences they elicit, including experiences best captured by the uneasy label of "spiritual." As researchers David Yaden and Roland Griffiths conclude, "the subjective effects of psychedelics are necessary for their enduring beneficial effects and . . . these subjective effects account for the majority of their benefit."[2]

A rapprochement between the interior and exterior, subjective and objective, is long overdue. Our book is a continuing and ongoing exploration into the deep nature of the human brain/mind that attempts to achieve an integrative synthesis, a big-picture perspective. We bring together the traditional neuroscientific outside-in view (of the brain) with an inside-out one (of the mind). We mean to dive far below the surface of the familiar and ordinary states of the mind—what we experience when we are awake and going about our day. The brain/mind is more accurately depicted as consisting of layers upon layers of organization, with a "nested structure" that calls to mind those Russian dolls that fit one inside another.[3]

As for how nestedness manifests itself in human consciousness, the layers that are deeper and more interior can be understood as being more foundational or causal than the surface layers. In other words, what happens in those apparently surreal corners of the brain/mind that we draw nearer to in the worlds of dream, imagination, memory, and hallucination, as well as in the hinterlands of consciousness itself, seems to form something like a base, on top of which the more ordinary states of consciousness rest.

To complement what modern neuroscience tells us about the brain from an outside-in perspective, we will draw from the world's wisdom and spiritual traditions that have explored human consciousness from the inner perspective. We will call on insights from Buddhism and other religions of the Far East, but also from the richness of the Christian tradition. Our rationale for drawing from these ancient wells will become clearer after Chapter 1, where we explore how these wisdom traditions have, for millennia, made experiential and in-depth studies of the furthest reaches of the human mind, from the inside out. Among the yogis, yoginis, monks, and nuns, some of whose writings we will draw on, were masters of exploring non-ordinary states of consciousness.

Their expeditions through these deep interior terrains provide us with a way to complement the currently reigning frameworks for interpreting the human brain and mind—scientific investigation methods that are often exceedingly mechanistic and materialist. If modern science provides detailed descriptions of *how* the brain and mind functions, as well as how it breaks down in disease, it often remains silent on questions of value and our higher potentials (the *what for* and *why* questions). This knowledge is preserved in the world's great spiritual traditions and repeatedly encountered through lived experience. The wisdom contained in these traditions speaks to questions of ultimate meaning and purpose, addressing the further reaches of human consciousness in ways that confront existential themes that have perennially confronted human beings.[4] Building a bridge between the two realms—scientific and contemplative, quantitative and qualitative—is, we believe, not only intellectually fruitful and productive of unexpected synergism, but also crucial to gaining a holistic and integrative view of our humanness. It represents a cross-pollination of knowledge systems that will add depth and fresh vitality to a more optimistic brain and mind science of the future.

In the burgeoning field of "contemplative science," which has brought researchers and scholars into ongoing dialogue with contemplatives, Buddhism has been the predominant partner. This is understandable given the extremely sophisticated understanding of the human mind found in Buddhist schools of thought, as well as its preeminent concern with empiricism and its penchant for scholarly analysis. Buddhism's dominant involvement, however, has meant that other wisdom traditions have been substantially neglected—traditions that have also developed highly sophisticated knowledge of different states of human consciousness. Here, across several chapters, we will draw substantially from the wells of ancient Christianity, knowing that these sources will be less familiar to many readers.[5] There is, of course, more collected wisdom on these subjects than any one book could explore—but an effort to rebalance and increase the diversity of traditions that are treated as respected partners in contemplative science, we believe, can only bear fruit.

Our hope is that this book will excite readers to further explore the complementarities between the outside-in and inside-out perspectives on the human brain/mind, and to rethink some presuppositions they may not have questioned previously in their scientific or personal worldviews. If our readers become more interested in exploring their own and others' consciousness, and learn ways to cultivate greater inner silence and a restful quieting of agitated thoughts, we will have achieved our goal.

We also recognize, however, that any divide-spanning book stands in a precarious position, with the potential to alienate people on both sides. So we should at the outset acknowledge the fault lines: scientists may find the alliance being proposed an uncomfortable one, while the contemplative community may object to the inevitable oversimplifications that will be introduced in the chapters that follow.

Our aim is not to produce a work of comparative religion or an extended polemical tract comparing the various contemplative strands' lists of specific claims and judging which are most and least consistent with modern scientific evidence. Readers should not expect to find those arguments in the chapters to follow. Neither will we engage with complex theological subtleties, although we are aware that debates over these exist. We want, instead, to sketch a grand view that aims at synthesis and integration—the full story of the mysteries and rich potential of the human brain/mind and the wonders of consciousness. So while we do not attempt to advocate for any particular religious truth or practice, it should be said that we do have a very positive evaluation of the broad aim of living a contemplative and reflective life, understanding one's mind, and acquiring interior peace and stability.

Neither do we believe that the great spiritual traditions stand in need of the imprimatur of science or a "seal of approval" from modern laboratories. This is because, in many cases, the contemplatives within these traditions have access to a knowledge that they have painstakingly gained for themselves, from the inside out. The ability of these various traditions to produce awakened, enlightened, or saintly men and women speaks for itself.

A major theme of this book concerns a distinct way of knowing that contemplatives have refined: a way that is different from the ways in which scientific knowledge works. The contemplative way of knowing is more incarnational than it is representational; that is, more direct and nonconceptual than concerned with finding expression through linguistic, mathematical or logical symbols. These different ways of knowing are not contradictory so much as distinct but potentially complementary.

The scientific study of contemplative masters and how their brains/minds differ from the "average," can surely be illuminating, but it is quite distinct from the fruits that a lifelong dedication to a contemplative path can bring. So, although many of the techniques and disciplines for stabilizing the mind practiced by generations of contemplatives are surely highly valuable to contemporary scientists of consciousness, we should restrain an urge to become overenthusiastic in reducing these religions to a "mere" science. There is, however, certainly much that can be a source of inspiration for contemporary scientists.

Throughout this book we will use, when appropriate, insights drawn from scientific studies of psychedelics and other non-ordinary, altered states of consciousness. It might be obvious, but deserves to be stated nevertheless, that we personally do not endorse or encourage any reader in the irresponsible use of any illegal/pharmaceutical or other profound ways of entering into non-ordinary states of consciousness (like days-long extended retreats in darkness or meditative silence). Regarding psychedelics in particular, we agree with Stanislav Grof's characterization of such substances as "non-specific amplifiers" that can amplify both light and shadow in one's mind.[6] We repudiate the wildly oversimplistic but still popular misconception that non-ordinary or altered states are necessarily "higher" states of consciousness. While it is certainly possible that non-ordinary states of consciousness can have integrative, healing effects, especially with adequate preparation and support, it is also true that they can promote dis-integrative tendencies.[7] Precisely because of their power, a pre-mature encounter with non-ordinary states of consciousness can become highly disorienting. By contrast, a lifelong dedication to spiritual practices with strong ethical foundations, an emphasis on the cultivation of virtues like compassion and humility, and being under the skilled guidance of wise elders, can yield depths of personal transformation and insight that promote a renewal of consciousness far more enduring than passing states induced by other means can hope to afford.

1

The Matter of the Mind

In 1995 in Tokyo, a new kind of exhibit called *Body Worlds* was making waves, drawing crowds of visitors who were simultaneously captivated and repulsed by what they saw. On display were human cadavers available for close viewing and, in some cases, touching. The specimens were preserved using a process called plastination, which replaces the body's fluids and fats with synthetic polymers. Plastinated bodies are dry and odorless, retaining a fleshy texture without decay. The show went on to become one of the world's most popular traveling expositions, attracting many millions of people across North America, Asia, and Europe. A decade later, the second edition of *Body Worlds* premiered at the California Science Center in Los Angeles including a new exhibit focused on a particular piece of anatomy: "The Brain—Our Three Pound Gem."

Controversies have trailed *Body Worlds,* ranging from concerns about the donors' consent to religious misgivings and to accusations that the display of cadavers for public consumption is tantamount to pornography. The exhibit's illuminated cases in otherwise dark showrooms bring one face to face with the most intimate physical details of the human body. Seeing the intricate meshwork of nerves and the pinkish pulp of brain matter that previously cradled within its folds the memories of a breathing human being, it is impossible not to be awed. Medical and neuroscience students who have ever held a human brain in their hands, with its frayed rope-like spinal cord dangling below, will know firsthand the brutally honest self-reflection such contact produces. Suddenly, all of those anatomical drawings, enmeshed in a web of obscure Latin names, become an earthy, three-dimensional, and palpable reality.

The Quest for a Unified Theory of Consciousness

Imagine a parallel universe in which the evolutionary process resulted in human bodies with skin and bones made out of a diaphanous material. Let us call it the *Body Worlds* universe. On the inside, the biology is identical to our terrestrial biology with one crucial difference: no barrier prevents light rays from penetrating and bouncing off your interior cavities. Your skull is invisible, and you can see your brain as a pulsating, gelatinous mass, continually bathed in blood and cerebrospinal fluids.

A founding figure of neuroscience, Sir Charles Sherrington, bequeathed to us a lasting metaphor of the brain as "an enchanted loom." One can almost believe that Sherrington was inhabiting the *Body Worlds* universe as he described it:

> The great topmost sheet of the mass, that where hardly a light had twinkled or moved, becomes now a sparkling field of rhythmic flashing points with trains of traveling sparks hurrying hither and thither. . . . It is as if the Milky Way entered upon some cosmic dance. Swiftly the head mass becomes an enchanted loom where millions of flashing shuttles weave a dissolving pattern, always a meaningful pattern though never an abiding one; a shifting harmony of subpatterns.[1]

Now imagine that you can see this cacophony of illuminated sparks emanating from your brain, all happening as you are thinking, perceiving, feeling. When anger arises, you see a chaotic dance of neuronal points, and when you are feeling calm, the glowing embers settle and their heat dissipates.

In this parallel universe, the connection between two realms—the world of flesh and physical matter, and the inner world of experience—would perhaps be intuitively obvious. Scientists might even have long since derived a final theory of consciousness that allowed for a perfect mapping of cellular processes occurring in the brain to the private, conscious experiences of individuals.

In our own universe today we remain very far from having anything resembling this full-fledged science of neurophenomenology (combining the *neuro* focus on the brain, and the *phenomenological* study of the very stuff of experience—its interior fabric, so to speak). The modern, scientific understanding of consciousness is, at best, incomplete. Between the neural and the phenomenological there is a yawning chasm. We have increasingly detailed images of the brain that reveal the workings of single cells to whole brain networks. And we know that, at some level, the physicality of brains must relate rather closely to the interior nooks and crannies of the subjective worlds that form our interfaces with the world and people around us. But the nature of the bridge that connects these two realms remains hauntingly evasive.

The hope here is to look at both the neural and phenomenological dimensions of reality without losing sight of either one. This book argues for stepping back to take in both the objective facts of brains and the lived experience of the human mind, and integrating these into a comprehensive, stereoscopic vision. To do this, however, we must first examine some unresolved issues that stand in our way.

Brains and Minds—Never the Twain Shall Meet?

The emergence of cellular structures in our evolutionary history, with a membrane separating the inner compartment from the external world, served as a foundation for a radically novel feature of reality: the appearance of interiority. It was a development arguably as consequential as the Big Bang itself. The current evidence suggests that the first bacterium-like organism appeared approximately 3.7 billion years ago, and with it, this new feature of interiority on our planet. With the origin of the cell membrane, there was an organized body, with a division between a private "inside" and the world "outside."

The cell membrane provided a porous compartment that separated the interior cavity, which housed the molecular networks necessary for the energy and maintenance of life, from the world of primordial ooze that buffeted it from the outside. Some scientists hypothesize that the primitive ability of a cell to maintain a self-disclosing boundary while moving into the world, sensing and responding to events outside of its interior, provided the initial bio-"logic" that evolution would go on to exploit across subsequent eons, leading eventually to the full-fledged subjective consciousness that we now enjoy.[2]

The appearance of nerve cells, organized first into distributed and decentralized nets, but then into more complex nervous systems and eventually brains, heralded the next major landmarks in the universe's potential to create portals into deeper recesses of interiority. Brains, just like other bodily organs, are composed of physical matter. But unlike spleens, hearts, and lungs, the physical stuff of brains became endowed with a completely novel dimension. This was that "marvel of consciousness," as the writer Vladimir Nabokov described it in reverent awe: "that sudden window swinging open to a sunlit landscape amid the dark night of non-being."[3] The underlying nature by which this inner cosmos comes to be is a deep and abiding mystery.

An electrode tip that applies minute amounts of electricity to a section of the brain not only causes a transient disruption of cellular activity; for the owner of that brain, it can open a portal to an entire realm of subjective experiences. This other, interior dimension of reality is the very universe of consciousness that contemporary science is trying to account for. There are two realms: a subjective inside and an objective outside. Every conscious creature is another embodiment of a deeply personal point of view on the world. Everything you will ever know, from the cradle to the grave, will take place inside this bubble of consciousness so deeply and indelibly connected with your very life.

Where does the mind, this delicate center of subjectivity, come from? Is it simply a product of biochemical processes unfolding within the body? If so, then how does the teeming world of cellular activity "secrete" a mind? Are brains somehow more important or more "real" than minds? Or, on the contrary, is the mind necessary for the world of matter to appear at all, since without a mind in the first place, there would be nothing and no one to whom the world *could* appear?

These kinds of questions have puzzled untold generations of humans. Although widely varying answers have been proposed, they have generally cleaved into two lines.[4] One camp, epitomized by the French philosopher René Descartes, says that there are really two kinds of stuff in existence. There is the world of solid objects, composed of physical things that extend into and take up space, and then there is the world of immaterial, mental phenomena, which are not to be found anywhere in space. There is a cleavage in the very structure of reality, according to the Cartesian view, between these two. It is not only that they *appear* different—they are fundamentally different types of things which *really* exist, according to Descartes. Somehow, in some yet-to-be-discovered way, the world of invisible, mental stuff comes to interact with the visible, spatial world of physics. This answer to the puzzle of consciousness is known by philosophers as *substance dualism* (matter and mind being the two distinct kinds of substances).

In direct contrast to dualists, the *monist* camp claims that only one kind of stuff (one substance) actually exists: physical matter. The subjective, mental world, according to these materialists, is a kind of immaterial froth that arises as a by-product of the biophysical processes occurring in cell populations. Since conscious experiences (or *qualia*) appear to be nonphysical, a materialist worldview sees them as being passively subservient to the world of brains and bodies. Sort of like rainbows, they may be pretty to look at but are completely impotent to do anything besides serving as ornaments.

Different flavors of materialism vary substantially in how strictly they insist on the primacy of the physical. But, at bottom, materialists agree that minds exist (or seem to exist) only insofar as they are rooted within a physical substrate. For them, the arrow of causality *always* points down to the realm of matter, which is presumed to be more real than conscious qualities. Maybe the most brash and succinct summary of this perspective comes to us courtesy of the late Francis Crick, co-discoverer of the structure of DNA, whose "astonishing hypothesis" says that we are "nothing but a pack of neurons."[5]

In one flavor or another, materialism has dominated the intellectual framework within the scientific world at least since the Scientific Revolution.[6] Despite doubts voiced by some of the historically most prominent neuroscientists, materialism has been the reigning paradigm, influencing the kinds of questions researchers ask and the tools used to answer them.[7] As they sought in part to emulate the successes of physics and chemistry, a mechanistic and reductionist flavor of biological materialism came to define the study of mental phenomena in Europe and America. The consequence has been that the inner world of experience has held a kind of second-class citizenship in scientific thought—when it was not excluded altogether. Most self-respecting neuroscientists steered clear of the study of conscious experience and the dynamic interrelationship of brain matter and inner experiences.

Yet there have been influential voices arguing for the primacy of experience in any scientific investigation of the brain and mind. At the dawn of experimental psychology, in the late nineteenth century, William James placed the phenomenon of consciousness at the forefront and held that recorded observations of it should constitute a primary database for researchers.[8] And although his emphasis on interior experience as the primary datum of psychology did not take hold at the time, and was overpowered by approaches that minimized subjectivity, it was not completely driven out. Lately, new cracks have started to appear in the strict materialist foundation.

Recent decades have witnessed breathtaking technological developments in our ability to observe, prod, and study brain matter. Brain-scanning fMRI machines, with magnets more powerful than the Earth's magnetic field and costing millions of dollars, are a staple feature in virtually every modern neuroscience research center. But the tools that scientists have developed to probe the subjective side of the mind-matter equation are embarrassingly crude. In the twenty-first century, we understand how to use the spin of hydrogen atoms to reconstruct structural images of the living human brain—but when a researcher slides someone into a brain scanner, the amount of research data being collected on the rich subjective experiences presumably parallelling those spectacular brain images is usually zero or close to zero. A typical study might have the subject use a button to signal the perception of simple shapes, or look at pictures of human faces with frozen emotional expressions as the scanner captures which blobs in the brain "light up" in response. In some cases, subjects are handed questionnaires that require them to compress the subtle varieties of human experience into the form's prescribed vocabulary.

The result is that, although we know in astonishing detail about almost every bump and groove on the brain's surface, when it comes to charting the space of inner experiences we are still navigating with a cribbed map poorly drawn on a paper napkin. The philosopher John Searle bemoans this state of affairs, writing that "it would be difficult to exaggerate the disastrous effects that the failure to come to terms with the subjectivity of consciousness has had on the philosophical and psychological work of the past half century."[9] More than anything, he believes, "it is the neglect of consciousness that accounts for so much barrenness and sterility in psychology, the philosophy of mind, and cognitive science."[10] A mature science of consciousness will eventually need to discover how to pin its hard-won discoveries about brain physiology to maps of phenomenological landscapes.

Navigating Mind Spaces

Also getting in the way of a comprehensive account of consciousness is the major impediment that our scientific theorizing has been much too provincial. Nearly all neuroscientific research in this fledgling field has proceeded from the perspective of waking consciousness.

One need not be an expert psychonaut to appreciate that human consciousness extends to staggering depths. Simply observe the changes in your own experiences over a single day-night cycle. During waking hours, waves of electromagnetic energy hitting the back of your eye help fashion the images of an external "reality," while every night you sojourn through dream worlds that are no less "real" and interactive for you, even though your sense organs are quiescent.

In the traditional scientific consensus, the waking world maps onto reality and the dream world is a *mere* hallucination, of cursory interest perhaps, but in no way of central interest. And if in scientific treatments subjective experiences have been neglected, then subjective experiences *in non-ordinary states of consciousness* have been outright spurned. A growing mountain of evidence, however, suggests that this consensus misses a much more interesting truth: during dream time, the constraints that operate to shape subjective reality shift their center of gravity, from the world "out there" to an "inner world" that is equally resplendent in its riches.

At this point, it is helpful to introduce the concept of a *state-space*.[11] A state-space encompasses all the possible states available to a given system. To make the idea of a state-space more imaginable, picture a three-dimensional cube with breadth, length, and depth. Next, visualize a small, incandescent sphere that surfs through the internal volume of the cube. This represents your bubble of consciousness. When you are awake and going about your day, your consciousness bubble is traveling within a relatively small portion of this enormous cube. As you shift into dreaming, your sphere of consciousness is able to drift into other regions of the state-space, where it bobs along until your reawakening pulls it back within the more constrained space. We can look forward to a time when it will be known in much greater detail how different types of trajectories in mind space translate into fluctuating clouds of dynamical brain activity.

A unified account of consciousness will need to be broad enough to contain our imaginary cube, mapping it from corner to corner. And yet, large stretches of these experiential territories are *terra incognita* as far as contemporary science is concerned. The bulk of research initiatives have been focused on shining light on a few small pockets of the state-space, primarily the broad, sunny avenues of waking perception and thought. As famously stated by William James, "our normal waking consciousness . . . is but one special type of consciousness, whilst all about it, parted from it by the filmiest of screens, there lie potential forms of consciousness entirely different. . . . No account of the universe in its totality can be final which leaves these other forms of consciousness quite disregarded."[12]

Waking and dreaming have been the two major centers of gravity for contemporary consciousness researchers. This leaves out a vast amount of territory. Consider the phenomenon of lucid dreaming. Lucid dreaming is a fascinating mix of waking and dreaming consciousness in which the dreamer is fully aware of being inside a dream. Given that they do not fit very comfortably within our traditional understanding of experience, lucid dreams have been a topic of considerable scientific controversy, with some researchers even denying their existence. Despite an abundance of anecdotal accounts, some researchers explain them away as brief awakenings which subjects subsequently misinterpreted and misreported as instances of waking awareness infiltrating a dream.

It took an ingenious study by Stanford University psychologist Stephen LaBerge to convince the scientific community of the veracity of lucid dreaming.[13] Having selected a small group of subjects with a learned "ability to have lucid dreams on demand," he hooked them up to equipment that monitored their brain electrical activity and other vital signs as they drifted off to sleep. The subjects had been instructed that, if they slipped into a state of lucid dreaming, they should perform a predefined set of dreamed eye movements (left, right, left, for instance) and fist clenches to signal that occurrence. Remarkably, the experiment was a resounding success. Subjects were able to communicate to the researchers using the agreed-upon muscle codes, while the recording equipment attested that every such signal was received while the subjects were in fact sleeping, not awake. Lucid dreaming was scientifically established as a *bona fide* class of conscious experience.

The tendency to lavish scientific attention on normal waking conscious experience betrays an underlying assumption that this specific type of consciousness is the default anchor point for our explorations. Much as it was once believed that Earth occupied the center of the known universe, waking consciousness is often assigned the central position in the conscious state-space, around which other states of consciousness cluster as curious offshoots.

In principle, this is an entirely reasonable stance, since waking consciousness presumably evolved to allow animals, including humans, sophisticated ways of interacting with the world around them. What is more problematic is that there is often another, more treacherous assumption lurking in the background—namely, that our waking consciousness provides the "correct" perspective by disclosing a sole reality, departures from which are dubbed *altered states*.

Introduced in the 1960s, this term refers to a variety of states that have been relegated to marginal roles (when not completely ignored) in contemporary theories of consciousness.[14] The noted neurobiologist J. Allan Hobson, for example, considers waking consciousness "normal," and other variations "artificial."[15] As anthropologist David Lewis-Williams puts it, the phrase *altered states of consciousness* "is posited on the . . . concept of the 'consciousness of rationality'. It implies that there is 'ordinary consciousness' that is considered genuine and good, and then perverted, or 'altered,' states. But . . . all parts of the [consciousness] spectrum are equally 'genuine.'"[16]

To avoid pejorative connotations, we use the more neutral term *nonordinary states* to refer these other parts of the spectrum. To be clear, by calling attention to them we are not simply trying to add some spice—a dash from studies of lucid dreaming here, a pinch from studies of psychedelics there—to the consciousness stew. The world of consciousness, we suggest, springs forth from the inside out, in an organic mushrooming or ballooning-out process of deep beauty and elegance. The waking state does not necessarily stake a stronger claim to a more real, ontological status within the conscious state-space. So-called non-ordinary states can, under the right lens, reveal processes unfolding at the depths, those through which our everyday consciousness differentiates from a more primary potential.

Let us state a thesis that will be developed much more thoroughly in the chapters to come. Rather than the commonsense view that the real world impresses itself upon our mind as a brass stamp does on sealing wax, it is actually the other way around: an internal set of processes impresses itself upon what we experience as surrounding us. What lies in the interior precedes and is more foundational than what we experience as the world "out there." In fact, the stuff from which the waking state is built is precisely the same material out of which hallucinations and dreams are constructed. Penetrating through the layers of presumably non-ordinary states, we argue, brings us to deeper recesses toward the very core or interiority of consciousness itself—the source from which the entire phantasmagoria emerges.

Unfurling the Consciousness Spectrum

Let us unpack this idea further. Picture the spectrum of visible light, which runs from the brightest red to the deepest purple. Now imagine consciousness as a similar spectrum that spans from the bright daylight of consciousness to the pitch dark of deep sleep or anesthesia. Just as it is hard to pinpoint the exact place on the visual spectrum where red becomes orange, so too is it difficult to zero in on a transitional point on the consciousness spectrum. The moment when one state gives way to a different type of conscious event becomes even less evident to us because we are, in fact, continuously shimmying along the continuum's length. A twenty-four-hour period may involve a full transit of the spectrum, but when we zoom in on specific borderlines we see how the transit across them is complicated by mini-cycles of moving back and forth between states that might last mere minutes, or even mere seconds. The mental life constantly oscillates between being attuned more to the outer world and more ensconced in its own interior.[17]

One end of the consciousness spectrum corresponds to the states that most of us see as our "baseline" state—namely, when we are awake, alert, and engaged with our environment. In this mode of consciousness, the nervous system and the outside world are tightly interlocked, one with the other. Neural processing in alert wakefulness is strongly constrained by signals originating from the outside world, and our subjective state is anchored to specific coordinates in space and time. We are potentially open to being influenced and instructed by the outside world, when occupying this segment of the spectrum.

At this extreme end, consciousness is dominated by an abstract and highly conceptual style of cognitive processing strongly structured by language. Our thoughts are largely formulated as a kind of inner commentary, which helps to ensure that thought sequences are linear and organized tightly around a specific task. This is the state of consciousness we *need* to be in when solving a particular problem, whether writing lines of computer code, working out an equation, composing an email to a colleague, or operating heavy machinery. Strong upsurges of emotion, uncontrolled emergence of memories, spontaneous wandering of thoughts—all of these are controlled, and kept to a minimum, when walking this tightrope of consciousness.

Occupying the tip of the consciousness spectrum requires discipline, and it feels like hard work. Over time, our cognitive resources get depleted. Eventually, our hold becomes slippery, and we slide down the spectrum. Shimmying a few steps below, mind wandering starts becoming more frequent, with recurring forays between focusing on the external world and interior imagery associated with memory, feelings, and fantasy. This middle portion of the consciousness spectrum is a kind of default setting, the groove into which our mind spontaneously settles throughout the day, and where most of our lived reality happens.

You might enter this state when gazing out of the window, mid-afternoon, or when driving back home and reflecting on what you will have for dinner. In scientific terminology, this mode is known as spontaneous or task-independent thought. Most of our so-called resting-state thoughts exhibit a structure that can be roughly parsed into different domains of self-relevant processing (thinking about yourself or your feelings), theory of mind (thinking about other people), and awareness of inner bodily processes, like breathing and heart rate.[18] Much of these thought sequences, which are now increasingly suffused with different degrees of emotional charge, are the phenomenological counterpart of ongoing patterns of brain activity, particularly that occurring within anatomical regions known as the default-mode network.

Even further down the spectrum, consciousness starts to become increasingly introverted, as the interlock between the brain and the outside world becomes much looser: the inner world starts to loom larger and larger while the outer world becomes muted. A boundary is now approaching, past which our subjective reality is thoroughly interiorized. As that boundary gets closer, the subjective world becomes more unpredictable and freewheeling, less anchored to verbal thinking and richer in imagery. Imagery is increasingly drawn from deeper levels of the mind's interior, becoming less fixed in space and time. Fragments from memory are liberally intertwined with completely fantastical, imaginative elements, and this vast interior world blooms with vibrant currents of emotional charge.

The most common time to visit this twilight zone of the consciousness spectrum is while navigating the spaces between wakefulness and the light stages of sleep. In technical terms, this state is known as *hypnagogia*.[19] Just as a doorway can lead either from the outside in or the inside out, so too we traverse this passageway again when re-emerging from sleep to wakefulness, when it is called a *hypnopompic* state. Although many people are oblivious about the existence of these exotic neighborhoods of their own mental life, ancient contemplative schools, as well as modern psychology, have extensively documented hypnagogic realms. With training, it is possible to prolong the time spent here and to systematically observe how consciousness operates this far down the spectrum.

As absorption and enchantment with hypnagogic imagery increases, the border is finally crossed and our reality is sourced from deep within the mind's interior. We are now in the land of nocturnal dreaming, where we inhabit worlds illuminated by a light entirely from within the mind. In the sensory seclusion of a cocooned, sleeping brain, there is still an entire world (or worlds) to explore. Here, our thoughts are organized into narrative sequences. Memory and fantasy are unconstrained, and the very fabric of experience is often intensely emotional. In most contemporary theories of consciousness, the spectrum bottoms out in the dream. We will, however, strongly challenge that claim in

Chapter 12, where we consider puzzling examples of so-called min-imal phenomenal experiences (MPEs), seemingly empty of all content except for a pure and boundless subjectivity.

Two more points need emphasis when discussing consciousness as a spectrum phenomenon. First, as noted by David Gelernter, the upper (extrovertive, opened to the outside world) portions are oriented largely around *doing,* while the introvertive depths are defined by simply *being.*[20] In the 1970s, the psychologist Arthur Deikman re-ferred to these two modes of consciousness as an active and a recep-tive mode.[21] The former is characterized by engaged activity and high levels of control. The latter mode is the one we enter into when we allow ourselves to become absorbed, entranced, or when we are over-whelmed by powerful waves of feeling or made to witness a procession of enchanting images. Our modern, technological civilization lays heavy (maybe even exclusive, and in any case, almost certainly un-healthy) emphasis on the active, doing mode—efficient performance and problem solving carry the day for many of us.

The raw vibrance of sensation, the pulsatile nature of feeling tex-tures, the sheer experience of being alive, including ethereal qualities that dominate states of hypnagogia and dreams. All these are states of receptivity, pure being, which are not in the immediate service of spe-cific and delimited tasks, and are consequently much more likely to be ignored by scientific styles of thinking and analysis.

Yet it is precisely these most interior fragments of the consciousness spectrum—states of nearly pure being—that form the very substratum of our experiential world. The consciousness spectrum originates from the primordial depths and then unfurls toward the surface of our bright ordinary consciousness. From a developmental perspec-tive, dreaming consciousness precedes the waking one by several months *in utero.* There is also extensive psychological evidence that sleep deprivation, and in particular lack of rapid eye movement (REM) sleep when dreaming is most likely to occur, has multiple negative ef-fects on waking experiences. As the dream researcher J. Allan Hobson has concluded, "the integrity of waking consciousness depends on the integrity of dream consciousness."[22] The experience of reality appears to be organized in layers upon layers, and as we shed the surface, we increasingly move toward the primordial, evolutionary ancient depths. Whether it is through the use of tools like profound sensory deprivation, lucid dreaming, deep meditative states of stillness, psychedelic-assisted forms of psychotherapy, or otherwise, as we travel down the spectrum we get progressively closer to the root and source of our lived reality.

A Path Forward

This book envisions a radical approach to overcoming some of the impediments standing in the way of a mature science of consciousness. At this point, it is informative to step back and reflect on the lopsided nature of scientific theorizing that has been strongly shaped by historical strands dominating European and American worldviews. To this extent, the science of consciousness that we have inherited has an element of ethnocentrism—and, just as crucially, *conscious-state centrism,* to coin a term—baked into it. Our research programs are stamped with cultural attitudes and presuppositions, which inevitably influence the types of questions scientists ask as well as the answers that are most likely to be provided. These implicit biases have often privileged one mode of data-gathering at the expense of another: brains (physical) over minds (immaterial), and waking consciousness over other pockets of a vast conscious state-space.

Our historically and culturally entrenched view of what constitutes the fundamental basis of reality—our ontology—has directly shaped and constrained the methods we have used to acquire new bits of knowledge—our epistemology. Epistemology in turn delimits ontology, as the way we gather data shapes what we consider to be real. In this sense, all observations made by scientists presuppose a specific worldview. The data gathered are, as philosophers would say, theory-laden. Once they have been set in motion and calcified through long historical precedent, the patterns of exchange between a dominant ontology and epistemology become mutually reinforcing, and by their very nature, they function to place boundaries around our understanding of natural processes that are intrinsically dynamic and multifaceted. It is not so much that this strategy has not borne its fruits–it surely has–so much as it remains incomplete.

In the remainder of this chapter, we will sketch a way forward that begins to bridge the existing gaps in contemporary scientific theories of consciousness. In a surprising turn, the clues for this synthesis can be discovered in the world's numerous religious traditions, which have explored the human mind from the inside out, rather than the outside in. In a very real sense, they have developed and refined a sophisticated body of psychological observations of the human mind across the entire consciousness spectrum, especially in their inner, mystical dimensions. Unfortunately, most scientific researchers are unfamiliar with much of that knowledge. Gerald May, a psychiatrist who was also deeply conversant with contemplative themes, once wrote that spiritual giants like St. Teresa of Avila and St. John of the Cross often had "clearer insights into the dynamics of consciousness and attention than most modern neuroscientists."[23]

But consider the doyen of empirical science, Roger Bacon, who recommended a system of scholarship that combined the gathering of "outer" knowledge of the world with a systematic exploration of "inner" experience. For Bacon, it was precisely those following a dedicated spiritual path, with an intensive focus on studying the mind's depths, who would be responsible for charting these interior boundaries.[24] Today, we can only wonder about the alternate trajectory that our modern scientific tradition, rooted in European thought, might have followed had the excessive mechanization of the natural world that occurred in wake of the Cartesian split been offset by heeding Bacon's advice for a complementary mode of knowledge gathering.

In a nutshell, those who mastered these religious traditions were, first and foremost, experts in the rewiring of human consciousness that, in many cases, today's science is only beginning to appreciate. Of course, they were not probing people's brains with technologically advanced sensors, nor were they obtaining delicate recordings from living or sliced brains. Nevertheless, they most certainly *were* developing sophisticated and profound methods for turning attention inward to focus on mechanisms and processes that structure the fabric of experience, including the very foundations of awareness. They were masters at systematizing consciousness, observing the mind's manifestations across states ranging from the paroxysms of rage, lust, and fear to profound relaxation, meditative absorption, dreams (non-lucid and lucid), the nether regions of sleep, and even the borderlands between life and death. In fact, what we call Buddhism is expressed in Tibetan as *nangpa sangyepai chö,* perhaps best translated as "inner science."[25]

In the ancient Christian monastic tradition, the first stage of the spiritual journey, called *praktikos,* required gaining extensive insights into one's own psychological makeup, including the role of one's deepest desires and fears in constructing the most foundational types of subjective experience.[26] All of this elaborate knowledge of how to identify, work with, and transform different states of consciousness, and the consensually validated maps of the human mind obtained over many hundreds of years, repeated across many, many individuals is nothing short of a goldmine.

The religious traditions of the world, on which we can draw to complement our exploration of the human mind, do not respect neat geographical boundaries. Some of the traditions on which this book focuses trace their roots to the Indian subcontinent, while others come to us from the Near East and Northern Africa, and from regions that extend from the Greek peninsula throughout vast regions of Eastern Europe, the frozen tundra of Siberia, and across Central and South America.

The contemporary scholastic and medical-psychotherapeutic tradition has made valiant efforts to develop its own science of experience, primarily through the work of phenomenologist philosophers including Brentano, Husserl, Merleau-Ponty, and Heidegger, and the early twentieth-century structuralist school of psychology headed up by Wilhelm Wundt and Edward Titchener.[27] Unfortunately, these attempts were largely neglected. Moreover, the founders soon discovered that the effort required to obtain high-quality research data on conscious experiences was far from trivial and required extensive training of research subjects. Also complicating the psychoanalysts' attempts to probe into their patients' inner worlds were those patients' own subtle, or completely unconscious, defensive attempts to distort, conceal, and obscure the deeper reaches of their minds.

The exploration of the experiential is not a task that ought to be left to amateurs, as emphasized by Alan Wallace, the Tibetan Buddhist scholar.[28] No one expects those uninitiated in advanced quantitative and statistical methods, or the exigencies of obtaining and interpreting brain recordings, to be able to make serious contributions to the scientific body of data. Our standards for obtaining data on the microstructure and dynamics of experience ought to be analogously stringent. The pioneering efforts of the Chilean biologist and neuroscientist

Francisco Varela, one of the founders of neurophenomenology as a legitimate discipline, have been instrumental in fostering a rapprochement between subjective and objective sources of data in contemporary neuroscience, including numerous methods for improving the rigor of the former.[29] In Chapter 3, we will examine, in more detail, a sophisticated set of methodologies that researchers can now leverage to hone "experiential microscopes," including formalized protocols for this type of data acquisition, as they have been developed in the interdisciplinary field of contemplative science.

Different Ways of Knowing

How do we come to know things? Today's mainstream intellectual culture generally recognizes two dominant routes for acquiring knowledge. In the first case, what we know is what we can observe and measure. "I'll believe it when I see it," we say colloquially. This constitutes that body of data that science deals with through carefully controlled experiments.

The second route is through the exercise of rational, logical faculties. Mathematical proofs or philosophical syllogisms, for example, are worked out through logical deduction from certain principles. Consider the famous syllogism: *All humans are mortal. Aristotle is human. Therefore, Aristotle is mortal.* The conclusion of Aristotle's mortality is something we can know with ironclad certainty just using logic, without having to gather any additional data.

The edifice of modern science has been built largely on these two pillars of acquiring knowledge. An important feature of both methods is that they objectify the contents of knowledge. That is, both stand over and against the object that is known. It is straightforward to appreciate that this is true in the case of sense-based knowledge, as it really does aim to measure or quantify external appearances. The brain, as a physical organ, is clearly observable and measurable in its minutiae, but how about the mind and consciousness? Is your consciousness small, big, wet, dry, red or sphere shaped? Where does a thought occur? Such questions stump us. They sound nonsensical, because consciousness is a *prima facie* interior phenomenon, that only reveals itself in its very mode of being, from the inside out. Subjectivity seems to be an indisputable datum, yet it evades straightforward measurement. In contrast to inanimate objects like rocks, consciousness discloses itself to itself. Yet consciousness as such is not directly visible, even if its external correlates are quantifiable.

The spiritual traditions of the world have valued other, largely neglected, routes to gaining knowledge that complement the first two strategies. These less-traveled roads to acquiring knowledge seem uniquely suited to interrogating the subjective nature of the mind. They can go by different names, but as a group they feature what the philosopher-turned-mystic Franklin Merrell-Wolff called *knowing-through-identity*.[30] While other forms of knowing rely on a style of observation by which the individual objectifies and stands apart from what it observes, this is a knowing that is direct and gained through actually *being* that which is known. It is participatory knowledge rather than objectifying knowledge. For a primitive example of this form of knowing we might consider antibodies: How do they *know* about foreign antigens? They do so by literally conforming themselves to the shape of the invading antigen. To capture this direct epistemic strategy in a phrase we might say, "to know the antigen, become the antigen."

As applied to subjective experience, this would mean that no amount of reading, for example, about the sweetness of honey, or learning details of receptor physiology and chemistry, could ever allow you to know what honey tastes like. That can only remain unknown until you taste a spoonful. Consider another example. Lucid dreaming, as we saw earlier, was long considered a logical impossibility, while the subjective experience of dream lucidity attested to its very real status. In both of these examples, it is the inside-out, phenomenal perspective that is indispensable. This inside-out knowledge cannot be gained by only reading textbooks, listening to lectures, or taking examinations—it requires careful analysis of the experiential fabric. The great spiritual traditions of the world consist of elaborate, generational systems that preserve and transmit methods of deep inner contemplation that enable direct knowledge about the latent potentials of the human mind.[31] For too long, academic disciplines studying the mind and brain have neglected these traditional knowledge systems that deal with and principally interrogate subjectivity, but in principle, they stand to be enriched by entering into a dialogue with them. Today the field of cognitive neuroscience stands at the frontier of a far greater willingness to embark upon such an adventurous project than in decades past.[32]

Authentic contemplative schools provide an educational system of sorts for the production of experts in consciousness exploration and transformation. One of the contributions of the world's great spiritual traditions is that they have invested millennia in refining and perfecting knowing-through-identity. This endeavor depends on receiving supervision from a qualified teacher and many years of dedicated training, every bit as demanding as immersion in a modern scientific curriculum. The student often begins by learning how to settle the mind and develop a focus that is taut and supple enough to achieve stable equilibrium of awareness. Expert meditators accumulate tens of thousands of hours of training, qualifying them as the world's elite in observing the dynamics of the human mind.[33]

The viability of an experiential mode of knowledge is that the resultant findings can be compared with those of other practitioners who have also put in the requisite work to make similar observations. One can point to the many shared observations about the nature of mind that practitioners converged on. It is exactly this richly layered, consensus-based body of knowledge that offers the possibility of informing a proper understanding of phenomenology. Without such a convergent coordinate system, the prospect of a topography of inner spaces would be a hopeless, meandering enterprise of "navel gazing" and, in that case, rightly dismissed as a proper topic of scientific inquiry.

We should not lose sight of the fact that many generations of contemplatives and mystics were successfully replicating their effects for thousands of years, in ways that largely generalized across independent observers. This fact argues against what Alan Wallace has termed the "taboo of subjectivity," which has long characterized scientific explorations of the mind.[34] Certainly, scientists should limit subjectivity when it refers to unreliable and biased observations, but this does not mean that subjective data themselves are untrustworthy and incapable of being independently observed and organized into a consensual knowledge base. The ability to repeatedly visit particular pockets of the conscious state-space and use careful descriptions to achieve agreement across people concerning its most salient features makes these data no less credible than the information your optometrist uses when conducting your eye exam. The truth is that modern medicine often relies on gathering subjective data without blinking an eye or considering it unreliable.

There are reasons, then, to be intellectually open, and even eager, for the enrichment in perspective that a neuroscience of consciousness stands to gain by encompassing the discoveries of contemplatives. This is especially true as scientists today are increasingly making forays into deep spaces of the mind that have been charted as locales, and often pursued and prized, by religious traditions with millennia-long head starts. Explorers of these non-ordinary states of consciousness were able to bring back profound and healing insights that others could learn and benefit from.

As contemporary consciousness researchers are increasingly beginning to investigate experiences elicited by different types of psychedelics, various stages of sleep, and powerful states of so-called *ego dissolution,* mystical religious traditions have a critical contribution to make. Without prolonged and extensive training, introspective access to these non-ordinary states is especially challenging as the experiences often exceed our ability to adopt a reflective or witnessing stance. The computer scientist David Gelernter coined the term *consciousness burn* to refer to this phenomenon, whereby experiences of what we might term pure being are so all-encompassing, so concentrated and engrossing, that despite their prevalence they end up as so many blind spots dotting the conscious landscape. It is only with prolonged contemplative training that one's inner witnessing stance is established on a base sufficiently solid to keep consciousness from burning itself out in a bright flame of absorbed being as one inhabits such a state. Any prospect of building a database of these very real but very slippery terrains requires years of expertise. The Tibetan Buddhist tradition, for example, charts the full spectrum of mental cartographies and makes a crucial distinction between gross (or coarse), subtle, and very subtle dimensions of mental life.[35] The coarse levels are where most of our doing happens, and they are relatively straightforward to probe using commonly used methods of contemporary neuroscience and brain imaging—by, for instance, correlating specific patterns of neural activity to the perception of discrete visual stimuli. But ultimately, the coarse levels of the mind depend and rest on the subtle and very subtle layers, which are far more difficult to investigate and are characterized primarily by *being* and very little (or no) *doing*. And states of being are best known precisely by *being* in them—knowing-through-identity.

Beyond False Dichotomies

Before we draw this chapter to close, we need to revisit that old battleground between the dualist and monist camps. This is important ground to reconsider now, because if we are to move forward with bridging the chasm between brains and minds, we need to secure a defensible starting position. As long as one believes that the mind is really nothing other than what the brain does, then the only *real* insights about mental life are those that will be gleaned from dissecting, stimulating, and recording from brains. The discoveries of generations of contemplatives, viewed in that light, might be interesting from an anthropological perspective, but would tell us little about consciousness itself.

The framework we adopt is neither materialistic nor dualistic, and it firmly rejects the false dichotomy that these are the only two games in town. There is a third possibility, a unique mix that goes by the name of *dual-aspect monism*. This perspective helps to reconcile seeming contradictions, and it enjoys a venerable historical pedigree with independent strains emerging across both the western and eastern hemispheres of the globe.[36]

The *dual-aspect* part means that two complementary perspectives can be had. From the "outside" or third-person perspective, we can describe physiological processes in the brain which can also, when they are experienced from the "inside" or first-person perspective, manifest as a subjective universe of mind. Neither perspective alone is exhaustive, and neither one allows you to know the other. It is worth rereading that last sentence and reflecting on it. No matter how much you know about neuroanatomy, neurophysiology, and neurochemistry, purely physical knowledge implies no necessary connection to first-person experience. Physical events alone leave a crucial aspect of the natural world unaccounted for. Likewise, if you were to introspectively observe your experiences, you could do so even until the end of time and you would never gain any insight about the complexities of biophysical processes in your own nervous system. The mental is not reducible to the physical, just as much as the physical is not reducible to the mental. Rather, the physical and mental coinhere and enfold one another into a single, integrated reality. They are inextricably braided together into one underlying manifold. Coinherence rather than reductionism holds the key. A complete and exhaustive description of this underlying reality requires data gathered from both observational windows.

Say you are sitting in a room, captivated by a candle flickering in front of you. Your immediate experience is that of being in a space with four walls and a dancing point of light, but most definitely *not* one of electrical and chemical signals being passed between cells inside your skull. Yet, with the right recording instruments, it *would* be possible to simultaneously observe these cellular activities happening in parallel with all of your internal experiences. Crucially, each of the two observational stances seems to us to always displace its complementary counterpart, and each provides only one lookout point onto a more complex reality that underlies them. This point was emphasized by theoretical physicist and quantum pioneer Wolfgang Pauli and psychologist Carl Jung, who worked out many of these ideas in their decades-long intellectual exchange.

In other words, our perceptual systems can take either the first- or the third-person stance, but not both simultaneously.[37] The mental and the physical are like the two sides of the same coin. To make a discrete observation in time, let us imagine, the coin has to land either on one face or the other, but this does not mean that there are, in fact, two different coins. A single observation in time displaces the other but does not negate its existence. This mutual exclusivity would be true even in the imaginary world of transparent bodies that we considered at the chapter's outset. We would never be able to directly see a person's private experience of colors or tastes by peering into showers of their brain activity, even if we could more directly intuit an intimate association between the two.[38]

So, at least in our view, the brain is not somehow *more* important or real than the mind, nor is the mind more fundamental than the brain. Both matter and mind originate from a common ground, which is what puts the monism in dual-aspect monism. At a fundamental level, there is a unified base that contains no fragmentation between an "inside" and an "outside" or "mental" and "physical." Jung called it the *unus mundus* (One World)—a holistic unity that only later becomes fractured into subjective and objective poles. And that *unus mundus* is a dynamic, living, and relational process rather than a structural thing.

As to what constitutes this underlying "ground of being," out of which both matter and mind originate, that remains a deep and abiding mystery. It is possible that direct access to this underlying level of unity is simply unavailable to us humans. Many of the world's great spiritual traditions, by contrast, claim that it is possible for humans to come into direct contact with an ultimate reality that is behind the heavily constructed, and often dreamlike, realities that many individuals inhabit by default. Entering into communion with this ultimate reality is in fact the stated goal of many religious traditions, especially in their mystical aspects. This single ground is neither strictly physical nor strictly mental. Stated another way, it is *both* at the same time. If you want to get fancy about it, it is psychophysical in nature.[39]

In chapters to come, we will delve into the workings of the unified brain/mind across a broad spectrum of states, bridging the objective findings of neuroscientists with the subjective findings contained in the world's spiritual traditions. Inspired by the Tibetan tripartite division of coarse, subtle, and very subtle layers of the mind, our book employs a similar structure, starting from the familiar daylit zones of consciousness and then progressing down the spectrum, burrowing deeper into the interior recesses. As we shimmy down the spectrum, the experience of being a self in the world undergoes radical alterations, as one might expect when considering the deep hierarchical structure of the brain— an organ that ceaselessly attempts to build sophisticated models of the world in which it is embedded.

To complement the presentation of information, we have interspersed invitations (called "In-Sights") for readers to explore a first-person view on some of the topics to be discussed throughout the book. These are optional ways for readers to engage the material at a deeper level, through gentle examinations of their own experience. These In-Sights will open for a receptive reader the portal to their own interiority in ways that can be known uniquely through direct experience rather than simply reading *about* them.

2

A Brain That Talks to Itself

Imagine you are part of an experiment: you are awake and isolated in a hermetically sealed, dark chamber, with a team of researchers using an array of technologically advanced probes to record the electrical activity inside your brain. The neuroscientists conducting this experiment in midtown Manhattan have taken the utmost precautions to eliminate every photon of light in the chamber, and its walls are sound insulated. Beyond your own heartbeat and breath, nothing is audible. The floor has been engineered to eliminate even the most minute vibrations.

Once a record of your brain signatures has been obtained, the chamber door swings open, and you are invited to step out. Next, with the recording equipment still in place, you leave the room and take a walk through the throngs of people in Times Square. Your senses are exposed to a constant barrage of stimulation in many forms.

Later, when the scientists analyze the large amounts of brain activity data gathered under these two radically different experimental conditions, they make a discovery that initially surprises them. Apart from some differences in the overall "volume" of neural signals, the basic structure of your ongoing brain activity appears quite similar in the two settings. The cornucopia of teeming sights and sounds in Times Square seems to have merely nudged the ongoing chatter among neurons in your brain.

As they probe deeper into the data, the researchers discover that even the precise ways in which your visual and auditory brain was being modified while you took in the scene of a busy public space contain essentially no novel information, relative to what was happening in utter darkness. The complex electrical wave patterns recorded during your jaunt through the streets resemble the organized set of activations that took place as your brain explored that dark, silent chamber.

To accommodate these findings, the scientists realize they must revise their prior notion of the brain as a kind of hypersophisticated computer that remains largely inert until provided with the proper raw-material input for producing other kinds of output. They begin to conceive of the brain as instead engaged in a constant monologue, generating spontaneous neuronal chatter. This is a new recognition of a brain that never stops "talking to itself," although external events can subtly shape and sculpt its chatter. At the same time, the scientists realize that this different understanding of the brain might have broader implications; it could turn upside-down some long-established and cherished notions about the nature of everything around us.

A Brief History of Two Views of the Brain

Across the history of neuroscience, two different ways of understanding brains have developed. One views the brain in passive terms, as a computing apparatus activated by signals coming in from outside.[1] The other, which is the perspective we explore in this book, conceives of the brain as a *spontaneously active* organ whose activities are influenced, but not determined, by incoming energies. The former framework has been the historically predominant view, originating in the late nineteenth century; it has tended toward a conception of the brain as the reflex organ *par excellence*. A reflex consists of some defined output (for example, a knee jerk) consequent upon the right sort of input (for example, a sharp tap on the patellar tendon). The reflexological framework extends this logic to the brain, seeing it as a master reflex-like apparatus, with many chains of computation sandwiched between input and the output.

From this perspective, the proper approach to studying brains is to bombard them with specific inputs, controlled by scientists, and then to carefully record the resulting responses. By studying responses to extrinsic stimuli, scientists are able to make inferences about the series of processing steps presumably taking place in those long neuronal chains. Traditionally, the context within which these successive stages of processing occurs has been treated only as background of little consequence.

The analysis of the neural bases of vision offers a paradigmatic example. To study how the brain processes visual inputs, scientists would use cats, anesthetizing them (to obtain clean recordings uncontaminated by movement), propping their eyes open, interrogating their brain cells by exposing them to abstract and elementary forms of light stimuli (like doughnut-shaped rings and slanted bars), and measuring the responses. It was assumed that the pattern of responses, and their transformations across different levels of the brain, would not vary much between states of consciousness, even ones as starkly different as general anesthesia and wakefulness. This led to a persistent myth that it was possible to isolate segregated systems devoted to "reading in information," absorbing sensory signals irrespective of the broader motivational context or behavioral relevance of specific objects in the world. Here is how Maurice Merleau-Ponty, a philosopher and phenomenologist, describes this once-prevailing view: "Sense experience, thus detached from the affective and motor functions, became the mere reception of a quality, and physiologists thought they could follow, from the point of reception to the nervous centres, the projection of the external world in the living body."[2] More recently, theoretical neuroscientist Romain Brette has also argued against this isolating approach. "We speak for example of the visual system as a set of anatomical structures from the eye to the visual cortex," he writes. "But the visual system defined in this way is not actually a system, if it is disconnected from the elements without which it cannot have any function."[3]

Today, we have plenty of evidence that the spontaneously organized background state of the brain has a substantial impact in shaping neural responses to specific extrinsic stimuli.[4] Even relatively subtle changes in a waking state, like walking rather than remaining still, dramatically alter cellular firing rates in the visual brain.[5]

The second view of the brain referred to above—the "intrinsic" view—appeared in the early twentieth century, initially inspired by

T. Graham Brown's studies of the spinal cord.[6] Brown discovered that when the spinal cord of an animal was completely severed from incoming signals, neurons continued to produce organized, rhythmic patterns of activity on their own, contrary to what might be expected from a reflexive form of system organization. He realized that the neural circuits in the spine had an intrinsic source of activation that could incorporate and use external inputs when they were available, but did not require them. The so-called *central pattern generators* that Brown discovered were *autorhythmic*. They moved to the beat of their own drum.

It took another half-century before an appreciation of this autorhythmic nature of nerve cells was extended from the spinal cord up into different regions of the brain, including neural systems believed to be critical in maintaining consciousness. These discoveries are important because they reveal that a substantial share of neural activity is entirely self-generated, making the brain very much *unlike* a standard computing machine that requires structured input coming from the outside to produce meaningful outputs. These observations require a substantial shift away from treating the brain as an organ being passively driven by extrinsic inputs and toward understanding it as a system that sustains its own rich, interior set of reference frames.

From the intrinsic vantage point, the brain is understood to be a self-organizing, self-referential organ whose operations are largely spontaneous and related to predicting, interpreting, and responding to the world. Self-organization means that brain activity is spontaneously ordered: its order emerges from the numerous interactions of its own component parts rather than being supplied by outside sources. And by self-referential, we refer to the curious observation that everything the brain can possibly know about can be known only through configuring (and reconfiguring) internal properties and states of the brain itself.[7]

Imagine, for example, a mysterious black box that generates very convincing and sophisticated simulations of the outside world—yet is situated in a world that remains off-limits to it. While some admittedly intricate cabling allows for responses to be emitted and for consequences of those responses to be received, nothing directly crosses the boundary between the box and the world. No lid can be lifted to give the workings inside the box unmediated access to how things *really are* out there. Everything the box can ever know of things outside its boundaries is its own interpretation of how its purely internal activity is changing. Now consider, as philosopher Andy Clark argues, that this is no fictional scenario.[8] Your brain is a real-world version of that black box. Brains have evolved to simulate and emulate the world, not to faithfully translate or have a direct window onto the objective world as it presumably *really* exists.[9]

In sum, viewing the brain as internally oriented is a far cry from seeing it as a standard computer transforming inputs to outputs. It demands that we can no longer treat the background brain activities constantly unfolding in the absence of external input as mere "noise" that muddies a proper study of responses to extrinsic inputs.[10] It means restoring the organism to its status as an intact, meaning-making agent that spontaneously explores its environment, energized by sets of expectations that were sculpted by prior interactions and that are continuously being updated.

Not surprisingly, an understanding of the brain as internally rather than externally oriented creates a substantially different context for discussing the nature of perception, feeling, cognition, behavior, and consciousness. A great deal of this book is about figuring out what this fuller picture of the internal (intrinsic, endogenous) brain looks like.

The Self-Organizing Brain

As we turn to the intriguing work of developing that picture, we must anticipate that our readers come to this book with varying levels of prior knowledge about neurobiological structure and function. We devote the rest of this chapter, therefore, to providing the baseline of knowledge required to understand the story we will go on to tell. This will necessarily be a brief primer, but sufficient even for readers who have made no prior forays into the literature of neuroscience. For those well-versed in the science, its highly selective coverage will also serve as helpful orientation, since it places strong emphasis on the brain's spontaneous, self-organizing, and endogenous activities—and asserts that the details of brain organization are best appreciated by a dynamical framework that honors these properties.

Brain Structure

Like every other organ of the body, the brain is a massive collection of microscopic cells. Brain cells are generally speaking of two types, *neurons* and *glia,* with many different subtypes of each. Glia are as numerous as neurons, but less famous than their counterparts. It was mistakenly believed, not very long ago, that the chief function of glia was simply to provide structural support to neurons. Scientists are now realizing the limitations of this old conception of glia, and recognition is growing that these cells are involved in higher-order functions like cognition.[11] Biologist Douglas Field has referred to glia as constituting an "other brain" that has been underexplored.[12] It is likely that there will be plenty of surprising findings concerning glial function in the coming years, but our story now turns to neurons.

The status of neurons as the basic building blocks of the nervous system dates to the late nineteenth century and the work of a charismatic artist and scientist, Santiago Ramón y Cajal. A true visionary genius, Cajal used his formidable powers of observation to deduce the cellular wiring diagram of the nervous system on the basis of a cell-staining method he perfected.[13] His efforts led to the formulation of what is known as the *neuron doctrine,* the idea that neurons are the fundamental structural and functional units of the nervous system.

Neurons are elaborate cells that can be subdivided into lower-level compartments. Among these, *dendrites* are treelike extensions that sample the extracellular environment and serve as the neurons' main sites for sensing their environment. These signals are integrated at the level of the main *cell body* of the neuron. Cell bodies cohere to form the brain's *gray matter.* Protruding from cell bodies are long, branching processes known as *axons* that serve as conduits for outflowing electrical impulses. Many axons are covered in insulating layers of fatty tissue, which speeds up the rate of neural signaling. This fatty wrapping is responsible for the term *white matter,* the moniker given to bundles of axonal fibers.

In short, the signaling flow within a neuron moves from the dendrites to the cell body, where signals are integrated. In some cases, a discrete electrical signal might then be relayed down the length of the axon. Sacs filled with chemicals known as *neurotransmitters* are released when the signal reaches the tip of the axon. Some of the most

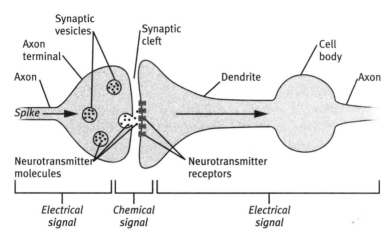

An electrical signal (spike) invading the axon terminal of a pre-synaptic neuron releases neurotransmitters into the synapse. The neurotransmitters bind receptor proteins on the post-synaptic neuron, thereby initiating an electrical response in the receiving neuron. © 2023 Joan M. K. Tycko.

well-known neurotransmitters are serotonin (critical for the mind-altering effects of psychedelics) and dopamine (necessary for energizing us to go out in search of rewards). Neurotransmitters diffuse across minute gaps between neurons known as *synapses,* influencing the dendrites of neighboring (postsynaptic) neurons by binding to sets of specialized proteins in those cells' membranes. At a synapse, the electrical signal sent by one neuron elicits an electrical response from the receiving neuron through the intermediary step of neurotransmitter binding.

Neurons tend to form connections with other neurons, and most of these connections are reciprocal.[14] So if cell A connects to cell B, there is likely a return connection from B to A. The complete physical-wiring blueprint, known as the organism's *connectome,* ends up setting the boundaries of what that great society of neurons can accomplish.

Having reviewed the basic cellular inventory list, we can now turn to the broad layout of the human nervous system. The *peripheral nervous system* is an intricate network of nerve cells connected to the body (not only to the body as seen from the outside, but also to its visceral, jelly-like interior, as sensed through interoception) and to receptors that capture energies from the external environment. The *central nervous system,* consisting of the spinal cord and the brain, is responsible for integrating all of these signals and using them to refine the sequencing of actions directed out into the world. The brain is intimately attuned to the body, as well as to the broader world within which the body moves. Its fundamental embodiment means that one's sense of being a self in the world is, quite fundamentally, anchored in bodily processes. The body is linked to the world in two main ways: first, by the skeletomuscular system that provides the means for directing actions into the world; and second, by the various environmental energies that stream through different sensory portals—hearing, vision, smell, and so on.

Motor action is our primary way of interacting with the world. This way of approaching the nervous system departs from the story that is usually presented in the traditional, external or outside-in perspective, which views the brain as a complex input-output device. That perspective tends toward depicting the body as a machine that merely executes commands issued by the brain, the master computer. To think of the body as a slave to the brain gets it the wrong way around. It is much closer to the truth to describe the brain and nervous system as a specialized system of nerve tissue that is there to serve the rest of the body.

From the intrinsic perspective that we endorse, creatures cannot help but ceaselessly generate self-initiated, spontaneous motor explorations that originate deep from their core and extend out into the world. The main role of sensory signals is to provide cues for "error correction," causing a creature to constrain and prune its endogenous nervous activity in ways that improve its fitness to thrive in its environment.[15] As it acts on the world, the sensory consequences of those actions feed back into its nervous system, creating the constant, ongoing loop of action and perception that is the basis of learning. Each perceptual act entails the operation of such a loop, and these loops collectively specify the "rhythms of our being." For example, in the case of vision, the eye is constantly making micro-movements.[16] When an image is completely stabilized on the eye, stimulus uncertainty and variability drop to a near vanishing point and there is no error correction that is required. The result is that we lose all perception of the stabilized image and it rapidly fades out of consciousness. A good way to personally experience this phenomenon is described in In-Sight 2.1.

In-Sight 2.1

In 1804 a Swiss physician named Ignaz Paul Vital Troxler realized that, if he kept his gaze fixed on a small portion of an image, the areas peripheral to that focal point would soon lose their intensity and fade from view. Only later did neuroscience explain that, as he forced his eyes to avoid scanning those areas, his brain quickly learned to discount them as unvarying stimuli and tuned them out. To experience the Troxler Effect for yourself, go to a version of it, with instructions, at https://mymodernmet.com/troxler-effect-optical -illusions. You may be surprised, as you stare fixedly at the center of the image, how quickly the palette of colors fades nearly to white.

It is not the motor system that is enslaved to the sensory system, as the classic view would have claimed. Instead, the motor system makes use of the sensory system to gain the input it needs for its next motor loop, as it relentlessly explores the environment in which the body finds itself. Originally, this view was held by a minority of neuroscientists, but it is a way of understanding the nervous system that is gaining traction. In deep evolutionary time, the components of neural tissue dedicated to motor and sensing functions were integrated.[17] Unlike plants, which in most cases can extract all they require while remaining firmly rooted in one place, animals can only satisfy their needs (for example, thirst, hunger, sexual desire) by moving about the world in which they are immersed, seeking out resources and arranging better circumstances for themselves. But *moving* (which could mean walking, running, swimming, flying, or some combination of those, depending on the body plan) is a feat that requires massive coordination. The entire value of having a dedicated nervous system was that it provided for intelligent guidance—with "intelligent," in this context, simply meaning guidance that helped the system's owner survive and produce offspring.

As the spinal cord enters the skull, it extends into the *brain stem*, a region along which all incoming and outgoing signals are trafficked. As a general rule, if we view the brain from the top of the head, the further down a structure lies, and the closer it is to the core or midline,

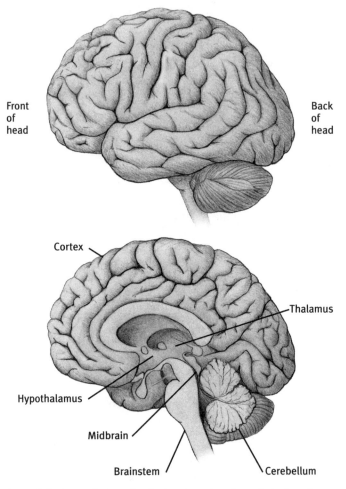

Front
of
head

Back
of
head

Cortex

Thalamus

Hypothalamus

Midbrain

Brainstem

Cerebellum

A side view of the intact human brain (top), and a view of the interior portion of the brain with one hemisphere removed (bottom). The bottom panel highlights a number of notable brain structures. © 2023 Joan M. K. Tycko.

the more evolutionarily ancient it is. The brain stem is the oldest part of the brain, and its integrity is absolutely critical to the maintenance of life. Damage to this area, only about three inches in length, is devastating, and often fatal. Groups of cells found in the brain stem regulate breathing, control of the heart, levels of alertness and awareness, and voluntary as well as involuntary behaviors.

A tiny cluster of cells in the brain stem can exert powerful effects on the arousal state of the entire brain, as happens every night when we become tired and our consciousness shifts from waking to a stage of light sleep, and then to progressively deeper stages. This complex of cells bears a very high level of internal organization and it goes by the name of the *ascending reticulothalamic activating system.* Projections from these cells ascend up throughout the entire mantle of the cortex, regulating the brain's background tone of arousal.[18]

Crowning the brain stem is the *midbrain,* an area responsible for triggering reflexive eye movements, among other functions. The midbrain is the first point in the nervous system in which there is an integrated, virtual map of the body, including its surface as well as the insides. This may mean that the midbrain provides the bottommost foundation for a unified sense of selfhood or agency.[19]

At the top of the midbrain, and no longer considered to be a part of the brain stem, is the *diencephalon.* The two main brain structures that are worth remembering here are the *thalamus* and, slightly below it, the *hypothalamus.* With the exception of our sense of smell, the thalamus shuttles all other sensory signals up to the cerebrum in a series of radiating, fanlike connections. Activity cascading along this superhighway of connections is critical in shaping attention and supporting consciousness.[20] The hypothalamus is involved in sensing and maintaining delicate control over the body's viscera, and in coordinating the emotional behavior of animals, including humans.

In close proximity to these diencephalic structures, and distributed throughout the middle, inner region of the brain are numerous neural sites, sometimes called *limbic,* that play a pivotal role in motivated behavior. As we will see in Chapter 5, these structures are some of the main drivers that give the "oomph" to one's emotional life.[21] The limbic zones, including structures like the hippocampus and the amygdala, are involved in highly sophisticated and integrative functions that defy any facile divisions between intentionality, feeling, memory, and reasoning. According to Walter J. Freeman, "the limbic system is the principal architect of action in space-time."[22] In other words, the wellsprings of intentionality and motivated, purposeful behavior are to be found in these deep brain nuclei, and they pattern the creature's most primordial explorations of the world.

The *cerebrum* wraps around the inner core of the brain and forms the most recent evolutionary layer. By far, the most visually impressive part of the cerebrum is the *cerebral cortex,* consisting of the left and right hemispheres. The cortex forms the wrinkled surface of the brain and is linked to some of the highest functions of the mind, including long-term planning, abstract thinking, and language. Each hemisphere is divided into a set of four lobes.

Generally speaking, the posterior half of the cortex serves as the site for receiving, analyzing, synthesizing, interpreting, and storing sensory information.[23] The occipital lobes are heavily involved in vision, the temporal lobes in hearing, and the parietal lobe in touch sensations. These sensory functions do not operate in isolation. As mentioned earlier, they steer and direct intelligent action. The frontal half of the cortex is largely implicated in planning, controlling, and initiating motor acts that are expressed out into the world. The frontal lobes integrate sensory inflows from the outside world and connections from evolutionarily ancient brain regions that monitor the internal states of the body and regulate overall levels of arousal. Regions along the midline of the cerebrum, such as the cingulate gyrus, are integrally involved in an embodied sense of selfhood.

The Cortex as an Imaginarium

Delving further into the structural layout of the cortex, we can appreciate how the brain becomes increasingly introverted—dedicated to its own workings—and less constrained by immediate circumstances. Instead of being a slave to the senses, or an impersonal recording device, the cortex is closer to a vast imaginarium—a site where ephemeral, fantastical creations are let loose and can coalesce in protean ways. In the jungle canopy of its numerous neuronal interconnections, the cortex gains its freedom from the sensorimotor contingencies of the present moment, and simulates a wide range of possible, as-if scenarios. Locked up in the human skull is the world's most convincing virtual-reality system.

We can consider the brain's complex connectivity patterns at the level of synaptic neuronal connections, and then further up, at the level of larger cortical regions. First, in terms of synaptic wiring, between 95 percent and 98 percent of white matter fibers connect the cortex to itself rather than to incoming projections from the outside world.[24] Consider the example of the visual brain. The responses of these neurons are mostly determined by connections with other parts of the cortex rather than by direct projections from the external environment.[25] In fact, the influence of input from the outside world is miniscule in comparison to the chatter originating in the cortex itself. These findings are startling and have long been a source of puzzlement. As early as 1906, that early neuroscientist Cajal, who had just been honored with a Nobel Prize, speculated in his acceptance speech about the role played by the overwhelming number of fibers reaching down from the topmost cortex into the optic relay cells of the thalamus. His student Rafael Lorente de Nó subsequently devoted his career to studying how their massive connectivity enabled feedback loops giving rise to closed, "self-reexciting" chains of neurons in the cortex sheet.[26]

Even at the level of the thalamus, which is a major relay station for receiving optic fibers originating from the eyeball, the descending connections that come down from the cortex are far more numerous than the ascending connections, by a factor of three to one.[27] To put it bluntly, most of what you see is really a story that your cortex is telling itself. As Chapter 7 will explore, when you dream, this internal story is *all* that you are left with, and yet it is enough to build stunningly convincing self-in-world simulations.

At the level of the cortex as a whole, a distinction worth mentioning is between its primary *projection* areas and its secondary *association* areas. The term *projection cortex* refers to those aspects of cortical real estate allocated to housing an orderly sensory (or motor) map—and in using the word *map* we are speaking quite literally. For example, in the case of vision, the optic nerve leaving the back of your eyeball projects in an orderly fashion onto neurons of your primary visual cortex. In other words, adjacent light receptors in the eye eventually stimulate neighboring neurons in the primary visual cortex to create an organized map of the visual world. The same is true for the primary auditory and somatosensory cortices, which handle hearing and touch, respectively. This orderly organizing principle is known as *topographic mapping*.

In contrast to the strips of projection cortex, *association cortices* integrate and pool activity from multiple sense modalities. This allows for far more nuanced and meaningful interactions with the world. For instance, to the primary visual cortex, an apple is little more than an agglomeration of points of contrasting light and dark spots with particular orientations. Neurons beyond the primary visual cortex are critically implicated in being able to identify the object as an apple, with a particular texture and color. Moving outside of the primary zones, as the flow of neural activity becomes more multisensory, you are able to recognize an apple not just from sight, but also by touching, smelling, tasting, or hearing someone bite into one. In fact, merely imagining a juicy apple can be sufficient to evoke the many sensory riches that apples afford.

A striking feature in the evolution of the cortex in mammals is the expansion of its association areas. In human brains, the projection areas are tiny outposts scattered across vast stretches of association territories, where brains can entertain fantastic, multisensory combinations of stimuli and abstract concepts that do not exist in the world. This shift in cortical organization reflects a long, evolutionary arc by which the brain's processes became progressively less tethered to immediate exigencies, less caught up in instinctive reactions to stimuli directly impinging on sense receptors.[28] No doubt the increased liberation from the particulars of immediate space and time was what allowed the enormous capabilities we now enjoy to be spawned in the human brain. Regions of the frontal cortex are the most unmoored from the outside world, freeing them up to entertain increasingly abstract and hypothetical scenarios, like planning a retirement projected to span decades into the future.

Brain Function

The structural wiring of the nervous system, called the *connectome,* forms the anatomical framework upon which electrochemical fireworks are ignited. Understanding how this architecture of the brain comes alive starts with delving into some elementary aspects of brain function.

As basic cellular units, neurons are electrically excitable widgets. Any given neuron's charge is constantly fluctuating, sometimes becoming more electrically negative and other times more positive with respect to the extracellular space. If a neuron becomes transiently more positive in charge, it may reach a certain threshold value at which it emits a *spike:* it fires as a single, all-or-none impulse that's somewhat like a binary digital code. Conversely, anything that makes a neuron more negative in charge will push it further away from initiating a spike.

A neuron can be excitatory—increasing the probability that neurons connected to it will spike—or it can be inhibitory, with the opposite effect. Whether a neuron fires a spike or fails to do so is the outcome of its pooling thousands of such inputs, analogous to the ballots cast in a political election. If a swell of excitatory "votes" swamps the inhibitory ones and the firing threshold is exceeded, the cell will spike; otherwise, it remains silent. A spike event can precipitate the expulsion of neurotransmitters into the synapse. Neurotransmitters then glom on to their corresponding binding sites on adjoining neurons, initiating a further sequence of signaling cascades.

A fascinating property of certain neurons is that they are not dependent on synaptic inputs to determine whether to spike. Evolution has endowed some neurons with membrane proteins that confer on them a substantial degree of autonomy. As a result, such neurons can emit spikes or maintain ongoing fluctuations of electrical activity in the complete absence of incoming signals.[29] Brown's above-mentioned discovery of the motion-producing central pattern generators in the spinal cord, more than a century ago, provides an example. Other neurons with similar properties were later discovered in regions that control breathing, swallowing, and some other functions. Today we know that very many neurons in the cortex can sustain rich patterns of neural output even if they are isolated from all input. When single cells with autonomous rhythmic fluctuations are wired up with others, quite complex patterns result.

In isolation, neurons are notoriously unreliable parts and they often spectacularly fail to get their message across to their neighbors.[30] Fortunately, your brain has about eighty-six billion of them and redundancy makes up for the mediocrity of individual units. When neurons coalesce into larger groups known as *neuronal ensembles* they can function more efficiently by synchronizing their respective activities, forming what has been called a "brainweb," similar to patterns of information exchange on the Internet.[31] There is strength in numbers, and the joint action of many hundred neurons is able to accomplish feats that an isolated neuron simply could not pull off. For this reason, some scientists have argued for revising the standard neuron doctrine.[32]

Although the neuron is clearly the basic structural feature of the brain, the neuronal ensembles are the more salient unit to focus on if you are trying to understand the development of brain functions. In the language of complexity science, these functions are *emergent properties* of the brain—properties exhibited only at the combined level, as the extremely simple actions of the many "agents" in a system add up to a larger pattern none of them individually planned. (The patterns created by swarms of bees and flocks of starlings are classic examples of emergent properties.) From a strictly biological perspective, the brain, compared to other organs, seems uniquely suited to support emergent functions, owing to its complex connectional patterns.

Much like human societies, brain cells are also organized into societal arrangements at multiple levels. Streets, neighborhoods, and boroughs like Manhattan all have their analogs, up to the entire states that are the different lobes of the cortex. An orderly sequence of arrangements is evident in the cortex: one to two hundred neurons are bundled into a *minicolumn,* and about fifty such minicolumns get aggregated into a larger *macrocolumn,* many of which in turn are nestled within a *local cortical area network.* The intertwining of different spatial scales also entails a highly complex admixture of neural activities unfolding over distinct time scales, so that space and time become inextricably linked in the brain as they are also in the universe at large.[33] For this reason, we will often refer to the brain's combined *spatiotemporal* activities.

Neuroscientists have found ways to spy on the activity of neurons at all of these different levels of organization. Extremely thin microelectrodes can pick up the firing rates of single cells, whereas multiple-unit recordings with larger electrode tips are sensitive to the combined action of many neurons. Still larger electrodes can be placed on the scalp surface and used to pick up the fluctuations of electrical activity that are strong enough to be conducted through the skull, reflecting the combined action of thousands to millions of cells. This is the technique of electroencephalography, which can be used to capture neural processes down to thousandths of seconds. Other machines, dedicated to functional magnetic resonance imaging (fMRI), can be used to record much slower changes in the flow of oxygenated blood within the brain.

Across this extensive monitoring of neural activity, the presence of *oscillations* has been detected at every level, from single cells to the entire brain.[34] Oscillations shape your perceptions, thoughts, feelings, and your sense of who you are. In fact, oscillations are not only replete throughout your brain and body, but also outside of you and are crucial to linking your inner and outer worlds. An oscillation can be defined as any rhythmic or cyclical fluctuation between the states of a variable. That may sound abstract to begin with, but consider some concrete examples. The day-night cycles corresponding to the Earth's rotation about its axis, the shifts in consciousness between waking and sleeping, and the beating of a heart are all instances of oscillatory phenomena. Neuronal oscillations are rhythmic fluctuations in excitability over time.

Frequency and power are two important aspects of an oscillation. *Frequency* refers to the speed of an oscillation: the amount of time consumed by one whole cycle of a fluctuating process. Going back to our earlier examples, the Earth takes twenty-four hours to complete a full rotation around its axis. A heart can beat once every second. A single neuron, with its complex assortment of membrane proteins gating the traffic of electrically charged particles into itself, is capable of generating a wide range of different frequencies. Since the electrical activity of neurons fluctuates very rapidly, frequency is measured in the unit of *Hertz (Hz)*, the number of cycles occurring per second. The second descriptor, *power*, is the amount of energy or strength contained in an oscillation. Over time, power waxes and wanes as the system under consideration becomes either more or less active. During the waxing phase, the height of an oscillation becomes large, and in the waning phase it shrinks.

Of most interest to us are the oscillations that emerge in populations of neurons. Oscillations abound whenever excitatory and inhibitory neurons get wired up together. The spectrum of frequencies covered by rhythmic brain activity is quite large, ranging from the infraslow (less than 0.1 Hz) to the ultrafast (200 Hz and higher). The slow and fast frequencies interact with and modulate each other in numerous ways.

An intuitive way to think about brain dynamics is by picturing a vast ocean of brain activity. Earth's oceans naturally sustain a rich repertoire of waves. Some waves are incredibly large, spanning huge distances and reaching enormous heights. These large waves gain momentum and move on a very slow time scale, but they can exert profound effects. Other waves appear as successively smaller ripples on the water's surface. The waves lapping at the shore are locally occurring, rapidly oscillating water molecules.

In a similar way, neural oscillations are related to the spatial distances over which they integrate activity. Slow oscillations reflect the recruitment of large networks of neurons, while fast frequencies tend to be confined to smaller neuronal patches.[35] Much as, in the ocean, tinier waves appear as crests atop more massive waves, faster frequencies in the brain are regulated by the ongoing up-and-down phases of slow oscillations. For example, high-frequency oscillations (around 40 Hz) in a very local region of the brain alternate in power, becoming very large when a global and much slower wave (around 4 Hz) happens to be tracing an upward arc. Then, as the slow frequency begins its dip into the trough of an ongoing cycle, 40 Hz power diminishes.[36] Nature has discovered ingenious ways to braid together complex spatiotemporal webs of brain activity spanning very different magnitudes. And lest the ocean metaphor seem too contrived or fanciful, the evidence demonstrates that it is precisely through wavelike dynamics that distinct brain regions communicate in one of their primary modes. These microscopic waves of neuronal activity, when they meet up with waves emanating from distinct populations, can either constructively combine to create larger wave patterns or suppress other surges of activity in a process called *destructive wave interference*.[37]

Just as single neurons are never completely quiescent, the brain as a whole also emits a continuous electrical hum that is a mixture of many different oscillatory frequencies. For a long time, brain scientists tended to dismiss this spontaneous humming as a kind of "brain noise" with limited functional significance, since it characterizes the brain at rest, when it is not responding to any specific processing demands. The brain noise, then, was believed to be somewhat like the "snow," or white noise, picked up by an old analog television set when its antenna is not receiving any broadcaster's transmission signal.

Scientists were instead focused on probing the brain's responses to different types of task inputs—performing experiments where, for example, they delivered a sensory stimulus across hundreds of repetitions and recorded the resultant ripples in the churn of the neural ocean.

This methodology was motivated by the observation that averaging across numerous trials of task-evoked activity minimizes the contribution of supposedly random background noise. The belief that this was the proper approach rested on the view of the brain as a complex input-output information processing device, corresponding to the reflex-based framework described above. Whatever role spontaneous, self-generated neural activity might have played, it was left largely out of the theories spawned by this reigning research paradigm. In sharp contrast to this near-total neglect, the past two decades have witnessed a major shift in focus on the meaning carried by the brain's internally generated background chatter.[38]

A number of important lessons can now be shared about the supposed brain noise.[39] The first is that, when it comes to energy expenditure, the brain spends about twenty times more resources on maintaining its ongoing monologue than on responding to specific task-based demands. It is far from the case that this spontaneous churn of neural activities is swamped by the introduction of outside stimuli. To the contrary, phenomena like the conscious perception of the objects around you consume a meager 5 percent of your neural energy budget, constituting a mere blip on the radar.

Most of the glucose and oxygen guzzled by neurons (which, by the way, absorb a whopping 20 percent of the body's overall resources) goes to supporting their ongoing communication cascades. Neuroscientist Marcus Raichle has dubbed this activity the brain's "dark energy," by analogy to the mysterious stuff believed to comprise most of the mass-energy content of the universe.[40] Whatever role is played by that silent majority of neuronal activity, ceaselessly operating in the background, it is the bulk of what brains actually do. In a prescient statement, the neurologist and psychiatrist Kurt Goldstein anticipated this discovery in 1934. "The system is never at rest, but in a continual state of excitation," he wrote. "The nervous system has often been considered as an organ at rest, in which excitation arises only as a response to stimuli. . . . It was not recognized that events that follow a definite stimulus are only an expression of a change of excitation in the nervous system, that they represent only a special pattern of the excitation process."[41]

The second lesson we can share about brain noise is that it is not truly random at all. There is, in fact, meaningful spatiotemporal structure in spontaneous brain activity. At rest, and without anything in particular to do, the brain self-organizes into a highly coherent set of areas that share neural chatter with one another within specific frequency bandwidths called *resting-state networks*.[42] These networks have been extensively investigated using fMRI technology, starting with the discovery of the *default-mode network,* composed of a number of cortical and deeper structures running along the middle of the brain. This set of brain areas actually becomes *more* metabolically and neurally active when the brain is quietly resting and disengaged from external demands.

Later, researchers found that the default-mode network was far from unique: there turn out to be about a dozen other resting-state networks, and even sensory areas of the brain that directly receive stimulation from the outside world exhibit spontaneous patterns of coupled low-frequency oscillations (less than 0.1 Hz). The structural connectome constrains but does not entirely determine the precise pattern of synchronous activity. In a sense, the patterns of broadly distributed and synchronized network activity constitute a more subtle dimension, invisible to the naked eye, that is coinherent with physical anatomy. Using electroencephalography—and a related method known as magnetoencephalography that detects magnetic instead of electrical brain activity—scientists have corroborated the widespread prevalence of resting-state networks fluctuating across a range of faster oscillatory frequencies. These two types of diagnostic equipment, despite their sensitivities to very different orders of spatiotemporal brain activity, produce a convergent narrative about the intrinsic structure of the background neural hum.

The final lesson important to pass along here about brain noise is that this massive internal chatter actually directly shapes what we become aware of when there are signals coming into the nervous system from the world. For anyone hanging on to the notion of the brain as a giant machine being passively fed information from the outside in, this fact is probably the most damaging to that model. A standard machine, performing according to spec, can be expected to provide consistent outputs when repeatedly presented with unvarying inputs. This is not the case with the brain, which exhibits a substantial amount of trial-to-trial variability in neural responding, even when probed with the exact same stimulus. A large share of this variance from one presentation to another is explained by the underlying flux of autono-mously generated activity.[43]

There is strong evidence from both human and nonhuman animals that varying amounts and strengths of spontaneous neural signaling have a clear and direct effects on how one responds to external stimuli, and whether one detects or misses incoming data.[44] In technical terms, there is a *rest-stimulus interaction* that causes us to become aware only of those externally occurring signals that mesh with or get entangled in the brain's self-created webs of spatiotemporal activity. Incidentally, this process also works in reverse (as a *stimulus-rest interaction*), so that recent events, after passing through, leave their unique traces to continue reverberating or echoing through spontaneous patterns of neural activity. Developmental change, over the course of an entire life-time, is a process of imprinting records of events and shaping the fabric of the brain's spatiotemporal repertoires. All of our experiences leave imprints, some subtle, others less so, on the web of brain-body activities.

By way of grand summary, the brain is wired up and spends an over-whelming majority of its energy talking to itself. Encased in the skull's bony shell, it continuously spins stories to itself, choosing to incorpo-rate outside influences when it deems them expedient, but otherwise carrying on a ceaseless neural monologue. Drawing out the implica-tions of this strange set of affairs will be the central theme of the remainder of this book. As we move to Chapter 3, we shift our atten-tion to the deep, subjective core of experience, which so far has been completely neglected by the objective, third-person approach to the study of brains.

3

Looking Directly at the Mind

The brain exhibits an intricate, hierarchical organization from top to bottom. Single cells gather into larger and larger aggregations and eventually into complex regions and networks. For convenience, scientists sometimes describe these as forming at three levels of scale: the micro, the meso, and the macro. In fact, these categories are the subtle artifact of the limited instruments used to observe and measure the brain. The brain is really an integrated organ that constantly embroils all these levels, from micro to macro, in its uninterrupted and holistic functioning. Think of droplets of moisture self-organizing into a hurricane vortex, the dynamics of which are simultaneously shaping the behavior of individual droplets. In the same manner, single neurons influence the shape of the aggregate, population-level activity that in turn commands their individual firings.

While the brain is a physical organ whose structure and function can be measured using objective (third-person) neuroscience tools, as discussed in Chapter 2, there is also an interior dimension of lived experience (a first-person perspective), the depth of which is unknowable to a third-party observer. This dimension cannot be gauged with devices, but it is no less real for that. At the risk of sounding glib, let us just say that everyone has a mind and not just a brain.

Chapter 2 adopted an outside-in perspective on the brain, but here we take an inside-out approach, exploring and dissecting the experiential (phenomenal) dimensions of the mind. How do we investigate these? There are a number of methods, some of which are relatively new. As will become clear, however, the contemporary science of consciousness stands to benefit enormously from techniques that are very old, developed across millennia by contemplative practitioners dedicated to observing the coarse- and fine-grained workings of the human mind.

Space and Time: The Common Currency of Neuro-Mental Events

As outlined in Chapter 1, phenomenal experience is not *caused by* brain operations, in the way that one billiard ball hitting another causes the second to move. Neural activity and phenomenal experience are rather two manifestations of the same underlying psycho-physical reality that reflects a dynamical *process* that continuously unfolds in space and time.

As psychiatrist and philosopher Georg Northoff explains, space and time form a bridge between the neural and mental, the objective and subjective.[1] Chapter 2 described how the brain spontaneously "talks itself" into existence by weaving webs of spatial activity (from single neurons into mini- and macro-columns, into local cortical areas, and into entire cerebral hemispheres) and combining rhythms of temporal activity (from infraslow neuronal oscillations to ultrafast ones), using the anatomical connectome as a scaffold. Since space and time are fundamentally entangled, we refer to *spatiotemporal configurations*.

A given spatiotemporal configuration is only a specific instance drawn from many possible configurations that are ever-changing in a dynamic flow. William James famously coined the phrase "stream of consciousness" to describe the mind's flow of shifting perceptions, feelings, and thoughts.[2] Some of the oldest Buddhist texts from the Pali canon likewise teach that consciousness is a confluence of two streams—one material, the other mental—in continual flux.[3]

Space and time, then, provide the mechanism that accounts for the correspondence between the neuronal and phenomenal levels. Scientists, among them Selen Atasoy at Oxford University, are now doing work to show that underlying these spatiotemporal processes are *harmonic waves*.[4] Such waves, which are ubiquitous in the natural world, are a central component in scientific theories of heat, light, sound, electromagnetism, fluid mechanics, and gravity. In the neurobiological context, harmonic waves emerge from the excitatory and inhibitory cellular dynamics playing out on top of the anatomical backbone. These harmonic modes constitute basic neural building blocks. Different qualitative states of consciousness, such as waking and sleep, correspond to the engagement of different combinations of harmonic waves, which influence the spatial and temporal distribution of brain network activity.

On one end of the consciousness spectrum, there is a relatively impoverished repertoire of spatiotemporal activity in the brain. From the objective, third-person perspective, this is evident in patterns of neural activity that are low on a quantity known as *entropy*. In mathematical terms, entropy is a metric that captures the level of uncertainty exhibited by a system. Neural activity with low measured entropy is highly predictable, regular, and low in complexity, therefore yielding little information. These restricted spatiotemporal activities are experienced as states of consciousness with very minimal content, as is presumably the case during the depths of dreamless sleep or anesthetic sedation.

By contrast, in alert wakefulness, spatiotemporal activities are diverse. They are reflected in increasing neural entropy values and richer, more numerous, contents of conscious experience. As the brain's spatiotemporal webs expand and reverberate throughout the cortex, especially when they reach the frontal lobes of the brain, an individual is not only conscious, but capable of verbally reporting on their experience. This represents a sharp contrast to sedation or dreamless sleep, when patterns of neural activity shrink into local islands, without reaching an ignition threshold necessary to sustain a brainwide symphony of activation. Periods of dreaming, for example, evidence more complex patterns of brain network engagement, with activity patterns higher in entropy, than periods of dreamless sleep.

Non-ordinary states of consciousness, such as an acute episode of psychosis or a high-dose psychedelic state, are associated with enriched and diverse spatiotemporal activities relative to ordinary waking experiences. A psychedelic like LSD increases entropy in brain signals during a spontaneous resting state, over and above ordinary wakefulness. This heightened irregularity (or information content) and complexity of neural activity patterns is evident as a multiplicity of alterations in perceptual, cognitive, and emotional functions. Robin Carhart-Harris, a leading psychedelic researcher, has synthesized a wealth of these findings into a comprehensive entropic brain theory.[5] His theory stipulates a direct link between the quality of any conscious state and the amount of entropy assessed using different measures of brain function. Relative to ordinary waking experiences, the non-ordinary states induced by psychedelics are marked by enhanced entropy patterns.

The tools and techniques of today's neuroscience are brimming with fresh possibilities for bringing the outside-in (brain) and inside-out (mind) perspectives into alignment. In other words, mathematical measures like entropy can help to bridge the link between the quantitative and qualitative aspects of neurophysiology and this bridge is reflected in space-time. Delivering on this promise, so necessary for a full account of the brain and mind, depends on being able to measure phenomenological experiences in ways as sophisticated and detailed as those used to measure the activities of the brain itself.

The Neurophenomenology Project

Bridging the gap between measures of neural activity and conscious experiences is the aim and agenda for a research frontier known as neurophenomenology. In Chapter 2, we took stock of the basic makeup of the brain and nervous system. Here, we turn to the subjective, phenomenal dimension.

To start with, a distinction must be drawn between the *levels* and the *contents* of consciousness. The *level* or *state* of consciousness is a continuum ranging from unconsciousness or minimal consciousness on the one end all the way to alert wakefulness on the other. Christof Koch, a pioneering consciousness researcher, describes it as encompassing "conscious experiences in their entirety, irrespective of their specific contents."[6] His reference to the *contents* of consciousness, by contrast, points to what we are conscious of, including our perceptions (sight, sound, taste, touch, smell, and bodily awareness), memories, feelings, thoughts, hallucinations, and sense of selfhood. In the Tibetan lineage, all of these mental arisings are known as *namthogs*.[7] One way to think of the mind is as a highly active *namthog* machine. In the context of contemporary science and philosophy, individual instances of subjective, conscious experience are known as *qualia*. Each quale is very distinct: for example, the quale of smelling fresh coffee is quite distinct from the quale of seeing the design of the coffee cup.

At a higher level, the variety of contents within consciousness can be bundled into aggregates corresponding to a world "out there" and a self to whom the world appears.[8] This self seems to be "inside" looking out onto the world, but it can also experience a variety of more subtle mental contents and phenomena—a topic to be considered in depth in Chapter 11. The body is in a unique position: it belongs both to the world outside, as when you see yourself in a mirror, while it also has an internal aspect, as when you get butterflies in your stomach.

To understand the mind, it is also vital to distinguish between forms of consciousness.[9] The first of these, alternately called *primary, phenomenal,* or *anoetic,* corresponds to raw, immediate, subjective experience that is not subsequently reflected on or elaborated in language. Many nonhuman animals, and certainly all mammals, are capable of having perceptual experiences of the external world that are imbued with various feeling tones, and this form of consciousness allows them to adjust their behaviors to survive and avoid danger.

By contrast, *secondary, extended,* or *noetic* consciousness moves beyond the here and now and is associated with possessing the ability to report on having a particular experience. This more mature and more complex form of consciousness largely depends on language and the ability to form higher-order mental representations. It culminates in being able to become aware that *I am aware,* and that *you are aware that I am aware.*

Contemporary science is making inroads in linking levels/state, contents, and the form of consciousness (primary or secondary) to separable neurophysiological mechanisms. These distinctions are still fairly coarse, however, and they produce only fairly abstract maps of consciousness. What is needed to go beyond this are techniques for conducting more fine-grained measurements of conscious experience that are ever changing and always in flux and can parallel the instruments for the objective quantification of neural activity.

In the 1990s, Francisco Varela, Evan Thompson, and some others initiated what they called their "neurophenomenological research program" to try to produce these sorts of measurements.[10] To Varela, the justification for such a program was twofold. First, lived experience cannot be reduced to a purely physical, neuronal explanation; in any study of consciousness, the phenomenology itself is the primordial datum and must not be discarded. Second, by leveraging disciplined methods, the rigorous first-person study of conscious states can uncover stable dimensions of experience that are universally shared among different observers. Such techniques offer "a clear methodological ground leading to a communal validation and shared knowledge," Varella writes.[11] They allow a truly trans-subjective database to be assembled for consciousness research. As Varela notes, this requires that one "attain a level of mastery in phenomenological examination."[12] Just as no aspiring sommelier can become competent by only hearing about wine-tasting, mastery in this field extends well beyond fielding questionnaires, which can provide only limited or, as some have called them, *thin* descriptions of experience.[13]

Varela and his colleagues' research program was placed firmly in the tradition of phenomenology, a movement that Edmund Husserl inaugurated in continental Europe at the beginning of the twentieth century. At the heart of this tradition were techniques for the systematic exploration of the felt immediacy of lived experience. Varela saw this tradition as in sync both with the attention that William James paid to consciousness, during the founding period of psychology as a scientific discipline in America, and with much older spiritual traditions. Most consistently, Varela drew from Asian contemplative traditions.

The practice of *epoché* is key to systematic phenomenological investigation.[14] In this method, one approaches the flow of consciousness *exactly as it appears in the moment* while suspending (or bracketing) extraneous assumptions. Only a description of the bare givenness of a conscious act is left—its immediacy—while interpretations of the experience, or the ways in which that experience is subsequently reflected and refracted through multiple conceptual lenses, are bracketed. For example, seeing a chair, one would attempt to bracket away the concept of a chair and instead attend to the immediate sensory experience of the chair, including the texture of its wood, its color, and so on. Detailed descriptions, more than explanations, of experience are what phenomenology is looking for.

The philosopher Natalie Depraz worked with Varela to analyze the phenomenological epoché as a set of subtle, interior shifts in attention unfolding organically across three interconnected and mutually reinforcing phases.[15] First, one must suspend habitual judgments, prejudices, and ways of interacting with the world. Our default tendency is to be immediately "glued" or pulled into all of our conscious contents. Philosophers refer to this as the *natural* or *unexamined attitude* that must be broken to enable an inward turn of attention, away from its passive capture by external stimuli. Sometimes a startling occurrence can accomplish this disruption spontaneously, by introducing a shock to one's routine, but contemplative techniques teach practitioners to accomplish the break voluntarily.

The second phase, which grows out of this, is a "reversal of attention" away from passive absorption into the alluring spectacle of the world to the raw act of becoming aware, the experiencing itself. This is a fairly subtle shift, but it can also be expressed as switching from our usual near-total focus on the "what" to the "how" of any given moment of experience.

For example, in hearing a sound outside of the window, rather than instantly rushing to categorize the experience ("it is my neighbor's dog barking again"), one would become attuned to all those mental acts that bring an experience up to the surface of our awareness. This could include the initial, inchoate registration of the raw sound itself, its gradual localization (for example, "it is coming from behind me, and to my right"), and nuanced flickers of emotional feeling that accompany the background of one's experience, such as frustration that one's attention is being distracted by the barking.

The final phase involves entering into a state of quiet receptiveness—a "letting go." In this phase, one adopts a diffuse mode of attention that allows for experiences to spontaneously complete themselves, rather than attempting to interfere and redirect their flow. Clearly, all phases of the phenomenological epoché run counter to our habitual ways of engaging with the world and our experience of it. For this reason, learning the skill requires extensive (and patient) training of observers.

A barrier that has impeded detailed neurophenomenological studies is the difficulty of acquiring the skills needed to provide high-quality descriptive reports. One solution to this problem is to rely on subjects who already have expertise in phenomenological examination—

namely, advanced meditators. Short of this, however, even minimal training can equip subjects to describe their inner experiences, often with the same precision they would achieve in describing external objects. Different methodological tools can help to shepherd observers toward this level of precision. One such tool is experience sampling. Here, an external cue such as a beeper prompts a person to describe events and processes in their internal experience at *that very instant in time.* Other techniques, like the experiential method that Donald Price and James Barrell developed in the 1970s and 1980s, rely on interviews designed to unpack the details of experiential acts.[16]

Claire Petitmengin developed another, more refined technique called the *micro-phenomenological interview,* which has been likened to a "psychological microscope."[17] This method can be used—usually within a short time—to produce descriptions of experience at levels of granularity approximating the richness of data gathered from modern brain imaging. The interview uses a stepwise set of questions to uncover the bare structures of lived experience. It takes effort to tease out these structures and bring them to full awareness. When the subject digresses from the here-and-now particulars of experience, the interview is redirected to present-moment awareness through gentle prompts. It deepens descriptions of experience along a temporal dimension and the nontemporal "stuff" of experience. For example, here is a subject describing the earliest stages of a thought erupting into conscious awareness: "Not so much an image, but a felt sense that something arises. Like a little movement . . . a perturbation. It's not a thought yet. It's just a kind of a stirring. Something is about to happen."[18]

This grain of detail reveals that experiences do not come as a bolt from the blue, but typically unfold through an organic series of developmental stages, from diffuse and inchoate to more differentiated and specific. The interview method can illuminate the germination of consciousness as a kind of mental version of time-lapse photography, showing how phenomenal acts evolve into their more mature expressions.

On most subjects, we know far more than we are capable of verbalizing. Psychologist Eugene Gendlin coined the term "felt sense" to refer to a holistic experience prior to its maturation and differentiation at later phases of organization.[19] The felt sense is direct, immediate, and nonconceptual.[20] In addition to this temporal dimension, the interview method also fleshes out spatial and intensity qualities of an experience and might include prompts to characterize a conscious experience as "fuzzy," or "clear," for example. Mental imagery might be two- or three-dimensional, vivid or dull, and a sound may localize in various locations. A thought might be experienced as happening in the head, a feeling might be in the gut, the perception of an evocative piece of music might be experienced as occurring neither within the body nor in external space but in some in-between dimension.

Contemplative Approaches

The most sophisticated and expensive brain imaging tool would remain useless in the hands of an untrained user. When it comes to phenomenological measurements, by contrast, there is often a widespread and naive assumption that obtaining descriptions of conscious experience is trivial. This is only true if the quality of the data gathered is disregarded.

Thanks to the existence of various meditative disciplines and techniques, across various contemplative schools, there are tried-and-true approaches to a close-up study of the mind. The Dalai Lama, for example, depicts Buddhism as an empirical tool for the study of consciousness:

> [A] disciplined use of introspection would be most suited to probe the psychological and phenomenological aspects of our cognitive and emotional states.
>
> What occurs during meditative contemplation in a tradition such as Buddhism and what occurs during introspection in the ordinary sense are two quite different things. In the context of Buddhism, introspection is employed with careful attention to the dangers of extreme subjectivism—such as fantasies and delusions—and with the cultivation of a disciplined state of mind. . . . [A]n untrained mind will have no ability to apply the introspective focus on a chosen object and will fail to recognize when processes of the mind show themselves.[21]

The need for highly disciplined methods of mental observation is made poignant by the fact that here the measurement "instrument" (the mind) is being used to study itself. In other words, there is no separation between the object of knowledge and the instrument used to obtain the basic datum of consciousness, as there is when an fMRI machine obtains images of the brain. First-person perspectives therefore face very real challenges that are less pronounced with the use of third-person technologies.

Different Buddhist schools have spent millennia collecting rich descriptions of human phenomenology, often with detail and precision that outstrips what is currently available in the contemporary scientific literature.[22] There are at least two ways to engage with these traditions. One is simply to draw from the deep treasury of well-documented mental cartographies that they already possess and that have been validated across practitioners within a specific lineage. These observations can then be used to generate and test novel hypotheses, confirm or disconfirm existing scientific theories, or simply compare against detailed knowledge of functional brain activity. Another way is to recruit contemplative masters within those traditions as practiced and seasoned research participants who can provide first-person reports that exceed the precision and detail available from naive study participants.

Barriers of Entry to First-Person Explorations of Consciousness Identification

A person's unconscious immersion in what the phenomenologists call the natural attitude is probably the first and strongest obstacle to any principled, first-person investigation of consciousness.[23] The natural attitude involves an unthinking acceptance of things appearing in consciousness as disclosing an objectively existing reality. The unreflective fusion with the contents of experience is known as identification. The clinical psychologist, John Welwood, offered the following definition: "Identification is like a glue by which consciousness attaches itself to contents of consciousness—thoughts, feelings, images, beliefs, memories—and assumes with each of them, 'that is me,' or, 'that represents me.'"[24] Think of a rapt audience at a cinema, jaws agape. Each audience member has strongly identified with the unfolding movie narrative to the point of becoming incensed by the actions of fictional characters. The example is apt, as we are often riveted to an even more enchanting inner cinema that we call life.

Any systematic investigation of the stream of consciousness demands, at the least, some ability to stand apart from the contents passing through consciousness, freely witnessing them without becoming absorbed. Think of an example where you sit down with the intention of quiet meditation. No sooner do you settle into a comfortable position then the thought occurs to you that you forgot to send an important email. Identification with the thought means that you then lose sight of your original intention to meditate, and you get up from your meditation cushion to send the email. Simply noticing the passing thought about the email would be an example of being able to stand apart and observe the thought without getting carried away by it.

When we engage in identification, we get stuck to the surface of consciousness, passively pulled this way and that by the rushing flow of thoughts, feelings, and perceptions. It then becomes difficult to make high-quality observations of the nuanced dynamics of conscious processes. By definition, reflection upon experience requires some Archimedean point—an interior lookout from which the experiencing itself can be perceived and studied. With strong identification, there is no such point. The observer is completely mired in the natural attitude. For this reason, the phenomenologists advocated for the practice of epoché. By the time one gets to the experiences that occur on the murkier sides of the consciousness spectrum—the deep regions of psychological space dominated by dreams, hallucinations, and boundless feelings of selflessness—there is a sore need for the most elite class of psychonauts. It is there, especially, that the contemplatives are able to furnish consciousness studies with high-quality observations.

The Buddhist tradition frequently uses a mirror metaphor to convey the concept of identification. Being identified means a profound level of involvement with reflections appearing in the mirror, while ignoring that they are in fact just that: passing reflections. Our life, most of the time, however, is lived *in* the reflections, rather than *as* the clear mirror. Living in the reflections translates to seamless and continuous fusion with the medley of contents that fill waking and dreaming experiences.

In the ever-changing flow of our mental life, a spatiotemporal configuration (or a dynamic sequence of configurations) corresponding to sadness appears, and we quickly identify with it by proclaiming "I am sad" or "I am depressed," rather than realizing it as a transient phenomenon. Identification, then, goes hand in hand with a process of *reification*—imposing a certain fixedness and rigidity onto the flow of mental life, attempting to treat temporary aggregates in consciousness (for example, a temporary feeling of sadness) as independently existing. Sadness then hardens into a sort of mental solid that feels ponderous and overtakes us.

To appreciate this, it might help to imagine aggregates of thoughts and emotions forming like clumps in the freely flowing conscious stream. Once this clumping (reification) occurs, then identification, the process of entanglement, follows so instantly, and so habitually, that we do not even notice it until afterward, if at all. A major portion of psychological suffering originates from identification. Buddhism says that even moments of ecstasy and intense joy can end in suffering if we have identified with and become attached to them, since those moments, too, are only so many soap bubbles bound to burst.[25]

Disturbances of Attention

The instability of attention is a second major barrier to systematic exploration of inner states. We could think of attention as being like an adjustable scope that allows for zooming in or out on subjective phenomena as well as maintaining a stable focus of view.[26] Being able to break with identification opens up the possibility of zooming in and out on experience. But without substantial effort and training, the "mind" scope constantly darts back and forth. Scientific studies indicate that, without realizing it, we lose contact with present reality and spend at least half the time we are awake getting lost in replaying scenarios from the past and running through possible future expectations, all the while running a commentary on the entire show.

To extend the metaphor, our scope may be not only unstable but also have a grimy lens. And without further training, various distortive lenses may obscure the view of mental phenomena that are strongly associated with emotional habits, wishes, desires, thoughts, biases, or compelling sensory perceptions. For a test of your own skill at observing the mindstream, try exploring In-Sight 3.1.

In-Sight 3.1

Assume a sitting position, with your spine straight, your hands in your lap or palms down on your knees, and your feet flat on the ground. You can do this exercise wherever you are, though ideally you will want to be in a semiprivate place where no one will try to disturb you for the next ten minutes. This will be a timed exercise, so set an alarm for ten minutes.

Now take a moment or two to come into the present moment. Inhabit your body, and really allow yourself to feel that you are sitting in your chair, with your feet on the ground. Feel the gravity that pulls you to the earth.

Now, with eyes closed, try to focus your attention on your sense of hearing. For the duration of this timed exercise, do your best to be aware of the sounds in your immediate environment. You can begin with the most obvious ones—perhaps the hum of passing cars, voices, laughter, bird calls, or whatever is most audible. Just be present with the sound, really being aware that you are hearing fluctuating sound waves as they arise, pass away, and evolve into other sounds. Your goal is to stay present with the undulating energies of sound. Slowly, now allow your awareness to encompass other sounds, including the more subtle, background ones that escaped your attention before. You might allow the sound of your own breathing to enter the same space of awareness that is holding the more salient sounds from the environment.

After a few moments, open your awareness out even further, and try to merge with the most subtle sounds that you can hear. Maybe you can start to hear your own heartbeat or pulse. Or you might notice that there is a pervasive silence in the background, and that the cacophony of various things you hear is distinguished only by virtue of their rising from and returning to that great stillness, like cresting waves.

Whatever happens, do your best to stay entirely with the sounds that are present. Allow your sense of hearing to register the changing play of energies that appear to it. Do not think about the sounds or concern yourself with labeling or classifying them. If you hear a dog bark or a wailing siren, just surf the sound waves, rather than thinking, "I wonder where that fire truck is going?" or getting lost in your feelings about the sounds ("How annoying, I wish the neighbor's dog would stop distracting me already.") If you find yourself drifting off into a daydream, thinking, or feeling lost or dull, gently remember what you are doing and come back to the naked sensation of hearing. Continue with the ceaseless arising of sound for the next ten minutes.

Most contemplative paths understand two major forms of attention disturbance—excitement and dullness—that interfere with making high-quality phenomenological observations. As with learning to interrupt spontaneously occurring identification, attention training also requires dedicated, prolonged practice.

Excitement

Reflect on your experience of engaging with In-Sight 3.1. Unless you happen to have extraordinary powers of focus, the sounds you perceived might have seemed something like pollen grains suspended in a turbulent liquid, buffeted about, this way and that—or like the sparks of an exploding firework tracing many paths of collision. With sound energy, no sooner is naked contact with it established than it ricochets off along a stray sequence of associations, thoughts, and fantasies. Or maybe you got distracted by a persistent itch—or in any of the other ways one gets knocked off course by the continuous agitations impinging on the mindstream.

Dullness

The other major impediment is turbidity, dullness, and inertia in the mindstream. If you tried doing In-Sight 3.1 after a heavy meal, it might have been a challenge merely to stay awake. In states of dullness, an initial focus on an object of awareness rapidly loses its sheen and clarity and ultimately sinks into torpidity and drowsiness. Even at milder levels of dullness, as the focus dims, you might have found yourself sinking into a half-daze, ensnared in a shimmering web of interpretations and ideas *about* the sounds you were hearing, or *about* the practice you were doing, or *about* how long before it was over or some other daydream. Like a muscle grown flabby from lack of exercise, for most of us, our attentive focus falls far short of what an inner science demands. In the absence of a consistent, contemplative training routine, over the course of a day we usually oscillate between uptight, distracted agitation and insensate dullness.

An Example of a Contemplative Curriculum

Next, we will consider an example of a contemplative training path that seeks to ameliorate the difficulties just outlined, through investigative methods for looking directly at the mind. We are now going to speak of a specific mental technique known as *meditation*. A difficulty that besets any discussion of meditation pertains to the imprecision of that word.

Meditation refers to an extremely diverse range of contemplative practices, drawn from traditions that vary tremendously in terms of their historical development, training instructions, ontological and metaphysical claims, and background presuppositions. In some cases, definite correspondences exist between separate traditions. For example, at a very general (and perhaps superficial) level, mindfulness techniques from Indo-Tibetan traditions approximate the practice of watchfulness (*nepsis*) described in the Christian East.[27] In a similar way, certain forms of receptive meditation and yogic practices described in the Far East find their echoes in the Orthodox Christian practice of prayerful stillness (hesychasm).[28] Yet these presumed similarities also risk masking far-reaching differences, causing one to overlook distinctive aspects of each respective contemplative path and to ignore its unique contributions.

A sufficiently thorough, comparative study of contemplative practices across religious traditions is well beyond our scope and expertise to undertake here. Even within a single tradition, there are numerous and critical differences in meditative techniques. As one example, within the Tibetan Buddhist tradition, meditation can involve complex psychosomatic practices heavily rooted in Indo-Tibetan medical models, to highly complex visualization practices, to formless meditation styles with no specific instructions.

To minimize vagueness, the discussion that follows is anchored around the Tibetan Buddhist path, drawing particularly on the Mahamudra and Dzogchen traditions.[29] Moreover, given the heavy scholastic bent in Buddhism and its proclivity to systematize its phenomenological observations, it is a particularly apt model for our purpose of demonstrating certain basic tenets.

The Necessity of Purification Practices

At the outset, we mention an aspect of training that is ignored or deemed irrelevant for a secular science of consciousness—namely, ethical purification and altruistic intention on the part of the person making inner observations. This is emphatically *not* the case in the great spiritual traditions, where ethical purification is considered a nonnegotiable component of the entire endeavor, and indeed the very base of it.

Ethical discipline (in Tibetan, *tsultrim*), as a preparatory stage for meditative practice, has both a negative and a positive dimension.[30] The "negative" aspect means to refrain from engaging in harmful patterns of body (for example, killing, stealing, sexual misconduct), speech (for example, lying, slander, harsh, idle talk), and mind (for example, wishing harm to others, coveting, pride). The positive dimension entails cultivating virtues such as patience, kindness, and generosity.

To habitually engage in unethical conditions reinforces those same harmful patterns and smudges and dirties the lens through which the mind experiences its self and the world. Allowing more or less free rein to gluttony and coveting, for example, lends strong momentum to these tendencies, introducing turbulence in the stream of consciousness. In turn, this consolidates our degree of identification and strongly shapes phenomenological experience. Being able to gain distance and a more objective platform from which to observe the mental flux is difficult indeed. Doing so is particularly challenging if one comes to these practices after a long day of lying, gossiping, and hurting others, as the toxic effects of these patterns of body, speech, and mind continue to reverberate.

An altruistic foundation lies at the core of most contemplative paths. Advanced meditators accumulate tens of thousands of hours of meditation practice, often in partial (or more intense) seclusion from others. Accordingly, the practitioner's intention must rise to the challenge and be as magnanimous as possible. As an example of the aspirations one cultivates and reinforces in oneself on setting out on this path, consider this one: "May all sentient beings have happiness and its causes." Or this: "May all sentient beings be free from suffering and its causes." Or this: "May I achieve enlightenment for the benefit of all sentient beings, that I may bring them all to enlightenment."[31]

This wish to bring *all* sentient beings to enlightenment—to nurture them and protect them from coming to harm—is known as *bodhichitta*. Seemingly counterintuitive, *Bodhicitta* is the tendency to treat others as equal, and even to put them as ahead of and superior to, oneself. One example might be to mentally rehearse exchanging one's own joys and good fortunes with the pain and suffering of other beings. By constantly practicing and reinforcing *bodhichitta,* the practitioner very gradually begins to erode the deeply ingrained self-centeredness that causes painful fixations in the mindstream. The meditation practices discussed below are understood as having a cosmic significance rather than merely being an entertaining pastime or something done to satisfy an idle intellectual curiosity.

Training the Stability of Attention

The first stages of training usually involve learning how to stabilize the focus attention and make the mind more pliable. This is achieved through calm abiding (*shamatha*) meditation. The same word also denotes the end goal of this stage of training—*shamatha* as a state of achievement.[32]

By adopting a prescribed set of bodily postures (for example, straight spine, chin slightly lowered, body neither too uptight nor too relaxed), the practitioner is helped to avoid both agitated excitement and excessive dullness. Quiet environments, with minimal distractions are strongly recommended, which in traditional contexts meant a forest retreat, with reliable access to basic amenities.

The practitioner learns to use different kinds of objects to anchor attention. The mind gradually accustoms itself (the Tibetan word for meditation, *gom,* literally translates as "getting used to") to rest on its chosen object of support with gentle equipoise, neither grasping on to it too tightly nor with too much laxity.[33] Just as a guitar string is of little value when stretched too taut, and when relaxed too much, one strives for a perfectly tuned instrument. Practitioners receive detailed instructions about how to combat the negative effects of distraction and drowsiness.

Two mental factors are brought to bear in this style of meditation. The first factor consists of mindfulness and the maintenance of a stable focus of attention that is gentle and nonjudgmental. One is careful not to slip into a trancelike state or get caught up in thoughts or feelings about the object of attention, but simply to remain with the experience.

The second factor, alertness, is a form of quality inspection, which repeatedly turns around to examine the mindstream and ensure no deviation from the intended goal. If the mind begins to wander, then alertness brings it back. Mindfulness and alertness are like muscle groups that receive repeated exercise over many hundreds of hours of meditation. Mindfulness weakens the proclivity to identify with what arises in consciousness, while alertness enhances the vividness of the object on which attention rests.

As Thrangu Rinpoche puts it, "mindfulness and alertness make us strong, attentive, and well-disciplined so we cannot be robbed by disturbing emotions and discursive thought."[34] Initially, one practices in short (minutes-long) sessions, before extending the length of *shamatha* meditation. With increasing familiarity, the meditator is able to simultaneously maintain optimal stability (balanced between excitation and dullness) and clarity (crispness, vividness) of mental experience. The lens used to inspect the mind has become something like a magnifying glass, able to gather and focus the luminous rays of awareness and focus them like a laser.

Usually, one begins with using external sense objects as supports for awareness. For example, one might start by placing attention on a physical stimulus such as a pebble, a letter, or a stone disc. Sounds, textures, tastes, smells, really any sensory phenomenon, can likewise be an appropriate object. With continued practice, the support for awareness starts to withdraw from the external world, to become more and more interior. At the juncture between the external and internal, a traditional support object is the breath, with attention placed on the rise and fall of inhalation-exhalation cycles.

Moving further into the interior realm, a more difficult support for awareness consists of mental visualizations which are held firm in the mind's eye. Eventually, at more advanced stages, one moves to *shamatha* meditation without a fixed reference point. Instead of using external (physical) or internal (mental) phenomena as a crutch for attention, the practitioner becomes aware of awareness itself. Daydreaming about the past or future is explicitly avoided, as is any analysis of the present moment: the mind simply settles into a state of relaxed alertness.[35]

The purpose of *shamatha* practice is to maintain one-pointed concentration (think back to our metaphor of the magnifying glass gathering the luminosity of awareness to a fine point) of increasingly longer duration, without becoming easily thrown off balance by incoming distractions or sinking into a daze. B. Allan Wallace compares it rather to a kind of telescope equivalent for investigating the depths of the mind.[36] As concentration stabilizes, it becomes so clear and hyperfocused that the sense of a perceiving subject and a perceived object disappear into a single (nondual) experience. This is a degree of mental equilibrium, a taste of something like a perpetually moving stillness, that most of us neither have nor will ever experience. As Ngapkha Chogyam and Khandro Denchen note in *Roaring Silence*, perhaps the nearest approximation for a nonexpert meditator is to spend several hours in total sensory deprivation while remaining fully awake.

At this point, we can think of *shamatha* less as a practice and more as a state that is achieved. In traditional terms, *shamatha* is technically "achieved" when a practitioner is able to maintain unwavering one-pointed concentration on an object of awareness for a *minimum* of four continuous hours—a feat so astounding that most scientists would dismiss it as even being a possibility.[37] While the successful achievement of *shamatha* is associated with a high degree of well-being and numerous purported psychosomatic benefits, it is decidedly not the ultimate goal of the Tibetan Buddhist tradition. Rather *shamatha* is a preliminary step for being able to move beyond it to more fundamental realizations.

What Lies Beyond

Shamatha provides the requisite stability to serve as a launching pad for what follows next: contemplative insight, *vipashyana*, meditations consisting of deeply penetrative and thorough investigations into the basic constituents of lived reality.[38] Relative to *shamatha*, which is focused on training concentration through unswerving placement of attention, *vipashyana* is a more analytical form of meditation. Here, the practitioner begins to systematically probe and investigate the fundamental nature of the mind. At the beginning of this chapter, we indicated that the stuff of consciousness can be organized into aggregates corresponding to, first, a world "out there" and, second, a self to whom the world appears and which seems to be "inside" looking out onto the world. *Vipashyana* is a way of thoroughly investigating both of these aggregates to discover their ultimate truths and then allowing the resulting realizations to become assimilated into every moment of existence. Although *vipashyana* meditation is considered analytical, this refers to a direct, nonconceptual way of coming to knowledge concerning the phenomena being investigated and not a discursive form of argument.

We present here only the barest outline of *vipashyana* meditation. After first establishing oneself in the equipoise of *shamatha*, one might begin *vipashyana* practice by looking directly at the mind itself. The meditator poses penetrating questions concerning this supposed mind, such as *Where is the mind? Is it in the body? Is it outside of the body?* The meditator may then conduct a systematic search for the place where this elusive, seemingly independent thing called mind resides, while scanning throughout the entire body. Additional questions may be posed such as *Does my mind have a shape? A color? Does it have a location? Where do thoughts arise from and where do they disappear?*

It is vitally important that answers to these questions are not sought in previously stored encyclopedic knowledge, to be formulated as glib sentences. Rather, the meditator sits with the naked experience itself, shorn of its prior accumulated conceptual overlays. In Chapter 1, we called this knowing-through-identity. This is precisely why the stability and clarity offered by the prior achievement of *shamatha* is such an important prerequisite. At some point, according to Buddhist tradition, the meditator comes to have a powerful experience of the basic no-thing-ness of phenomena like the mind. Mind itself might come to be directly experienced as a kind of empty, expansive space—one that is inherently cognizant and sparkling with a luminous quality of fresh awareness.

In later stages of *vipashyana* practice, these deep inquiries are directed at the seemingly solid external world out there. Eventually, the meditator has a definitive realization that both of these poles, the subject and object, perceiver and perceived, are inseparable: both are part and parcel of the same fluidic medium of a more primordial awareness that is understood to be our "natural state," the true home of all phenomena in existence.[39]

One could say that the mechanism by which *vipashyana* works is through a kind of "undoing" process, which gradually relaxes, at ever deepening levels, the ceaseless clutching and grasping tendencies of our mind (the cycles of reification and identification) that lead to the apparent clumping within the mindstream. The meditator begins to learn undoing with *shamatha* meditation, in releasing distractions, for example. With practice, this process becomes second nature and gradually penetrates to more subtle aspects of experience. At this point there is a more direct encounter with reality itself, stripped from layers of subjectivity that tend to distort and cloud more "ordinary" ways of experiencing. The authenticity of the realization is traditionally checked through a kind of examination, performed by a more experienced meditative master, who might pose a series of questions to the adept, or simply observe how the adept breathes, moves, and reacts to changing circumstances.

Principled contemplative paths are not the unique prerogative of Hindu, Buddhist or any other tradition. Practices for stabilizing attention exist across most of the world's great religions and even under secular guises. In each case, we encounter a long and venerable lineage of masters who have preserved a living means of transmitting their knowledge down through the millennia.

We can now begin connecting the neurophysiological and phenomenological domains to the full spectrum of different states. The story will begin with exploring the most familiar example of waking consciousness, in which different sense modalities are integrated into a coherent, virtual reality–like simulation of the outside world that allows for constructing an evolutionary useful model of a separate self in the objective world. As we will see, the experience of mundane reality is not at all what it seems to be on the surface. Rather, this seemingly straightforward experience consists of internal neuronal configurations being construed as an immersive, "out of brain" world providing an arena within which an inner self acts.

4

The 500-Million-Year-Old
Virtual-Reality Simulator

Imagine looking out across a field of poppies on a cloudless summer day. From a naive perspective, the variegated shades of red and green, the sky's blueness, are simply all properties of an objective world, a straightforward disclosure of how things really are. It is a world that stands revealed, without any seeming effort on your part. The textures you see, the swaths of color, just as much as the gentle breeze that touches your skin, all of these sensory experiences seem to be transparent portals that disclose a single, unitary reality that is *really* there. *I'm somewhere "in here," and the world is "out there."* This belief is so foundational, so intuitive and implicit, that one would scarcely pause to question it. This belief has been with you from infancy and has served you very well. It even ensures your survival such that you do not end up as road splatter while crossing a busy street.

Yet, as will become clearer below, all of these perceptual qualities are deeply subjective phenomena, viewed and brought to life in a private phenomenal theater. "Not only are dreams experiences," writes the philosopher Antti Revonsuo, "but, in a way, all experiences are dreams."[1]

Looking out across the field, the shades of red are a kind of hallucination that gets painted onto the poppies. The grass, the plants, the field, all of these solids that seem so weighty and packed with matter, are rather the very specific congealing of radiant light forms that give specific shape to an elusive something we call consciousness. And the distant horizon does not truly reveal the edge of the physical world, but only the far limits of your phenomenal bubble—the bubble in which your life happens.

Every experience you will ever have will be poured, one might say, into the expanse of this sphere. But how does this bubble work? This chapter looks at the usual way in which it becomes organized, the typical architecture that it adopts. The characteristic phenomenological features of the bubble emerge spontaneously through the process of self-organization introduced in Chapter 2.

We will look into the evolutionary record to argue that the most convincing virtual-reality system ever known—the one that can never be surpassed, no matter how many "metaverses" Silicon Valley develops—is actually made of flesh and bone. And finally, we consider what not only neuroscientists and psychologists but generations upon generations of contemplative masters have gleaned from millennia of plumbing deep through the phenomenal bubble.

Before beginning the journey, you might wish to pause here and try a few ways to explore these topics from within the confines of your very own experiential bubble, responding to the invitation of In-Sight 4.1.

In-Sight 4.1

This demonstration is best done in an outdoor setting, at a spot that affords a relatively clear view about twenty feet in front of you.

Pause to gaze out, and take in the scene before you. You might be seeing natural scenery filled with trees, or a mountain range, or perhaps an urban setting populated with buildings and moving vehicles. Get a feeling for the texture of what is in front of you. Does it seem quite solid, definite, made up of physical matter? Now, extend an arm well in front of you and turn the palm of that hand toward your eyes. Focus your attention entirely on your palm. Keeping your eyes fixed on your hand, try to get a visual impression of how the appearance of the background has changed. After several moments, shift your focus to the scene beyond your hand. Shift your focus from foreground to background, from your hand to the environment, a few more times. Does physical reality alter depending on where you put your attention? How does your subjective experience change if you entertain the possibility that your experience is made entirely out of light forms instead of physical solids? Does it become lighter or heavier?

Here is another demonstration you can try at around dusk. Pause to locate the moon when it is first rising on the horizon. Does it look particularly large? Extend your arm fully and compare the size of the moon with the size of the fingernail on your index finger. The moon and your fingernail should appear to be roughly the same size. After a few hours, when the moon is approximately right above you, compare its size against your fingernail again. Has it shrunk? Does the moon change its size across its orbit? If your visual experience of the moon can change so much with context, what does this reveal about other experiences you have that seem so direct and immediate?

The Phenomenal Bubble Has a Shape and a Structure

The visual neuroscientist Steven Lehar has taken the idea of a bubble as more than a metaphor, indeed as corresponding quite closely to the very *shape* of our visual phenomenology. "The first and most obvious observation," he writes, "is that consciousness appears as a volumetric spatial void, containing colored objects and surfaces. This reveals that the representation in the brain takes the form of an explicit volumetric spatial model of external reality. The world we see around us therefore is not the real world itself, but merely a miniature virtual-reality replica of that world in an internal representation."[2]

You can verify for yourself that your experiences of the world around you have the shape of such a bubble, a "volumetric spatial void." Just look down a long, straight stretch of highway, its two sides of travel divided by a grassy median. Glancing into the near distance, you can readily confirm that the left and right side of the roads run strictly parallel to each other. Yet, if you cast your gaze out to the far horizon, you will also notice that these seemingly straight and parallel sides of the highway converge to a very definite point, approximately at eye level. The same point convergence happens as you study the road behind you, making it appear that the road is shaped like the rind of a melon slice.

As Lehar points out, this warping of perceived space is just what you would expect to find if you needed to model the world within the confines of a finite spatial volume.[3] This 3D spherical volume, which has an intrinsic dimension of spatiality, provides the fundamental phenomenal coordinate system within which your experiences of the world take place.

The phenomenal coordinate system is like a grid that gets filled in by specific contents, through the orderly, point-by-point topographic mapping process briefly reviewed in Chapter 2. This native coordinate space is not just used by your visual sense. All of your senses have a dimension of spatiality and are referenced with respect to a phenomenological coordinate system.[4] The sounds in a coffee house, someone calling your name, the sound of tires screeching—all of these are referenced to their respective spatial domains, some nearer and others further away. A tap on your shoulder happens in a very special part of the phenomenal coordinate system corresponding to your physical body. And the savoring of a delicious meal likewise seems to occur within some generally circumscribed portion of the phenomenal grid. The phenomenal grid is where your sense modalities cohere, the shared canvas on which they come to life. It is what unifies each moment of conscious awareness. Thoughts and feelings are a bit harder to pin down and localize spatially, although not always. Yet they too can only take place within an overall phenomenal bubble.

There are a few more general observations we can make. The first is that the entire phenomenal grid seems to be centered around a specific point of reference. If you close one eye right now, you will notice the blurry side of your nose. You have just found your center point, the first-person perspective. Your consciousness, at least most of the time, has this pronounced *perspectival* aspect.[5] The multimodal perceptual reality that you embody is organized around this perspectival point as a center of gravity.

Second, as we saw earlier, there is also a phenomenal horizon in addition to the center point (or at least a center nimbus). From your first-person perspective, the phenomenal horizon corresponds to those portions of the phenomenal bubble that are at maximum distance from

the center, from you. Things can look a bit unusual at the phenomenal horizon, as seen in earlier examples. The road is of course not
really a melon rind, and the skies are not *really* a dome. These examples simply reveal that our experiences occupy a spherical, bubble-
shaped volumetric expanse.

Our conscious experiences are shaped by where we focus our attention. When we pay attention to something, it becomes the center of
our experience, and other parts recede into the surroundings. As explained in Chapter 3, many contemplative techniques, at least initially,
are designed to allow us to gain some flexibility, voluntary control, and
stability of this focal spot of attention, so that our consciousness is
not simply buffeted about by constantly changing circumstances.
Contents falling outside of this spotlight tend to be experienced in a
way that is more unfocused, indistinct, blurred (as you might have noticed from doing the first demonstration in In-Sight 4.1). Thoughts
and feelings, which tend to be more amorphous and poorly localized
within the phenomenal coordinate system, might be particularly likely
to occupy the periphery. As will be discussed in Chapter 5, feelings
seem to stick to definite things in our environments. For example,
desire can become attached to the very specific sight of a chocolate
cake, so that the cake draws us toward itself like a magnet. In other
cases, a vague feeling can lurk in the background, subtly coloring everything we experience without ever quite materializing in the focus of
consciousness.

The combination of sensations, feelings, and thoughts, with the general layout described here, becomes organized as a totality (a *Gestalt*)
that encapsulates the sense of being a self-in-the-world. References to
the self-in-the-world model will recur throughout this book, as we
come to appreciate how it is organized across many layers of subtlety.[6]
This model typically becomes organized automatically, outside of voluntary control, with very little personal insight into its nature. As will
become clear, the world's various contemplative traditions inform us
that this self-in-world model is just that—a model—and that there are
in fact numerous other ways in which consciousness can become
organized. We will delve more into that later, but first, let us consider
just how faithful (or not) this self-in-the-world model is to an objectively existing reality.

Naive Realism Challenged

Put bluntly, naive realism is the philosophical stance that says the things you experience really *are* those things as they appear on the surface. We can go a step further and draw the metaphysical implication that, when you perceive a table in front of you, what you are perceiving *really* is the table as it fundamentally is, made out of dense, physical matter that you can touch and investigate for yourself. For many of us, this is simply the implicit, often unspoken, intuitive position we take for granted.

And yet, when we put this belief to the test, it quickly becomes obvious that, at least in its extreme version, it fails to withstand the weight of the evidence. Immanuel Kant, the Enlightenment philosopher, long ago claimed that knowledge of things-in-themselves (which he called *noumena*), or the world as it *really* is, remains forever inaccessible to us as limited creatures.[7] Modern physics since Einstein informs us that space has four dimensions, and some string theorists suggest that the world we see around us may actually consist of eleven.[8] Clearly that is not how things appear to us (at least not in our ordinary states of consciousness). All we can know, Kant said, is our experiences of the world, the realm of phenomena. Kant made a distinction between what we might call second-order and first-order reality, the former being the phenomena that appear to us, and the latter the mysterious noumena that are (according to Kant) out of bounds for us mere mortals.[9]

If you accept that the external world around you is actually the experience of an internal "volumetric spatial void," then you will naturally wonder about how accurately it mirrors the real world—you know, the one that is *really* there. One response, which arguably has enjoyed widespread endorsement within the neurosciences, is to say that the brain is in-*formed* by the outside world in a relatively straightforward way. The brain encodes and forms representations of the world in its role as a master computer. This view admits that your experiences are not directly those of Kant's noumena, but that there is nonetheless a fairly close mapping between noumena and phenomena. Even accepting that, under certain conditions, our visual system can get fooled by an odd optical illusion, one might still endorse a perspective that, most of the time, our sense organs are high-fidelity sensing devices, like cameras and microphones, through which the outside world streams in.

And yet, in this book, we depart even from this reassuring perspective. Instead, we will review evidence that the self-in-world "light show" you are inhabiting bears numerous similarities to the process of dreaming a reality into existence. The world outside your body cooperates with your dreaming, sometimes closely and sometimes not, depending on what segment of the consciousness spectrum you are in. The entire experience of being a self-in-world originates from a process of hypothesis generation, creating a continuously refined and updated model of how things are.[10] From an evolutionary standpoint, the model need not be absolutely true; it only has to be good enough to allow the organism to survive and pass on its genes.[11]

With this understanding in mind, the world outside is not passively impressed onto the brain through recording devices. Rather, the world reaches in to constrain and refine the internally generated models.[12] The self-in-world model is indeed meant to interact with the actual state of things, the mysterious noumena. Yet a self-in-world model can sustain itself even when isolated from sensory signals impinging from the external world. You can, and do, continue to be a self-in-world even when you are asleep and dreaming (as Chapter 7 will show in detail) and when you are awake but in a state of sensory deprivation (as explored in Chapter 8).

Maybe this sounds strange if it is your first time hearing it, but in truth it is not an unusual or even particularly original interpretation. It is simply what the latest neuroscientific evidence suggests—and what is arguably, by this point, the consensus opinion of scientists. As Lehar states, "the phenomena of dreams and hallucinations clearly demonstrate the capacity of the brain to construct complete virtual worlds even in the absence of sensory input. Perception is therefore somewhat like a guided hallucination, based on sensory stimulation."[13] J. Allan Hobson and Karl Friston write that "the brain is genetically equipped with a model that generates a virtual reality during sleep and is entrained by sensory input during waking."[14] And philosophers Thomas Metzinger and Jennifer Windt concur that "what we call waking life is a form of online dreaming."[15]

Our experiences are constructed on the fly to a far greater degree than most of us ever appreciate. In fact, cognitive scientist Donald

Hoffman has run computational models simulating artificial "living" agents in a completely virtual world to demonstrate that evolution does not favor highly accurate sensory systems, so long as they are good enough to survive and procreate.[16] His simulations indicate that the agents—which placed a premium on accurate representations of their virtual worlds, as close as possible to how things really were—actually went extinct after a few generations.

Why? A completely veridical representation of reality may threaten to overwhelm an organism with information that is not necessarily useful for enacting adaptive behaviors. What ultimately matters is that the agents contain internal models that succeed in surviving and passing on their progeny, regardless of how faithfully they capture reality as it is. A sensory apparatus that experiences the outside world even with wild departures from reality (say, seeing all objects around you as if through a pair of distorting goggles) could still fill the bill, so long as it did not increase one's chances of being killed off before procreating. The model of the world needs only to be a viable interface for shaping adaptive behaviors—not a particularly accurate version of ultimate reality, but a heavily filtered and edited version of it.

A concrete example recommends itself, which helps to illustrate that an organism in a sense cocreates the reality that it experiences—a claim that is in keeping with many of the great spiritual traditions. The landmark paper "What the Frog's Eye Tells the Frog's Brain," published back in 1959 by Jerome Lettvin and his colleagues, has much to teach us on this score.[17] Lettvin and his team delivered a range of different visual stimuli to the eye of an immobilized frog. The stimuli were shown inside of an aluminum hemisphere. Small objects were attached to magnets, so that they could be positioned and moved at various places within the hemisphere to present different types of stimulation of the frog's visual pathways. A tiny recording electrode was inserted into the optic nerve of the frog to capture the electrical signals being conducted from the eye to the brain.

After bombarding the frog's eye with a bevy of different stimulus combinations and types, Lettvin and his team discovered that the frog's eye was responsive to only four different kinds of visual stimulation. In other words, the frog's internal model of reality carved up the complexity of the outside world into just these four limited categories. One category corresponded to the detection of sharp edges and changes in contrast required to create a rough outline of the surrounding environment. Another effective stimulus was the sudden dimming of light, as would happen when a larger predator closed in. The optic nerve was also sensitive to moving edges in its visual field, which would sufficiently capture the presence of movements around the head. Finally, arguably the most interesting was a set of nerve fibers that reacted optimally to small, curved dark objects moving intermittently with respect to their background. This, you might have already noticed, is how frogs in the wild are able to detect flies buzzing around them and earn their meal. What is particularly notable about this study is that it suggested that even the eye itself, which one might expect to serve most as a cameralike device, is engaged in a process of narrowly filtering and interpreting the contours of the frog's lived reality.

What should we conclude from this evidence? Much of the time, we are actively involved in interpreting and constructing our vision, even without realizing it. Our sense organs and associated perceptual apparatus—not to mention our past experiences, our desires and fears, our thoughts, attitudes and beliefs—shape, twist, and contort the phenomenal model of reality we inhabit. Another example of this concerns what are called *filling-in* events. Filling-in describes a perceptual phenomenon in which any number of visual features—such as color, texture, or brightness—of a surrounding area are perceived to exist in a certain portion of the visual field where they are not actually physically present at all.[18]

Filling-in is commonly observed among patients who have sustained damage to their visual pathways. Such patients can have focal scotomas, meaning that selected regions of their visual space are empty. The existence of these blind spots can be readily demonstrated on objective visual tests. These patients often do not consciously experience the existence of these blind spots, however, because they use background cues to fill in the gaps. Looking, for example, at a wallpaper with a floral design, a patient might experience not a blurred or empty spot but an unbroken continuation of the floral pattern, actively using the surrounding design to compensate for the missing pieces.

Filling-in is not an exotic phenomenon demonstrated only by patients with damaged visual systems. This very moment you have a fairly large natural blind spot roughly in the middle of your visual field. The natural blind spot is formed by the optic disc, the part of the retina where the optic nerve exits the back of your eyeball to get routed into the brain. Fortunately, you do not go around with a noticeable black hole in the middle of your visual field. Instead, you perceive an interpolated best guess of what's in the blind spot, which forms the content of your experience. The end result is to preserve the apparent seamlessness of visual consciousness. To explore this phenomenon in more detail, you can try out In-Sight 4.2.

In-Sight 4.2

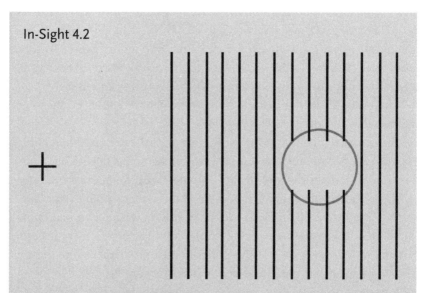

An illustration of texture filling-in at the optical blindspot. Marianne Maertens, "The Neural Representation of Illusory Contours," detail of figure 4.2, Open Archives Initiative.

With your left eye closed, firmly fix your right eye onto the plus sign on the left. Do not move your right eye; keep it focused on this plus sign. Now, slowly adjust the distance between your eye and the image (you will probably need to experiment with slowly bringing the image closer to the eye). Note that at some point the gray circle on the right disappears from your conscious awareness and the bar gratings that were within that circle become continuous, eliminating the gaps that were evident earlier when you looked at the image on the right. This is the point at which the circle is wholly in your natural blind spot, which is being filled in by an extrapolation of a surrounding pattern that is visible to you.

There is yet another example of filling-in, where the filling-in happens in time rather than in space. Think about your visual experience when you are crossing a street. Your visual awareness is that of a smooth and continuous world. As you look at other people strolling in a public park, you perceive everyone around you as engaged in smooth and fairly predictable motions rather than as a series of static snapshots. Yet the best evidence, drawn from studies of visual illusions, perceptual testing, and recordings of brain electrophysiology, indicates that the adult human brain periodically samples the visual world at a rate of approximately seven to thirteen times a second. Many areas in your brain are spontaneously taking samples in discrete cycles that oscillate roughly at this alpha-bandwidth frequency.[19] Rather than having a conscious perception of flickering in your visual field, your brain seems to smear out its modeling efforts in time, to create the convincing illusion of unbroken continuity. In rare cases, brain damage or certain mind-altering substances can cause this filling-in effort to fail, resulting in conscious recognition of the discrete nature of visual experience. A drug like LSD, for example, can result in a phenomenon known as visual trailing, in which a moving object results in a series of discrete, stationary scenes.[20]

Nature's Magic Trick

I'm somewhere "in here," and the world is "out there." We started off by claiming that this is one of our most foundational beliefs. It is also one of the most impressive of all of nature's tricks. The phenomenal bubble does not feel like it's located inside your skull. The stars up in the sky, the other cars on the road, the person you are having a conversation with—none of these seem to be woven into your body, or conjured inside your skull. They are very convincingly *out there,* whatever that phrase might mean.

This clever trick is known as *referral* or, to use Sir Charles Sherrington's term, *projicience.*[21] Projicience accounts for your mental experience of a visual object as being extended in external space rather than experiencing it as a topographically mapped object residing on the back of the retina, in your visual cortex, or somewhere in between. Instead, our experience of the world is pushed out, projected outside the confines of the brain. One philosopher-neuroscientist coined the term *out-of-brain* to refer to this phenomenon whereby sensory experiences, while presumably occurring deep within the brain, seem to take place in an external space.[22] This is true across a range of distances. When you are holding a fork and knife, and manipulating them to eat your food, you attribute the tactile sensations of the utensils to the fingers that are handling them, not to your brain's somatosensory cortex. When you experience a sharp pang of pain from a fall on the ice, the pain is referred to the injured joint. In the moment of orgasm, the intensely pleasurable sensations seem to be happening in . . . well, *not* in your brain.

Under specific circumstances, you can more directly see the dynamics of your own brain as seemingly happening in the outside world. Strange to think, perhaps, but engaging with In-Sight 4.3 is a fun way to see the inside of your own brain.

In-Sight 4.3

As you look at the image below, fix your eyes firmly on the dot to the left. Note that when the pinwheel is relegated in this way to your peripheral vision, its center seems to flicker.

An illusory flicker whose frequency matches the visual brain's oscillations is perceived at the center of the pinwheel. Rodika Sokoliuk and Rufin VanRullen, "The Flickering Wheel Illusion: When α Rhythms Make a Static Wheel Flicker," *Journal of Neuroscience* 33, no. 33 (2013): 13498–13504, figure 1.

This flicker is of course illusory but at the same time you are seeing something real. As demonstrated by Rodika Sokoliuk and Rufin Van-Rullen, the speed of the flicker is a fairly close match to the speed of alpha-band oscillations in your visual brain (which, in adults, reach a peak somewhere between 7 to 13 Hz).[23] These oscillations are believed to correspond to the rate at which you are visually sampling the world. We do not normally perceive our brain's internal oscillations, but in this case we can consciously experience them as occurring in the center hub of the pinwheel.

The topic of projicience raises something of an existential conundrum. Raise a finger to tap your temple and ask yourself: Does the visible world, which looks so convincing to me, exist outside or inside my skull? Is one's brain in the world, or the world in one's brain? Now consider this mind-twister posed by Steven Lehar. If you agree, he writes, with "the statement that everything you perceive is in some sense inside your head," then:

This could only mean that the head we have come to know as our own is not our true physical head but is merely a miniature perceptual copy of our head inside a perceptual copy of the world, all of which is completely contained within our true physical skull. Stated from the internal phenomenal perspective, out beyond the farthest things you can perceive in all directions (i.e., above the dome of the sky and below the earth under your feet, or beyond the walls, floor, and ceiling of the room you perceive around you), beyond those perceived surfaces is the inner surface of your true physical skull encompassing all that you perceive, and beyond that skull is an unimaginably immense external world, of which the world you see around you is merely a miniature virtual-reality replica.[24]

The philosopher Max Velmans offers another perspective, disagreeing with Lehar's proposition of a world completely contained within one's head; Velmans challenges a staunchly physicalist interpretation of brains and worlds. For him, the three-dimensional world we see around us is fundamentally inseparable from consciousness itself.

It is as much a world woven out of a fine luminous awareness, as it is composed of physical matter. In agreement with the dual-aspect monist position outlined in Chapter 1, this means that the underlying reality—the first-order, noumenal one—is psychophysical at root.

As for the question of precisely how nature is able to pull off this magic trick of projecting experience beyond its boundaries (creating out-of-brain sensations), that remains a mystery. Virtual reality and holography both afford fruitful analogies. Here is one comparison that Velmans draws out, entertaining the possibility that physiological processes inside the brain form a type of neural projection hologram:

> A projection hologram has the interesting property that the three-dimensional image it encodes is perceived to be out in space, in front of its two-dimensional surface, provided that it is viewed from an appropriate (frontal) perspective and it is illuminated by an appropriate (frontal) source of light. If the image is viewed from any other perspective (from the side or from behind), the only information one can detect about the image is in the complex interference patterns encoded on the holographic plate. In analogous fashion, the information in the neural "projection hologram" is displayed as a visual, three-dimensional object out in space only when it is viewed from the appropriate, first-person perspective of the perceiving subject. And this happens only when the necessary and sufficient conditions for consciousness are satisfied (when there is "illumination by an appropriate source of light").[25]

One of the necessary biological conditions for consciousness is an intact *ascending reticular activating system* (ARAS). You can think of the ARAS's activity as being like the electricity that comes from a wall outlet to power other devices.[26] The thalamus and its projections to the primary projection areas throughout the cortex then helps to populate the actual sensory contents that occupy the phenomenal coordinate system. The topographic maps of the sensory world provide the hypothetical "projection hologram" that Velmans describes. When the ARAS sends a volley of activity to wake up the great mantle of the brain, this helps to light up the putative neuronal projection holograms in a very convincing simulacrum of a 3D world.

An Evolutionary Timeline

When in the history of life on this planet did the phenomenal bubble appear, and why? It is impossible to answer these questions definitively. We can venture hypotheses, but the phenomenal bubble does not remain preserved in the geological record. We can have access to the fossilized remains of animals that populated this earth across eons, but there is no way to confirm the first-person perspective that those bodies might have been privy to. Neurologists Todd Feinberg and Jon Mallatt, in *The Ancient Origins of Consciousness: How the Brain Created Experience,* review a large body of evidence to suggest that the first glimmers of consciousness likely appeared between 560 million and 520 million years ago. This period marked an event known as the Cambrian explosion, a so-called biological Big Bang. This was a time when animals evolved faster than ever before. From ancient worms with relatively primitive nervous systems there appeared nearly all of the phyla of animals that have ever lived. Then came an increasing complexity of neural organization, evolving to exploit the incredible number of signals the animals received from their surroundings: an explosion of sights, sounds, smells, vibrations, and tastes. An intense biological arms race between predator and prey species led to a premium on evolving novel models of motility, and the sensory apparatus required to absorb the rich diversity of physical energies present within novel oceanic environments.

The distance senses (vision, hearing, and smell, for example) allowed for biologically useful simulations of the outside world. Prey animals could now visually spot, hear, or smell their predators and thereby escape being victims. They could also exploit available food and mating opportunities. Emotional circuits, which gave subjective depth to the environment, so that it was not just a cinematic show, but allowed for a prioritization of behaviors–for example, do not bother with eating when a predator is hot on your trail–conferred enormous survival advantages.

Animals, in Sir Charles Sherrington's memorable phrasing, came to wear a "shell of their immediate future" around their heads—their heads being the most forward-looking component of their bodies as they steered and carved new paths into physical environments.[27] The coupling of sensory-perceptual functions to motor actions then greatly expanded the ability for intelligent action choice. Instead of blindly tumbling about, and ending up as food for a predator, animals that were conscious, by virtue of nature's evolved virtual-reality simulations, gained an incredible advantage in the game of life. The content of the simulations corresponded (and continues to correspond, for us as well) to the most convenient possible set of predictions available to a self-in-world for successfully navigating through extended space. Of course, the immediate experience consists of this rich sensorial field, a highly immersive and realistic form of virtual reality. No one ever experiences a model as a model. The model as such is transparent, as philosophers say, to the subjective point of view.

The *Umwelt* and the Mandala

In the early twentieth century, the German biologist Jakob von Uexküll coined the term *Umwelt* to refer to the lived, experiential world (the life-world) of a biological organism.[28] The *Umwelt,* he hypothesized, is formed by the coupling of a sensory-perceptual apparatus with a motor-effector system. The abilities to perceive and to act on the world combine to form an integrated *Umwelt,* and each species of animal occupying a given ecological niche has its own. The lived worlds of bats, bees, dogs, and humans are all very different—as different as their bodies. We can only begin to imagine what it might feel like (from the inside-out perspective) to sense the world, as bats do, through echolocation, or to detect, like bees, the electrical fields of flowers, or to experience the overpowering and rich world of smells that our canine friends enjoy.

There is no *one* veridical way to experience the outside world. In a sense, the flower is equally an arrangement of electrical fields, an explosion of olfactory cues, and the experience of colors that we see on our stroll through the garden. Each *Umwelt* is uniquely suited to provide maximum fitness to a given species. Ultimate, first-order reality is multifaceted and ceaselessly overflowing in the mysteries of how it discloses itself.

The notion of life-worlds indicates that our experiences of the world are an intimate interpenetration of objective factors (there really is a world of noumena, a first-order reality) and intrinsically subjective ones.[29] As we will explore in Chapter 5, the self-in-world models are imbued with depths of subjective feeling and meaning that lend a hypnotic allure to our consciousness.

The protean nature of our phenomenal bubble, our *Umwelt,* is a factor of deep importance—sacredness, in fact—in all religious traditions. The positive message that these ancient spiritual schools offer us is that human experience possesses an enormous potential for flexible reorganization. More than that, they see the central question of value and meaning of human life as intimately tied in with penetrating more deeply into the nature of first-order reality by stripping the gripping illusions of self-in-world models. The phenomenal coordinate system, which constitutes our private interface for moving and acting within the world, is malleable and flexible, made not out of unyielding prison bars behind which we are stuck, but of the same radiant luminosity that irradiates our nightly dreams.

An especially vivid and visual recognition of this truth is discovered in the symbol of the mandala. The mandala is a circular form, adorned with elaborate geometric patterns and colors. Some of the most ornate mandalas (*khyil-khor* in Tibetan) come from the Tibetan Buddhist tradition, where they are more than just creative artistic depictions, but symbols of the deepest significance conceived from the most subtle reaches of the mind experienced by advanced meditators. They are not, however, unique to Tibetan Buddhism. The mandala is among humanity's most ancient symbols.[30]

Visually, the mandala is an all-encompassing sphere containing a structure with mathematical precision. A rich vocabulary of symbols, often ones depicting opposites, are grouped around the core of a nucleus right in the center of the sphere. In Tibetan Buddhism, in the very center of the mandala is an empty point, traditionally symbolizing the infinite potential of awareness itself and its capacity to manifest as an endless variety of forms. Embedded into the very structure of the mandala is the notion of a world that is projected out from this center point of radiation, the still point. They are images that directly reflect the constant intermixing of the cosmos with energies emanating from our own psyches. In the center-surround shape of the mandala we can recognize the phenomenal coordinate system of our own perspectival view of the world, as it dynamically organizes in every moment of our life. Here, then, we have a powerful and immediate expression of lived phenomenology, as captured in a quote from the Tibetan Buddhist perspective:

> Wherever you are is an aspect of khyil-khor. Whatever you are doing is part of the energy of khyil-khor. You are simultaneously centre and periphery of this experience known as khyil-khor—wherever you are. You are the centre of your own and in the periphery of the khyil-khor of others. It is a totally interpenetrating energy . . . a wonderful dancing energy. It is not possible to exclude anyone from your khyil-khor or to be excluded from anyone else's. Even if someone dislikes you, you remain within their field . . . [and] if someone feels very strong negative feelings toward you, you would figure even more potently in their field . . . [so that] ultimately, every being is part of your khyil-khor.[31]

The mandala offers an apt symbol that allows us to understand and hopefully transcend the dangers of falling into the suffocating and lonely forms of solipsism that talk of phenomenal bubbles and virtual-reality models alone might steer us toward. As embodied mortals, we are inextricably woven into the great web of life that extends beyond our own bodies. A serious discipline of contemplative practice and spiritual transformation gives to its practitioners repeated first-person realizations of the deep extent to which so much of ordinary experience of reality is a dreamlike construction. Insight into this "magic trick of consciousness" reduces the likelihood of our becoming trapped inside excessively rigid models of second-order reality. We are reminded of a sense of responsibility and duty for tending to and honoring the gift of our mandala, and for doing what we can to foster nurturing influences on the mandalas of other beings.

Metaphors, New and Ancient

We will now explore in greater detail the nature of the virtual-reality metaphor, which has gained traction in contemporary scientific studies of consciousness. Arguably the most consistent and systematic exponent of this metaphor of consciousness is philosopher and neuroscientist Antti Revonsuo. Many prominent figures in the science of consciousness have made similar arguments, including J. Allan Hobson, Karl Friston, Thomas Metzinger, Jennifer Windt, Anil Seth, Steven Lehar, and Donald Hofmann—among many others.

As we said earlier, to compare our life, across waking and dreaming, to a virtual-reality simulation is no longer an especially controversial claim for contemporary neuroscientists. If it will soon begin to sound clichéd, then this will mostly be for lack of serious introspection that appreciates the metaphor's full implications. What does it really mean for us if everyday consciousness shares many features with, and indeed functions as a highly convincing form of, virtual reality?

Virtual reality refers to the use of computer technology to generate simulated environments through which the user can move. Users can interact with the environment by manipulating different virtual objects in various ways. There are a few distinct versions of virtual-reality implementation. An especially common one involves fully immersive virtual reality using a head-mounted display. The user wears a headset which delivers realistic visual stimuli, covering a full field of view, combined with soundscapes. The visual and auditory stimuli are adjusted based on the user's motion, and a set of motion controllers are held in the hands or worn as gloves to allow the user to manipulate virtual objects. Haptic technology, consisting of vibrations and manual forces, can be used to augment the virtual-reality environment with the sense of touch.

A key concept in virtual reality is that of presence.[32] Presence is difficult to define, but easy to understand when experienced. It refers to the compelling sense that the user has that the virtual-reality environment is given or presented to them. It is the subjective feeling of being *in* and fully inhabiting the virtual world. Presence is facilitated when all of the things encountered in virtual reality have a specific "address" within one's phenomenal coordinate system. This address has both spatial and temporal properties: experiences occur at specific locations and exhibit their own defined durations within the coordinate system. The presence of entering a virtual room is defined by definite virtual objects that are there for you at definite locations and are encountered in an elusive "now." They appear to exist in a seemingly external space. When you are outfitted with a virtual-reality helmet, you find yourself situated within an apparent 3D world instead of being aware of the underlying hardware and software operations that run the entire simulation.

Yet, during virtual-reality immersion, the things that one sees and experiences do not really originate from an actual physical environment. They are computer-generated "hallucinations." A user may have a compelling sense of walking through an Amazonian rainforest, but, in fact, they have never left a confined room. The virtual-reality metaphor says that this is not entirely different from how our brains operate even when we are not outfitted with a virtual-reality headset. Our head is itself a kind of "organic" virtual-reality apparatus that creates a sense of presence from an actively updated model.

Consider your dreams. The content of your dreams obviously is not computer generated. It originates from a story that your own brain tells itself as it receives impulses from the brain stem and other deep brain structures. The worlds encountered in dreams are locales occurring in your conscious state-space. The dream chair that your dream self sits on appears real to your dream self, but it is stitched together out of a "fabric of consciousness," rather than dense matter. Dreaming reveals that naive realism stands on shaky foundations.

If the virtual-reality metaphor is taken to its conclusion, it suggests that when you wake up from a nightly romp through dream worlds, you are still wearing the same organic "VR helmet." The simulations are now programmed differently, which makes a very important difference, but many of the core principles still stand. From the purely inside-out, first-person perspective, the seemingly solid and definite chair you are sitting on during waking periods is, to a large extent, still a sophisticated virtual interface. When one pauses to really engage with such thought-experiments, one begins to appreciate how unsettling it can be to challenge one's faith in naive realism.

We can push this metaphor even further. When you are wearing a headset and immersed in that virtual world, all the things you see around you are computer-generated. You, the person to whom they appear, are not. But in the case of our evolved, organic VR system, the self to whom the world appears is no less virtual than the world through which the self moves. We are talking now about the phenomenal entity that you identify as being *you*—sometimes concretely experienced as being roughly up in your head, behind and between your eyes. As Revonsuo writes, this is that ineffable sense of "I am here now," or *presence*.

This sense of self is rooted, at its deepest levels, within your body. The virtual-reality metaphor, taken to its logical endpoint, says that this sense of self, too, is just another phenomenally felt aspect of the overall self-in-world model. At the level of ultimate, or first-order reality, there is only the changing set of spatiotemporal processes, while everything else is a virtual solidification into a seemingly coherent, substantially, and independently existing thing. If you ponder these possibilities, they may be sufficient to induce a distinct shift in your momentary state of consciousness.

Later chapters in this book will explore whether it might be possible to step beyond virtual-reality metaphors by loosening the tight grip that self-in-world models ordinarily have, so as to enter into a more direct encounter with first-order reality. Our argument is that the mystical writings of many spiritual traditions are often concerned precisely with this question, seeing ultimate value in being able to discard illusory models that distort objective first-order reality so as to come into contact with the ineffable sense of universal connectedness that humans are capable of experiencing. This means much more than recognizing as an intellectual truth that all of life is an interdependent web. It means becoming consciously aware of and participating in this larger, more objective reality—feeling oneself to be continuous with a dynamical first-order reality rather than existing as a separated and isolated unit.

Longchenpa's Web of Enchantment

A recurring theme throughout this book is that contemplative masters often arrive at insights that resemble findings discovered by a scientific approach to the study of consciousness, but they also extend these observations in rather radical ways. This is to be expected, perhaps, since, the contemplative paths are concerned primarily with direct, experiential (first-person perspective) realization of certain fundamental truths concerning our experience of reality rather than in establishing an abstract, intellectual body of observations stated in a third-person perspective.

In this context, the case of Longchenpa, a fourteenth-century mystic poet and yogi-scholar of Tibet, provides an excellent example.

Longchenpa is one of the most venerated exponents of the *Nyingma* (Ancient Teaching) tradition, one of the four major schools of Tibetan Buddhism. So many of his observations have proved surprisingly relevant to the modern consciousness studies and discoveries that references to works by Longchenpa will be scattered throughout this book. Here of most interest is the one often translated as the *Trilogy of Rest*. In this classic manual, Longchenpa summarizes the core teachings of Tibetan Buddhism in captivating poetry, compressing an incredibly rich set of scriptural sources into crystallized, pithy lines.

The last of its three volumes, titled by some translators *Finding Rest in Illusion,* offers the most illuminating points of contact for unraveling the virtual-reality metaphors.[33] Here we find an incredible, centuries-old formulation of many of the same principles now being propounded by contemporary scientists and philosophers of consciousness. Longchenpa collected eight different "nature of mind" analogies that had already become entrenched in the Indo-Tibetan lineage, comparing the totality of our experiences to a *dream,* a *magic show,* an *optical illusion,* a desert *mirage,* the *moon's reflection* in water, an *echo,* a *city of gandharvas* (the gandharvas being a class of celestial, musical demigods), and a ghostly *apparition.* According to Longchenpa, the world we see around us, the seemingly objective state of things, is inseparable from the extremely subjective perception of it by our own minds.

In each poem, Longchenpa first evokes the details of the analogy, in hauntingly evocative language, then recommends a set of linked meditative practices to be carried out both during daytime "waking dreams" and in the nightly descent down the consciousness spectrum into the world of nocturnal sleep. Each analogy goes beyond an intellectual set of statements and begins to serve as an experiential portal, inviting the reader to a transformation of consciousness from the inside, first-person perspective. The poetry becomes itself a kind of incantation that disrupts the habitual, automatic tendency to be immersed in one's simulation of being a seemingly solid self in the world, which is the usual way that consciousness gets organized.

Pause for a slow, meditative reading of the following lines, perhaps taking a few deep and mindful breaths before you begin: "Tell yourself repeatedly—grow used to it—that/Outwardly, all mountains, valleys, regions, towns,/The earth, fire, water, wind, and space,/Sentient beings and the rest,/The five sense objects: form, sound, texture, smell, and taste, / And inwardly, the body, faculties of sense, and consciousness—/ Are nothing more than dreams."[34]

All that populates our consciousness, the teeming panoply of appearances, is apparent and vivid, but lacks its own, independent existence as a substantial entity—apart from being immersed in a ceaselessly fluxing and interpenetrating web of reality, which Longchenpa calls the primordial, immaculate *maya*.[35] Imagine this primordial, immaculate, natural state as a medium that is intrinsically empty, yet also cognizant and lucid. The phenomena that appear in consciousness are contingent, like bubbles forming and bursting on the churning ocean of ultimate reality.

The message of Longchenpa is that the seemingly segregated, boundaried phenomena arise only when the right causes and conditions are brought together. The causes include the physical energies, which are like electromagnetic waves permeating the actual physical world as well as our evolved sensory apparatus and nervous system. When the condition arises of these causes coming together (such that very specific frequency bandwidths of the electromagnetic spectrum interact with our senses), then we have the appearance of virtual phenomenal worlds.

Due to our genetic predispositions and previous habits, this magical display very quickly congeals into the dual polarities of self and other, apprehender and the apprehended, as noted earlier. The splitting into self and other happens quickly as the inherently luminous and flowing nature of consciousness clumps into the distinct aggregates of subject and object. The arising of self and world happens simultaneously (in Buddhism, it is called *coemergent ignorance*), as both of these clumps or higher-level entities arise from the fluidic medium of awareness itself.[36] Both the self and the world appear to the ordinary consciousness as quite dense, concrete, and circumscribed entities. And indeed, the appearance can be very vivid. But Longchenpa repeatedly points us to the fact that this appearance is the very miragelike illusion he refers to so constantly in the eight analogies. It is through our deeply ingrained ignorance (*avidya*) that we miss the cosmic play of energies that is reality and misperceive it as a dream (alternately good or bad, depending on how things are going at any given point in time), or alternately, that we become like actors so absorbed in their roles that they constantly forget the setting of their play. From this ignorance is engendered what Longchenpa terms the vast hallucination of *samsara*—humanity's endless wandering through the pain and suffering of existence.

So while his words are consonant with those of contemporary scientists describing consciousness as a genetically equipped virtual-reality model, Longchenpa hints at something far more grand and mysterious. In his formulation, human beings can gain direct, firsthand experience in this illusory charade and so escape the gravitational pull exerted by self-in-world models to achieve a sense of unity and compassion with all that is. With few exceptions, such states of mind and being, released from the ordinary sense of self-consciousness and self-centeredness, have gone unexplored by modern psychologists.[37] In pages to come, we will turn to discussing these broader topics. First, however, we need to examine in more detail how the brain and mind build elaborate reality models, by infusing our experiences with sentiment, feeling, and value.

5

Adding Feelings to the Mix

Chapter 4 outlined how the brain's construction of moving sensory images positions the body in a three-dimensional world. That process of world creation is impressive, but if the story were to end there, people would find themselves in a cold, empty world lacking all meaning. We need more than that, and here is where feelings come into play—our joys and sorrows, our frustrations and affections. Think of them as the glue that makes our inner and outer worlds stick to one another. As part of an entangled web of forces, they play vital roles in attracting us toward or repelling us away from nearly everything we encounter.

Our feeling states are the magnetic forces that pull us into chains of action and reaction. Feelings are a barometer we use to gauge our successes and failures and to decide between competing opportunities. Intense experiences of feeling seem to attach themselves to the objects and people of the world in which we find ourselves immersed. From the moment of birth, our lives become vivid, full of color and verve, as our feeling tones range from ecstatic peaks to the depths of misery. It can be extremely difficult to shake off the impression that things *really are* good or bad, independent of our involvement with them or our viewpoint. Reality itself seems to adopt the contours of their constantly fluctuating patterns.

This chapter will reveal one of the great tricks that nature uses to pull the wool over our eyes. Whether good or bad, our feelings are the residues of stories our brains spin to themselves, that then get painted onto and massaged deep into the fabric of our lived experience. Events in the external world echo tones that emanate from inner processes; it is a kind of enchantment process by which we endow them with emotion and meaning they would not otherwise have. We can catch glimpses of this in the surreal annals of neurology. For some people with certain kinds of brain damage, everything around them takes on the haunting quality of being unreal, drained of zest. Scientists applying pulses of electricity to brain nuclei the size of a match head can make subjects burst out in laughter or spontaneously tremble with the primal dread of a hunted animal.

From the experiences of generations of contemplatives, different traditions have amassed a wealth of similar observations about the strongly captivating force of feelings. Because, to a large extent, feelings are the vectors of contact between self and world, inner and outer, spiritual transformation operates primarily in their unseen, internal realm. For the sake of brevity, this chapter will draw largely on the Buddhist tradition, and Chapter 6 will consult Christian contemplatives as its primary sources.

The Internal Keyboard of Feelings

At the core of all the joys and sorrows you have experienced or will experience in this life are the built-in potentials of your nervous system, just waiting to be sprung into action.[1] Although these feelings are ordinarily evoked by situations that you seem to encounter in the external world (including your dream worlds), they are not formed on the spot. Those adventitious events in your world are orchestrating specific sequences of prepared simulations, encoded by genes contributed by your ancient ancestors. Throughout life, you have been accumulating (usually, in unconscious ways), a highly individual set of triggers for evoking these intrinsic potentials. Thomas Keating, a Catholic monk and contemplative, referred to these sets as the acquired "emotional programs for happiness" that over the years come to effectively steer the course of individuals' lives.[2] From the moment of your conception, the objects, events, and people you have encountered have been sculpting and shaping your emotional programs. Yet the essence of emotional energy emanates most deeply from an inner kernel you share with all others of your mammalian and human lineage. The experiences of emotions themselves—the joy, sorrow, anger, and the rest—are a genetically endowed birthright.

A rich body of evidence has revealed much about the brain bases of emotions. Laboratory studies conducted over the past century or so have uncovered several layers of genetically encoded neuronal circuits for expressing emotions, with roots in some of the most evolutionarily ancient brain areas.[3] Before reviewing a few of the key findings, we should mention that debate continues about whether animals other than humans have internal experiences that accompany the activation of these primitive brain structures. Given the close similarity in the anatomy, chemistry, and physiology of all mammalian brains, a large contingent of scientists are willing to concede at least some rudimentary form of emotional or emotionlike awareness, such as raw feelings of goodness or badness.

When some of the brain structures and circuits identified by animal studies are stimulated in human beings, the human patients can use language to express the feelings that result.[4] Patients with severe depression can report a sense of bright optimism and burst out in joyous laughter if the electrode tip delivers a miniscule amount of electricity to the appropriate cluster of cells. Or, if the electrode tip is moved to another particular spot, they can describe the feeling of impending doom leading them toward overwhelming panic.

At the innermost layer, neuroscientists have discovered the psychobehavioral potentials corresponding to what are called *homeostatic* (or primordial) emotions.[5] For preservation of life, the chemistry in the body's interior must be maintained within a delicate range, a state known as homeostasis. Deviations from the life-supporting zone trigger primordial emotions in proportion to the degree of deviation. With progressive levels of starvation, for example, or oxygen depletion or dehydration, come surges of increasingly unpleasant sensations, spurring a rapid effort to remedy the life-threatening situation. A different set of primordial emotions, such as feelings of savoring and satiety, mark the opposite condition, when no urgent remedial action is required because life's basic needs are satisfied.

One reason that subjective experiential states were especially favored by evolution as cues for maintaining physiological conditions favorable to life is that they arise rapidly and effectively predict the near future. Take the example of the need to quench thirst.[6] The gratification of drinking is instant, even though it takes many minutes for water in the gut to be absorbed by the body and lead to meaningful changes in blood chemistry. That quick switch to no longer feeling thirsty serves the drinker well by anticipating the change in biochemistry and removing motivation for further intake. In an environment where excessive time spent at a watering hole would increase one's exposure to predators, an efficient stopping mechanism would be an especially important adaptation.

Given the critical importance of primordial emotions to our species' early survival, it is not surprising that they would endure millions of years and remain fully intact in our modern-day brains. The most obvious examples are the intense pleasure of sexual orgasm and the terror that accompanies suffocation. The primordial emotions are most immediately concerned with ensuring the maintenance of critical life processes.

Very closely related to the previous examples is a broad category of sensory emotions that arise from the inner or external surfaces of the body. A partial list would include pain and extreme cold or heat, the discomfort of a distended bladder or full colon, the pleasantness of having palatable food in the mouth, and the intense disgust if that food is spoiled.[7] Any of these feelings can overwhelm our consciousness, with the effect of motivating appropriate behavioral responses.

There may even be feelings that go beyond (or underly) these situation-specific, self-protective emotions. Antonio Damasio, among other researchers, has argued that constantly operating in the background is a more pervasive feeling state that is fundamentally the sense of being alive.[8] Certainly such a feeling is not as urgent or even perceptible as those that arise in response to brief sources of stimulation or in response to more pressing sensations like hunger and thirst. Yet we can infer them through indirect means, such as observing body posture and overall energy levels. Frequently, other people pick up on these feelings based on very subtle cues, as when someone notes, quite accurately, that a colleague's energy is flagging before that person has even realized this is true.

Neuroscientists know more about the emotions that are triggered by stimuli encountered in the external world, having extensively mapped the brain systems responsible for these responses. Arguably the most avidly researched has been the system responsible for energizing animals, including humans, to explore their environments. Since a species' survival depends on its members' success in finding water, food, and opportunities for sex, evolution has long ago equipped brains with the capacity to motivate behavioral search strategies. Neuroscientists call this brain program the *seeking system* (or *wanting system*).[9] Its components have been located in the brain areas that are least different between humans and other mammals—again, not surprising given how rudimentary resource-finding is to survival. These areas are composed of cells deep in the brain stem and midbrain, with fibers that fan out and project widely from there, using the neurotransmitter dopamine to communicate their messages.

For many years, dopamine was known as the "pleasure chemical," but this is at best an oversimplification. It is far more accurate to say that dopamine helps to engender eager anticipation. While this state is certainly experienced as having a positive tinge—a shivery exhilaration—it is more like foreplay than it is like the moment of attaining what one wishes. In fact, in the blissful period that follows the consummation of a desire, the body *shuts down* the seeking system, however transiently, through the use of chemicals called endorphins, its natural store of opiates. This makes great evolutionary sense, too, because expending scarce resources to remain on the hunt after a goal was achieved would be a profligate waste of energy.

When the cells making up an animal's seeking system become activated, the animal starts to forage: it becomes intensely curious about its surroundings and engages in plenty of sniffing and touching. An interesting aspect of this neurological program is that it is intrinsically "blind" or "objectless." The animal conducts a frenzied search without necessarily knowing what it is after. If food happens to be close by, this means the animal will approach it quicker and perhaps ingest more of it. But if water is the only available resource, then that will become the magnet, and if a sexually receptive partner is within reach, then attempts at copulation follow.

In humans, recruitment of the seeking pathways also underlies intellectual hunger, the need to know and amass constant sources of information. Drugs of abuse gain their power by being able to hijack this very system.[10] Ingesting an amphetamine, such as cocaine, is probably the most direct way to arouse the seeking brain circuit. In humans, these drugs induce a high energy state in which the world and other people suddenly become endlessly fascinating, everything seems aglow, and one feels right on the verge of some ecstatic revelation.

Another major emotional brain circuit generates fearful states. Fear is a negative emotion that we typically strive to avoid. Yet in evolutionary terms it has been indispensable for survival, helping animals, including humans, to escape from predators and other dangerous situations. The small, almond-shaped clump of cells known as the amygdala is an important component of the fear system, although this particular piece of brain hardware is more involved in learning what or whom to fear than in the experiential core of dread itself.

A more primitive brain area called the *central gray,* which is further down in the brain stem, plays a critical role in orchestrating behavioral and visceral components of fear. In an experimental setting where the subject is a rat, a mild current of electricity delivered to the central gray will almost instantly evoke a "freeze," with the animal suddenly becoming immobile as if trying to escape detection by a stalking predator. If the amount of current is increased in intensity, the rat becomes increasingly desperate to escape, trying to jump out of its cage even if these attempts cause pain.[11] In a few, inadvertent cases, such effects have been observed in a human, when a neurosurgeon attempting to find opiate-rich cell clusters has accidentally moved a stimulating electrode over this region. Patients who experienced this described being "scared to death" and having a "fear of going crazy," and along with their cries for the doctor to stop came hyperventilation and sudden surges in blood pressure.[12] Other externally directed emotional systems in the brain mediate intense experiences of rage, or tender feelings of love.

The potential to experience all emotional states is, in fact, independent of interactions with the world.[13] Of course, in the normal course of things, the emotional systems are intricately tangled and intermeshed with happenings in the external world and with, in the case of us humans, the internal world of thoughts. We can easily think ourselves into experiencing intense fear or anger or lust. But even when a person you regard as an outright enemy does or says something to evoke a homicidal rage in you, that rage, in all its fury and power, is nothing more than the unleashing of a fairly specific and preexisting experiential script (a psychobiological potential) that your ancient

102

brain structures have stored inside of them. Your sense that someone is pushing your buttons is almost literally true: that antagonist is making your nervous system play a particular tune that you feel compelled to dance to. The anguished state of anger that then follows is a private event, one that only you experience.

Of course, while the tune is playing, your subjective experience is *not* a state of noticing how your stored internal simulations are being released and played out. You do not perceive the neuronal firings echoing deep down in your brain stem and then percolating upward, spreading and branching throughout the entire mantle of your cortex. Rather, your subjective experience is that you are confronting reality itself. You are wholly unaware of how the world "out there" has adopted the contours of your own internal landscape of feeling. This conviction on your part is just another part of the alluring cognitive and emotional illusion that is continuously unfolding before you, conforming itself to your current set of circumstances and your unique history of previous learning experiences.

Consider the words of a great, contemporary master of Tibetan Buddhism, Dilgo Khyentse Rinpoche: "When you feel hatred toward someone, your hatred and anger are not in any way inherent either to that person as a whole or to any aspect of him. Your anger exists only in your own mind."[14] The real-world situation of being enraged by someone and firmly feeling justified in one's reactions is, in fact, not so different from watching a very captivating movie and shouting at the characters on the silver screen.

Two Forms of Learning

When an infant is just born, there are only a handful of external triggers capable of playing its feeling keyboard. More will only come with further instruction. These few triggers are so persistent and common in the history of a species that the memory of them becomes lodged in the genetic code. One example is that loud sounds startle and provoke fear in human newborns despite their having no previous learning. Another is that even rats bred for use in science experiments, who have never encountered their natural predators, will freeze in response to the smell of cat hair or fox feces.[15] These are inborn triggers resulting from species-level, rather than individual, learning—a type of learning that unfolds over many, many generations. In scientific jargon this type of learning is known as *ontogenetic.*

Most of the things we get emotional about throughout our lives are things that we have had to learn about through direct personal experience. This individual (or *phylogenetic*) learning is a unique feature of one's life history, and it confers an incredible amount of nuance and flexibility to the brain's emotional circuitry. A newborn forms a deep, loving bond that is tied solely to the familiar smell and look of Mom. Once established, this type of learning is highly selective. Only a very particular conglomeration of sensory cues is associated with learning who *your* mom is, and the reason this unique signature is so precious is exactly because it is irreplaceable.

Although dogs have not been natural predators of humans, if it happens that you are chased by a pack of wild dogs, you surely will not forget the experience. As you continue to accrue experiences, virtually everything in your world acquires, to one degree or another, a layer of emotionality over it. Things that we usually ignore occupy no meaningful emotional real estate in our psychological lives.[16] The subjective experience of your reality then selectively edits out aspects of the world with little emotional charge, while amplifying stimuli that are invested with desire, dread, or other, more subtle, emotional qualities.

The Buddhist tradition, through its concepts of collective and individual karma, in many ways honors the difference between ontogenetic and phylogenetic layers of learning.[17] Thanks to our karmic propensities, we are predisposed to view the world through one or another set of subjective filters. Because most of us have never invested the effort required to remove these filters, however, we are by default oblivious to their implicit contribution in shaping perception. If you never attempt to question your tinted glasses, and you have never been apart from them, you lose all awareness that you are wearing them. And all the while, you assume that your experience of the world is just how things really are (*really*). In the absence of dedicated contemplative training and discipline, nature itself seems to offer us only two ways of confronting our powerful emotional lives: either passively react to what our emotional systems are compelling us to do, or make a valiant effort to suppress their external display.

To illustrate how our perceptions are relative and karmically conditioned, Buddhist teachings mention that different psychological states configure events in the world to disclose the appearance of entirely disparate realities. Buddhist mythology organizes the different psychological modes into six distinct realms of existence: those of the gods, demigods, humans, animals, hungry ghosts, and hell beings. These realms can be conceptualized as the six alternate ways of experiencing reality that are always open to us, with each realm serving as a unique filter on the world of appearances.[18] This means that the substance that humans perceive as water, for example, is perceived as blood and pus in the psychological state of the infinitely greedy hungry ghosts, and as ambrosia when inhabiting the psychological world of gods. The main point is that, whether something is perceived as pus, ambrosia, or water, none of these perceptions is closer to the "reality" of the substance—because nothing has an inherent reality of its own, independent of the observing entity.

A more familiar example is an excursion into the countryside. If we are intensely preoccupied with stressful circumstances at our job or have our resources drained after a lengthy convalescence, the scenery leaves us cold and untouched. The blooming, buzzing spectacle in front of us seems drained of life, as if contained behind a thick pane of glass. And yet, the very same scenery, on a day when we feel satisfied with life, is intensely joyous and we are moved by the lush, rolling hills and the scent of fresh spring flowers. Each step along the path unleashes joyous fireworks of sensations, as we fall in love with the nature around us. The scenery out there is identical, but our interior lives disclose completely different realities. At the heart of it, the observed and the observer are a dyad.

The Glue That Binds: Syncing the Inner and Outer

So far we have overlooked one of the most distinct qualities of feelings, which is that they carry within them an indelible stamp of "me-ness." Fears, joys, and sorrows always belong to someone, are always "mine" for someone, at some point in time. How can science possibly explain this inexplicable sense that feelings are always yoked to an experiencer who seems to be on the inside? Neuroscientist Jaak Panksepp has suggested that at the heart of emotional experience lies a primordial image of the body as an agent that can act out into the world and be acted on or "impressed."[19] The current hypothesis is that a group of cells deep down in the brain stem contains a virtual representation of a self (Panksepp, provocatively, calls it a *simple ego-like life form*). Think of it as a center of gravity for the sense that there is some*body* on the inside who is capable of making things happen.

The body is not a machine. There is a presence that seems to look out through the eyes. A unique aspect of this primitive, sensed presence is that it combines both the external skeletomuscular frame visible on the surface (what we see when we look in the mirror) and all of the stuff on the inside—the smooth muscles or viscera (guts, blood vessels, and so on) that are modified during emotional episodes. The overlapping of the "inside" and "outside" maps occurs first in fairly ancient nuclei of the brain stem.

According to Panksepp, the simple ego-like life form has an intrinsic rhythm which results from linking up a population of individual nerve cells, each of with its own timing parameters.[20] The beat to which this sense of self moves is continuously modified by entering into synchrony with other populations of cells. So, for example, when the seeking system gets activated, its cellular firings effectively entrain or change the rhythm in the primordial outside/inside body representation, shifting us into an upbeat, psychologically taut state, as these different brain regions enter into a dynamic dance with one another. Intense episodes of fear, too, have their own unique "beat" which then syncs up with our strongly embedded sense of self.

We can sync (and unsync) many times as our experiential world spans across possible feeling states, each with a strong sense that, at the center of it, these experiences are happening for, or to, us, at a very deep level. The resulting symphony, with exquisite timing, may be the trick that nature exploited to help link up events happening outside of us, with the intrinsic meaning and significance that they have for us. The experiencer and the experience are inextricably linked together in the same dance, although the tune can change depending on the circumstances.

Images or Feelings: Which Came First?

Up to now we have considered two distinct types of consciousness. First, the one corresponding to the sensory images of the world, and the second, what neuroscientists call affective consciousness that encompasses the many shades of pleasure and displeasure through which we filter *all* of our experiences. There has been much discussion about which of these forms of consciousness was first to arrive on the evolutionary scene.[21] As fascinating as these discussions may be, it is unlikely that the underlying questions will be capable of being settled on any firm, empirical basis. Unfortunately, it is impossible to rewind the tape of history into the deep evolutionary past so as to catch a glimpse of the first forms of subjectivity that swam, slithered, or walked the surface of our planet.

Regardless of which flavor of experience came first, once the two main kinds of consciousness came into existence, they did not remain insulated from each other. The tendrils of pleasant and unpleasant feeling reached out to infiltrate and suffuse all of our senses.[22] The chemical senses of taste and smell retain this connection with feeling most intensely—think of how the perfume of an old lover can instantly evoke a resplendent flood of memories—but while smell is most obvious, none of the other senses goes untouched. The intermixing of feeling and perception is evident in the brain's wiring.[23]

We now have scientific evidence that images depicting emotional events or objects are seen as being brighter and more vivid than scenes to which we are indifferent, like pictures of office equipment or furniture.[24] The color that feeling lends to perception reaches so deep that modern neuroscience has yet to resolve whether a gap between them is possible. It is exactly on this topic that spiritual traditions have much to offer, both in terms of sophisticated experiential techniques and a rich phenomenological mapping of consciousness terrains virtually untouched by modern scientists. A state of *direct,* or *pure,* or *naked* perception, unobstructed by craving or aversion or by layers of fabricated cognitive conceptualizations, is held out as the pinnacle of mind-training practices. As Buddhist practitioner Keith Dowman puts it, "Virtually the entire Buddhist canon . . . is concerned in some way with the mechanics of desire and sensory perception."[25]

A World without Feeling

Our perceptual world is so densely interwoven with feelings that we take their interplay for granted and scarcely notice it. Yet, there are well-recorded cases in clinical neurology, where the mechanism that attributes lived meaning to the people, events, and objects around us, breaks down, either temporarily or permanently. The phenomenology of these patients is rendered such that the depth added by emotion to one's experience of the world is absent. These unfortunate individuals suffer a disintegration of experience, since the bridge that glues the inner realm of feelings to the simulation of a self-in-the-world is impaired.

One of these is a condition that results from a brain injury that disconnects the emotion-rich areas from those parts of the brain responsible for vision. The psychological life of patients with this malady is marked by persistent feelings of unreality. In particular, the visual objects they encounter are perceived as drained of value: flowers lose their essence and look synthetic; a beautiful country landscape is perceived as inexplicably empty.[26] Although such patients can still experience feelings evoked through other senses, their sense of vision becomes lifeless.

Stranger still is a medical condition known as Capgras Syndrome, in which a person comes to believe that people they know have been replaced by impostors who look exactly like the individuals they resemble, yet somehow are not really what they seem to be on the surface.[27] In the normal course of events, as we recognize the face of a friend or a loved one, we simultaneously have an emotional reaction to that person, through a process we are not consciously aware is taking place. With the appropriate instruments, scientists can record the subtle changes in pulse rate or sweat gland activity or pupil dilation that occur when a subject's partner comes into sight, for instance, or is mentioned by name. This kind of emotional resonance with a familiar person provides an immediate, intuitive confirmation that this is someone who means something to you, for better (or worse). It appears that for those afflicted with Capgras syndrome, neuronal misfiring breaks the connection between the recognition of someone and its previously established emotional impact. Because this missed connection creates the impression that something is not quite right, it is a reasonable conclusion, no matter how bizarre it may sound, that this must not *really* be the person in question. She may look like Mom, yet she is not—some deceptive replica has replaced her.

In the related Cotard Delusion, the mismatch is between the sight of one's own face and the feeling of familiarity. The frightening conclusion then is that one is a walking corpse. Individuals with Cotard syndrome believe that they are already dead or nonexistent. Needless to say, this can be a terrifying experience, with strong sensations of dissociation and depersonalization. What these medical case histories demonstrate, in a startling way, is the extent to which our perceptual experiences are infused with emotional tendencies that bind us to almost everything that falls within our sphere of awareness.

Experience Is Not What It Seems to Be

How does the view of emotions and feelings, informed by the latest neuroscience, mesh with the view that has emerged from many generations of contemplative exploration? One way to begin to answer this question is by turning to a philosophical tradition in Theravada Buddhism known as the Abhidharma. The Abhidharma philosophers turned a metaphorical microscope toward the fine structure of experience itself. In a way, their methods were not all that different from that of modern-day psychologists and neuroscientists. They carefully recorded empirical observations and elaborated their findings into a rigorous, atomistic theory of mind.

By dissecting phenomenology down to its constituent elements, here is what the Abhidharma philosophers discovered: although our subjective experiences seem to be fluid and continuous, if we train ourselves to look more closely, we will find that experience actually consists of a granular texture. What is more, our experiences are conditioned by a combination of physical and mental factors. It is as if we perceive the world through successively extending and retracting a set of "tentacles" to sample the environment, and each such movement sculpts the interior structure of the next discrete moment of experience in a never-ending chain.

The constituent elements classified in the Abhidharma system are fairly complex, but in the Mahayana school, they can be grouped into five universal experiential factors that are with us from birth.[28] The first refers to the moment of initial contact occurring at a sensory portal, the birth of a new event in awareness. Take, for example, what happens when a light is reflected from a visual object. This initiates a sequence of events that interact with the visual modality capturing light rays and eventually transforms them into a visible form. This stage involves a confluence of an object, the sensory organ that receives it (for example, the eye, for visual objects), and the dawning of a fresh, raw, unprocessed moment of awareness. To grasp this idea, think of the moment of shocked startle that occurs when you hear a very loud sound before you have had a chance to categorize it or formulate any reactions.

The next mental factor is the feeling that attends the form, its emotional resonance of pleasant, unpleasant, or neutral feelings. According to the Abhidharma system, this primitive feeling clings to and colors (so to speak) every form that enters our awareness, endowing it with a certain visceral tone. Each object is simultaneously discriminated, or separated from others, which serves a parsing function, carving the space of awareness into distinct units. These conceptual categorizations—for example, "this is a chair," "that is an apple," "that is a dog"—involve calling upon a vast network of stored associations that we have been learning from birth to the present moment.[29] The factor of intention then is the evoking of intention or will, either toward or away from these perceptual groupings and determining whether or not to invest a specific object in awareness with attention. If the second factor has attached a resonant visceral response to the object, it will likely pull and capture attention, whereas if it is neutral your attention may well slide right off that object. After all, you would devote scant attention to investigating a dust particle on the table. Finally, consciousness is the factor that binds these various components and refers them to each of the forms. The unfolding of these five factors usually happens so automatically and quickly that we scarcely notice it happening, at least without advanced attention training.

A clear distinction can be made, then, between the endless variety of objects that the senses can perceive and how they *appear* to us, which includes the ways in which we emotionally appraise them (for example, whether they make us feel good or bad). The "appearances" are the mind's constructions, in principle having no more of a claim to a substantial and independent existence than dreamlike hallucinations.

This is where a clear point of agreement can be found with modern neuroscientific perspectives. As one example, the savoring of a piece of chocolate cake is not located anywhere in the molecular composition of the sugar or flour from which the cake is baked. There is nothing about culinary enjoyment that exists out there, even if it might seem like the cake is inherently imbued with a delicious cakeyness. Rather, that rich feeling of satisfaction is a quality that emerges from an interior process of layering value onto the cake, due to a complex interaction of factors operative at a particular point in time.[30]

There is a simple thought experiment to demonstrate this. Imagine a cake that lasts forever: if you keep eating it, it will slowly lose its pleasantness and, at some point, you will find it disgusting.[31] Nothing about the chemical composition of the cake has changed. Only your feelings about it have, and they have now transformed it into a highly undesirable object. How is it, then, that the dreamlike qualities of these phenomenal components exert such powerful influences over one's body, speech, and mind?

The Trap

From the Buddhist point of view, the five factors outlined by Abhidharma set into motion a reflexive "push" and "pull" dynamic that comes to mold all of our experiences. As biologist Stuart Hameroff notes, these experiences have evolved from even more primitive "yuck" and "yum" evaluations made by single-celled organisms.[32] The spontaneous way that this natural dynamic plays itself out is that the organism becomes a kind of plaything of these powerful and primordial energies, blown about in different directions. For us humans, the development of the forebrain perhaps endows us with some capacity to suppress the outward expression of some of these explosive lurchings, but our inner, subjective life is still profoundly invaded and colored by the same affective chords.

In the Buddhist understanding, our regular ways of perceiving are clouded over by two veils that keep us from being able to remain in the immediacy of the lived moment and stay open to the potential that is hidden in every instant of our life.[33] The first of the veils refers to conceptualizing our experiences instead of actually living them. That is, we are constantly busy classifying and categorizing the full spectrum of experience into "this" and "that" things, existing out there, separate from us. We perform this mental pigeonholing so incessantly that mind-created categories proliferate to the point of squeezing out the world as it actually is at any given instant. We spend much of our time interacting with conceptual representations, often mired in verbal networks of the world, rather than with the actual world itself.

The second veil is the emotional one that clings to the perceptual categorization and is either one of seeking to pull specific instances of "this" or "that" toward us, or push them further away: in other words, the ancient craving and aversion dynamic that has been operative probably since the beginnings of life on earth. At the most fundamental level, the dual polarities of desire and dread structure the field of experience into people and things we either want to move closer toward, or escape from.

In Sanskrit, the double distortions through which we receive the world are known as *kleshas,* or mental afflictions, and they are the juice that gets the wheel of misery turning.[34] Underlying craving and aversion is the klesha of ignorance. This ignorance causes forgetfulness of the fundamental Buddhist tenet that, regardless of whether appearances in consciousness are pleasant or unpleasant, they are all an illusion, like a virtual-reality simulation, and we too are part of this simulation. Ignorance begins in the fragmentary moment it takes to become distracted and hypnotized by the magical show, as we ourselves become fused into the unfolding narrative. In Chapter 3, we referred to this process of absorption or fusion as identification, or the natural attitude, as it was called by the phenomenological philosophers. The result is to become intoxicated by the seeming veracity of the simulation, constantly ferried about, this way and that, trying to choreograph an incalcitrant world so that it meets our changing wants and needs. The magic show works so well, so quickly, and is so well entrenched in us, that it often takes concerted mental training and discipline to even notice it orchestrating the events of our conscious life behind the scenes.

Because of this deep-seated ignorance of our true condition, our situation only gets worse from this point. Like an animal caught in a snare, we try to escape but only get more deeply entangled within self-world models. From here, the three initial afflictions expand into a larger set of the so-called *five poisons,* as understood by the Buddhist traditions: ignorance, attachment, pride, jealousy, and anger. These are destructive emotional tendencies that distort our ability to experience ourselves and others in an objective way.

The best way to learn about the Buddhist five poisons is by way of a concrete example, which will help to illustrate how they compound each other, like a snowball whose size continually grows. Imagine the following scenario: a potential romantic partner catches your eye. You begin to develop a fascination with this person, and you start to fantasize about how happy you would be if only your interest was requited. It gets to a point where you crave more and more, and this person becomes the center of most of your daydreams (the poison of attachment). Eventually, an opportunity arises and you make your overture. You receive the most pleasant surprise when your efforts are rewarded, and you find that this partner, on whom your every chance for happiness now seems to depend, returns your affections. Your world becomes magically transformed. As you start to attend social gatherings with your attractive new partner, you feel like a million dollars (the poison of pride). You become increasingly attached to your new partner and to the way you feel. It is like being intoxicated. Then, gradually, a strange new development begins to take shape. Your partner senses your growing dependency and neediness, and is discomfited by them. After all, being responsible for another person's happiness can be stifling. Maybe the partner withdraws a bit, putting a bit of emotional distance between the two of you. Perhaps, someone else comes along, an alluring stranger who seems to offer a hint of freedom, more breathing space. You notice that your partner's attention for you is flagging, and you start nursing suspicions of infidelity (the poison of jealousy). It does not take long for you to become convinced that something is wrong. At night, your sleep suffers as embers of doubt and jealousy are fanned toward a conflagration. One day you see your partner with the mysterious stranger and your jealousy swamps you. You are beside yourself, seething with anger that completely overwhelms any rational thought you might have at this point, and you cause a scene (the poison of anger). Your partner, shocked and disgusted with your behavior, decides to end the relationship with you.

Suddenly abandoned, you struggle to come to terms with your loss and this cruel change of fortune. Since you allowed yourself to be dazzled into believing that the happiness you were searching for and briefly felt was a substantial attribute of that partner—of someone *out there*—the loss now seems irredeemable (the poison of ignorance). Eventually, you will probably recover—but only go on to entangle yourself in other situations similar to this, with a predictable script. Perpetuation of this cycle is equated to a constantly spinning wheel of dissatisfaction, since the very nature of this process is its transiency—otherwise known by the Buddhists as samsara. Samsara, in fact, is to constantly go around in circles. Many of the world's great spiritual traditions provide systematic ways to overcome this pleasure-and-aversion cycle's power to enslave our behaviors in addictive patterns.

The emphasis so far has clearly been on negative emotions. The preponderance of attention to negative emotions is not because the existence of different types of positive emotions are unrecognized by Buddhist contemplatives. Rather, it is because they are largely concerned with defusing and transforming the destructive nature of the afflictive emotions.[35]

The Ingredients of Emotion

Comparing conventional neuroscientific accounts of emotions and feelings with the view from contemplative traditions such as Buddhism, perhaps the starkest difference is that the former have mostly focused on studying specific emotions, such as fear, disgust, anger, and so forth, while the contemplatives have honed in on the more basic building blocks forming the experiential texture of different emotional states. Where the primary data have come from meditative explorations performed by trained observers, there is much more detail available concerning aspects of the energetic qualities, momentum, and duration of feeling components.

As one example, within the Abhidharma taxonomy reviewed earlier, the feeling component consists of what we might view as basic "atoms" of emotional states, a graded variation along a wide spectrum of pleasure and aversion, rather than consisting of a full-fledged and differentiated emotion like fear or anger per se. The affective qualities that are understood to infuse every aspect of our perceptions are not specific enough to earn labels of specific emotion types corresponding to contemporary scientific taxonomies. In keeping with this more dimensional view, contemporary psychologists now recognize that each emotional experience may be entirely unique, shaped in the moment by often complex interactions of how people view the event, their unique physiological response, and the social context in which the event is embedded.

In Buddhist psychology, the differentiation of distinct and complex emotions is understood to emerge from a coalescence of primitive dimensions of experience—like general arousal—and a host of more complex associations, involving thoughts, memories, and images of people or objects.[36] Clinical psychologist and Buddhist practitioner John Welwood offers a useful analogy of a pyramid of experiential qualities that lie below and provide a base for the experience of different discrete types of emotion.[37] The base consists of a very diffuse sense that the organism has of simply being alive, which has mostly escaped the attention of modern neuroscientists, with a few exceptions. Antonio Damasio's definition of background feelings, mentioned above, captures it reasonably well—it is a global sense either of joy and exuberance, when life processes are operating freely and within the range of homeostasis, or of dull lethargy and malaise.

From this base, there is further differentiation into a sort of twilight zone of subtle felt senses that are very closely linked to their embodied signatures, like a constriction in the chest, a burning in the gut, or muscle tension in the face. At the top of the pyramid, we finally get to our more familiar and consensually agreed-upon emotions of anger or dread, after integrating across a broad mixture of signals percolating from lower levels and attaching a verbal label to the conglomerate of sensations and responses. Nearly all spiritual traditions share the notion that surface phenomena appearing in consciousness, the contents of which have been the focus of major research efforts, are anchored to more primitive processes and ride atop a massive swell of energetic momentum. In-Sight 5.1 invites you to explore the subjective texture of feelings from the inside-out, first-person perspective, allowing you to catch subjective glimmers of these more global and undifferentiated feelings that evade simple categorizations.

In-Sight 5.1

With this exploration, we aim to probe into the subjective texture of feelings with intention and guided attention. When feelings dominate our consciousness, they routinely erupt with great force, often catching us off guard. Feelings can very rapidly function like magnets within consciousness, attracting emotional qualia (which, at the very root level, are either pleasant or unpleasant) and exerting physical sensations (like butterflies in the stomach, or a racing pulse). They deeply influence conceptual narratives (reshaping a story to, perhaps, recast specific individuals as villains or heroes), and the applications of cognitive labels (perhaps causing one to characterize the entire agglomerate using a word like *rage, fear,* or *sadness*). In this demonstration we will try to successively strip away the many things that become magnetized to a feeling state, to explore the bare felt experience without the additional baggage of conceptual stories, labels, or external pointers (which might, for example, try to indicate that a feeling is *because* of a certain situational trigger or person in the external environment).

To begin with, take a moment to quiet your mind and relax into a comfortable position, ideally sitting in a chair or lying down on a comfortable surface. After a few moments, recall a recent emotional episode that hooked you in and elicited the kind of multimodal experience we just described. Avoid choosing an incident that was so upsetting or traumatic that bringing it to mind would overwhelm your ability to complete the necessary steps. Instead, select an experience that was moderate in its intensity.

Next, try to recall this experience in as much detail as you can, to the point of gaining a visceral sense that it has been reactivated at a level reasonably close to how it felt in the original moments of its occurrence. Now, begin to gradually remove the outer layers of this experience. Start by refusing to put a specific emotion label on the experience, but allowing the unnamable sensations simply to be present in your body. Next, begin to remove the narrative. Narratives often have some subject-object structure along the lines of "I felt *x* because of the actions of *y*, or occurrence *z*." Slowly dissolve each of these, beginning with the external referents (persons or external situations). Next, see if you can also dissolve the sense of a discrete subject who experienced the specific emotional label. As you dissolve these particulars, you could even imagine the relevant people becoming invisible. Just try to be with the continuously unfolding raw sensations of the energy that exists in your body, and note its likeness to an ever-shifting mosaic of experience. Enter deeply into the feel of the lived experience without trying to understand or organize it in any way whatsoever. Allow your attention to freely track all of the changing and flowing sensations.

Are you able to feel a qualitative shift in how the feeling is experienced? How is it different from when the physiology was being overlaid with a thick veneer of narrative and labeling? What is it that changes? Can you feel the more open-ended nature of the depths of the feeling? How does this relate to the pyramid of experiential qualities mentioned earlier in the text?

A rapprochement between contemporary scientific models of emotion, with their traditional focus on specific emotional systems, and the accounts offered in various contemplatively grounded traditions, focused on broader dimensions of feeling, is certainly not out of the question. The theory of *constructed emotion,* as advanced and championed by Lisa Feldman Barrett, offers numerous points of convergence.[38] Her argument, which is supported by considerable evidence, particularly from human neuroimaging studies, is that labels like *fear, happiness,* and *anger* are only best-guess interpretations or stories constructed by brainwide networks to try to fit the raw data. The raw data itself consist of global and continuous variations along dimensions of energy/tension and pleasure/aversion, core features of experience that are inseparable from the experience of being alive. These data communicate vital information about processes happening inside of us, and they change along a spectrum rather than being prepackaged into specific clusters, like fear or anger. In light of our extensive capacities for abstract thought and our linguistic capacities, we learn, over time, to clothe these kinds of data with conceptual overlays and so the classic emotions are phenomena that *emerge from* these numerous interactions. So, we may learn, in keeping with the cultural context in which we have been raised, that when we generally feel an aching, negative sensation that persists over extended periods of time, we categorize it as grief. Different combinations of lower-level building blocks may eventually obtain an identical conceptual label, so there is no unique and invariant cluster of physiological inputs that are always assigned to a specific emotion type or label. The theory of constructed emotion very explicitly *does not* claim that the construction itself is a conscious, voluntary process that individuals actually think about or do effortfully: the construction of an emotion concept happens virtually automatically, much as the five factors specified by the Abhidharma texts unfold.

Conventionally, the findings obtained from nonhuman animals that the electrical or chemical stimulation of different brain areas can elicit behaviors that resemble fear, rage, or separation distress are interpreted as disconfirming constructed theories of emotion.[39] Although there are fairly subtle and nuanced academic distinctions to be parsed here, we can ask whether it really needs to be the case that the two possibilities are irreconcilable.

A different way of approaching these strands of data is to think that certain well-defined brain structures and circuits, especially ancient ones concentrated along the brain's midline and in its depth, provide powerful sets of constraints that rein in and bias the kinds of conceptual interpretations that can best explain resulting changes happening within the body. From this perspective, emotions are conceptually constructed entities that emerge from more primitive and basic ingredients. Specific brain circuits merely offer up ingredients or building blocks that are especially prone to being categorized in a certain way.

The backbone of awareness is always, at least to some extent, involved in and about monitoring critical life processes occurring in the organism's interior (think of Welwood's pyramid, with its base is anchored in this basic aliveness and Damasio's concept of background feelings). From here, it is feasible that the brain structures identified as being strongly implicated in an emotion like fear are not quite the simple levers or buttons as originally understood. The metaphor of the brain that sees it as being prepackaged with ancestral programs which can be activated by the tip of an electrode is inappropriately deterministic.

To honor the sophistication of what is actually occurring, it is better to conceptualize that certain core neural circuits, identified largely in nonhuman animal studies, propagate strong constraints or nudges toward a particular interpretation of the raw data. These are physically realized as neurodynamic vector patterns that then orient larger ensembles of sensory, motor, and cognitive activities toward particular flavors. Each of the emotional systems imposes its unique, differentiated orbits of meaning. To put it another way, a region in the brain stem, such as the central gray, helps to filter the background and global sense of basic aliveness into more particularized expressions, by imposing very urgent interpretations of how the organism is related to its world. Electrical stimulation of such a neuronal structure could be acting like a strong tether that forms and shapes global lived experiential qualities in ways that make them increasingly well specified and increasingly likely to be categorized using a smaller range of conceptual labels (for example, panic or fear), especially in humans with elaborate linguistic capacities.

This more nuanced, and contemplatively informed, interpretation of the traditional neuroscientific data could help to soften and reconcile some of the seeming disparities between the view of emotions as discrete, hard-coded programs (à la Jaak Panksepp) and the constructivist perspective of Lisa Feldman Barrett. In Chapter 6, we consider what happens when emotional states and imaginary simulations congeal and get stuck in unhealthy patterns.

6

The Psychopathology
of Everyday Life

So far, we have delved into the nature of the phenomenal bubble, the 500-million-year-old VR simulator that situates a person within a three-dimensional world of settings, things, and other living beings. Each individual radiating point of awareness, each private phenomenal bubble with a point of view, is shot through with pervasive tendrils of feeling that make life worth living, as evolutionary ancient brain circuits exert the magnetic pull of attraction and the push of repulsion. We will now consider a species of mental phenomena that belongs neither to sense impressions of the world, nor to the chords of feeling that either simmer in the background of consciousness or rise to overwhelm us. This third category of conscious content refers to thoughts: an exotic hybrid form of conscious quale that melds elements of memory and imagination, reality and fantasy, image and word, allowing individuals to become mental time travelers.

Yet thoughts can also become an overgrown jungle that entraps a person in a maze of suffering—the forms of mental anguish that all of us are subject to, the everyday psychopathology that characterizes much of human psychology. In Chapter 3, we used the metaphor of clumps or aggregates that can form in the dynamic stream of mental consciousness. Thoughts correspond to an especially frequent instance of such clumping formations. When thoughts get very clumpy, they can command so much attention that we lose awareness of the present moment. We are then at risk of becoming completely entangled with the surface appearances of the mind, the constantly rearranging contents of consciousness.

The Christian contemplative and professor of early Christianity Martin Laird refers to these entanglements with thoughts as "mind-tripping on inner videos." For Laird, this mind-tripping has to do with how "a certain thought or train of thoughts quickly steals our attention and sets off a cycle of inner chatter and commentary. This inner chatter is something like a video that constantly plays in mind only to be rewound and played again and again and again. . . . The insidious thing about these videos is that they have a way of cultivating a psychological identification with them. . . . The more we watch or listen to them, the more we identify with them, the more we live out of them and live them out."[1]

Thoughts are something like rainbows or clouds: they seem substantial until you try to move closer and to get a better look at them. Some thoughts are more "viscous" and "sticky," commanding major investments of attention. Others are so flighty we barely notice them. They form as some intersection of word (language), image, and feeling—not clearly any one of these, but some convergence of them all. Although there is a commonsense view that thoughts are a kind of inner speech, detailed phenomenological studies suggest that this is only one, and not necessarily the most common, way in which thoughts can emerge. Beyond inner speech, thoughts can also be experienced as inner hearing, sensory impressions, feelings, or a wordless sense of thinking or knowing something.[2] Insight 6.1 invites you to gently explore the nature of thoughts from the inside-out, first-person perspective.

In-Sight 6.1

Many of us believe we know exactly what thoughts are, but we rarely attempt to examine thoughts from up close (so to speak). We now invite you to try to "catch" thoughts, as if you were hunting for exotic butterflies, and to examine their appearance to the mind's eye. Assume a comfortable position and set a timer for five minutes. For that duration, make a sincere attempt to closely observe the thoughts that arise. Do not think about thoughts, but stay with the

(continued)

In-Sight 6.1 (*continued*)

raw experience of thoughts, as they appear in your mind. Try to be as detailed and specific as possible concerning how thoughts appear to consciousness.

The following are inquiries that you can use to guide this exploration.

- Are most of your thoughts verbal, or are they image-based?
- Is there any separation between your observation of thoughts and the thoughts themselves?
- Can you control your next thought, or are your thoughts involuntary?
- If you try to locate them in space, are your thoughts coming from an "inside" or an "outside" space—or from nowhere in particular?
- Can you try to intercept the earliest arising of a single thought, or track where it disappears? What metaphors best describe the manner of their emergence and disappearance?

Phenomenological studies of this sort indicate that waking thoughts are far more strange, dreamlike, and insubstantial than we often realize or care to admit.

Spontaneous Thought, Internal Monologue, and the Default-Mode Network

Everyone is intimately familiar with some variant of the following scenario: in the morning, you jump into the shower, and even before the jet of water hits your body, you are internally off, bounding through a jungle of thoughts and associations. You start thinking about an impending deadline at work: that spreadsheet you keep putting off, but that you desperately need to finish and send off to your boss for a final approval. This might trigger a burning flash of emotion, as psychologically painful as an inflamed tooth. The spreadsheet reminds you of your perfectionistic boss, who is never happy with your productivity and who is sure to find plenty of faults with your latest effort. In turn, you might now recognize an echo of your father's disapproval, and you are suddenly struck by how your boss reminds you of some unhappy childhood memories associated with rejection and evaluation. The emotional charge of this memory is a bit too cutting, so your mind adroitly deflects the chain of thoughts onto a more agreeable trajectory: the upcoming social gathering this weekend that you have been anticipating for a whole month. And then it hits you—you are now five days late in getting your rent money in. You make a mental note to make the payment once out of the shower, but as you are toweling off you receive a text message from a friend who is in dire straits and needs your advice on a pressing matter. It is not until you are driving on the freeway, wondering about how late you will be and planning alternative routes, that you remember that your rent is unpaid.

In introducing the consciousness spectrum in Chapter 1, we mentioned that forays up and down the spectrum are continuous, with more or less tautness on a proverbial mental leash. When executing highly focused tasks, the mindstream is highly constrained and thoughts form orderly and logical queues, as in a military parade. Goal-directed thought is effortful, and we eventually fall into the spontaneous churn of the mindstream. In the absence of a specific task or behavioral demand, the mindstream loosens up and becomes more boisterous. It is now less like a parade and more like a children's birthday party. Spontaneous mentation happens very predictably whenever we step back from problem solving or active engagement. This happens frequently when we are reflecting on personal concerns or when distracted by TV sets at airports, magazine racks at grocery checkouts, or performing routine actions. At this point we are more likely to be *had* by our thoughts than we are to *have* thoughts.

The unconstrained flux of mental activity, known as mind wandering, lies at the crux of much of our conscious life. By some estimates, we average somewhere around 6,000 thoughts each day.[3] The prolix nature of the inner mental life is simply the subjective counterpart of having a brain that is physically wired to spend an inordinate amount of its energy talking to itself, as revealed in Chapter 2.

As mind wandering begins to wax, mental time travel begins to take off with gusto. The outside world gets quieter, while the inner worlds of reverie, fantasy and imagination grow in vibrancy. Amidst this inner cacophony we can gradually lose awareness that we are conscious at all, being so utterly absorbed with these shifting mosaics of experience.

The content of spontaneous thoughts falls into a circumscribed set of categories anchored around self-relevant processing (thinking about yourself or your feelings and behaviors), theory of mind (thinking about other people and their invisible mental states), and planning future actions.[4] Given free rein, thoughts seem naturally to gravitate toward obsessing over scenes from the past (memory) or entertaining all kinds of fictional what-if scenarios (fantasy and imagination). Thoughts pull us away from the present moment, until we are lost in a zoo of fanciful simulations. The most compelling thoughts, those that absorb most attention and resonate most strongly, are ones with emotional charge, competing with our ability to remain attuned to the present moment.

The grooves into which thoughts slip into have been laid down by a lifetime of experiences. As a person interacts with the world, the perceptual, emotional, and learning/memory circuits in the brain get stimulated in idiosyncratic ways. With repeated experience, personally unique patterns get laid down, obeying simple laws of association that are often involuntary and unconscious. Exerting effortful control over attention helps avoid being completely engulfed by chains of thoughts and related emotions.

In 2010, social psychologists Daniel Gilbert and Matt Killingsworth were intrigued by the philosophical and religious traditions that "teach that happiness is to be found by living in the moment" and that a "wandering mind is an unhappy mind."[5] Were they correct in that assertion? To probe this question, the researchers developed a smartphone app that tracked the happiness of thousands of volunteers as they went about their daily lives and responded to random prompts to report their thoughts and emotions. On average, people's minds wandered from current tasks nearly half of the time, regardless of what they were doing, and mind-wandering was "generally the cause, and not merely the consequence, of unhappiness." The volunteers were less happy when their minds wandered, prompting Killingsworth and Gilbert to conclude: "a human mind is a wandering mind, and a wandering mind is an unhappy mind." This is not to paint an exclusively negative role of spontaneous thoughts and mind-wandering. Psychologists Jonathan Smallwood and Jonathan Schooler list some benefits of mind-wandering during daily life including providing mental breaks to relieve boredom, planning for the future, imagining possibilities, placing experiences in a meaningful context, and inspiring novel and creative ideas.[6] Making a distinction between intentional and unintentional episodes of mind wandering is likely key to separating the largely negative from the positive outcomes.[7]

Our wandering thoughts sometimes surge like a cascading river after torrents of rain, such as in states of high anxiety, agitation, or uncertainty, whereas at other times, they slow to a mere trickle and cease entirely. Psychologists Adrian Ward and Daniel Wegner identified a mental state distinct from mind wandering, which they called "mind blanking."[8] About 18 percent of people they studied stated that they were thinking "nothing at all" when they were questioned about their thoughts. These numbers suggest that mind blanking is not nearly as prevalent as mind wandering. As Ward and Wegner put it, the blank mind is "neither here nor there but nowhere." Brain-imaging methods confirm that while mind blanking can be distinguished from mind wandering, both seem especially likely to occur when sleeplike states are present during wakefulness.[9] If you are curious about what a blank mind feels like from the inside-out, first-person point of view, try inducing such an episode by engaging with In-Sight 6.2.

In-Sight 6.2

Even when we deliberately try to do it, imposing control over our thoughts can be challenging, if not impossible. Assume a comfortable sitting position, and set a timer for five minutes. During those five minutes, exert every ounce of your willpower to have an uninterrupted stream of thoughts, with no gap or break between successive thoughts. In other words, try to head off and shore up any thought-free gaps, and have a continuous and unbroken succession of thoughts over the next five minutes. It is critical that you try your best to have no interruptions between thoughts under any circumstance.

Take a moment after the five minutes are up to review your experience and make a detailed record for yourself of what you discovered. Was your mind able to obey the instructions of having no interruption in thoughts or did it "rebel" by doing the opposite? Did you have any experiences of your mind going blank? If so, what was your experience like during these blank moments? There are no right or wrong answers in this exploration.

The Default-Mode Network

In 2001, the neurologist Marcus Raichle published a paper detailing the discovery of the brain's default-mode network (DMN).[10] DMN brain structures constitute a neural highway integrating parts of the cerebral cortex with much deeper and evolutionary ancient structures associated with emotional and memory functions.

The DMN's discovery was a classic example of scientific serendipity in action. By the early 2000s, neuroscientists had been busy running fMRI studies to obtain stunning images of the human brain in action for over a decade, by tracking changes in blood flow due to neural activity. As a research subject engages in tasks like reading, watching movies, or performing mathematical calculations, the colorful blobs that "light up" across different sectors of the brain indicate which neural areas are actively involved in a specific activity. Researchers assume that brain areas show reduced activity in the absence of a cognitive task during resting baseline periods.

Researchers observed consistently greater activation, however, in a certain constellation of brain areas during the resting periods. Brain activity in these regions actually *decreased* from its baseline levels when a subject was given an active task to perform. This decrease was observed across numerous experiments. This constellation of neural regions was called the *default mode network* because of its tendency to become activated when research subjects were waiting to perform an experiment, as they were quietly lying in the fMRI scanner, alone with their thoughts. It was called default, because this is how the brain spontaneously organizes itself when left to its own devices.

Neuroimaging findings suggest that the DMN is the brain network that is directly linked to what we have termed spontaneous thoughts.[11] When researchers asked research subjects to flag instances of their mind getting absorbed in internal monologues, these episodes were correlated with increased neural activity in the DMN.

In Chapter 2, we alluded to the cortex as a vast "imaginarium" owing to the brain's recurrent wiring connectivity. Outside of rather small islands of so-called *projection zones*, where neurons very faithfully react to signals from the outside world, there are expansive stretches of association zones, where the brain is mostly talking to itself. The DMN sits at the very top of this hierarchy of association areas.[12] This means that areas that constitute the DMN are provided with the weakest constraints from sensory stimulation and have a free license to concoct fantastic scenarios.

The DMN is like the brain's reclusive artistic genius, running wild with creativity, without having to worry about paying the bills and keeping down a menial job. During periods of high DMN activity a person is basically disengaged from the direct reality of the body and world, as they are in the here and now. Colloquially one would say the person is spaced out. It seems that during this time the DMN's job is to sort through recent and distant memories, their emotional qualities, examine associations between perceptions and outcomes, and elaborate on various autobiographical narratives. It assists in turning attention to the inner world of memories, fantasies, thoughts, and so constantly simulates a variety of scenarios that might happen in the future.[13] Such hypothetical simulations offer one way to prepare for uncertainties by rehearsing a gamut of response strategies. On the bright side, when it is functioning optimally, the DMN can also be a source of creativity and novel insights.

Imagine a scenario where something urgent requires your attention: a flying object is hurtling toward your head, and you need to duck quickly. In this case, DMN activity needs to be rapidly dialed down, so that other brain networks involved in responding to the external world can be recruited to execute a coordinated response. The ability to flexibly adjust between the inner and outer world, between the fantastical generation of hypotheses and the immediacy of the present moment, is handled by fine-tuning reciprocal interactions among these brain networks.

But there is a dark side to the DMN. In some cases, the structures within this network are *too* absorbed in their mutual message-passing, and the network becomes "sticky."[14] Clinical depression, for example, appears to be linked to an overactive DMN. This makes perfect sense when we consider what the subjective experience of being in the throes of a chronically depressed mood feels like. A person's ability to engage in and enjoy social conversations, for instance, is hampered because the DMN is stuck in an overactive state, and the person cannot break out of their negative internal monologue enough to respond to positive social cues. It is possible to be stuck in the past, as when we are unable to stop rehearsing embarrassing or otherwise unpleasant episodes that happened days, weeks, months, or even years ago. Such mental time travel depletes attention available for responding to other stimuli, and in severe cases, completely obscures our ability to be available in the moment and to respond consciously.

Rumination, also called persistent negative thinking, refers to being stuck in a rut of brooding, negative thoughts about the past, and constantly recycling them, whereas worry refers to compulsive internal monologues about possible future scenarios, in which a person rehearses some of the most unpleasant potential outcomes, even though the vast majority never materialize. Rumination and worry are linked to a hyperactive DMN, trapped in a vicious cycle where circular thought loops further amplify anxiety and helplessness.[15] Once sufficient momentum builds, these dark feedback loops become self-sustaining, taking on a life of their own, and eroding a person's capacity to voluntarily interrupt and redirect the negative sequence of thought chains into healthier tracks.

Toward a Universal Theory of Mental Suffering

Some thoughts can operate like animal traps. Both offer something that initially serves as a potent lure, drawing attention like a magnet. The stronger the lure, the greater the desire to approach and engage, but also the higher the risk of tripping a spring-loaded door. The end result is that once a spring is triggered, entrapment follows, closing off flexible response strategies.

In psychological terms this is called *capture*. Psychiatrist David Kessler defines it as the fundamental mechanism through which an appearance within the mind—whether a thought, a feeling, an impulse, a particular person, object, place, or idea, whatever—grabs attention and sculpts perception.[16] Kessler focuses on three important elements in the capture process: a narrowing of attention, a subjective sense of loss of voluntary control, and a change in overall feeling state. Capture is where the glue of attention and feeling gets combined into one sticky situation. Attention becomes hijacked, crowding out all other possible contents of consciousness. Pleasant or unpleasant feeling tones are rapidly evoked, calling up associations from memory or gearing imagination into overdrive.

Take an everyday example. A friend makes a passing remark that you perceive as derogatory. Rather than simply registering the spoken words, your attention gets sucked into the malicious intent that you spy behind them—even if this attribution is entirely false—as you experience a flare-up of rage. Meanwhile, the mind imagines ways to retaliate while remembering all the other people who have disrespected you. No matter how much you try, you cannot move on. You feel an irresistible urge to get back at the offending party, and you react with a cutting comment. This is the incredible (and at times frightening) power of capture.

Although this example was negative, Kessler understands capture to be much broader. We can be captured by things that make us miserable or by pleasant thoughts about an upcoming vacation. For Kessler, it is the misfiring of this mechanism that, to some degree, lies behind every existing psychiatric malady. In severe depression, a patient is captured by a negative mood, a self-limiting belief, or thoughts of suicide. In anxiety, vague worries about the future might be the culprit. A delusion about extraterrestrial mind control could be the driving factor in paranoid schizophrenia, while beliefs about being omnipotent might begin a spiral into a manic episode. Disorders of addiction are the prototypical examples of capture's dark side.

Shortly we will return to considering more details surrounding the neurobiology of capture, but first we will take a contemplative perspective on this topic. As it turns out, the earliest Christian monks, in fourth-century Egyptian and Syriac deserts, were intimately familiar with psychological capture, and indeed had already established a detailed understanding of it, dissecting its subtle mechanics and shedding light on its depths with scientific precision and accuracy.

The Ancient Science of Captivity

The origins of Christian monasticism lie in Egypt, in the early part of the fourth century. By the end of that century, Egypt was replete with monasteries, which had expanded into Palestine. Independently, another strain of Christian monasticism was associated with settlements around major Syrian cities. Among these monastics and solitaries, known as the Desert Fathers and Mothers, were countless numbers of spiritual masters and saints venerated to the present day. Far from the bustling cities of the Roman Empire, in some of the most physically isolated places, these men and women delved into depths of the human psyche. One of the first lessons they learned was that leaving the corruptions of the world behind was only the beginning—the real battle was interior, unfolding within the battlefield of the mind.[17] Their insights into the mysteries of the unconscious preceded the writings of Sigmund Freud and Carl Jung by centuries.

One colorful character in this setting, from an Egyptian monastic settlement in Nitria, was Evagrius the Solitary. Among the most brilliant scholars and theologians of his day, Evagrius fled the Roman capital of Constantinople following a supposed romantic entanglement with a prominent noblewoman who happened to be the wife of an influential aristocrat. He eventually found his way into the Egyptian deserts, became a reclusive monk, and received spiritual guidance from some of the most preeminent Desert Fathers.[18]

In the seclusion of the desert, Evagrius gave himself over entirely to intense asceticism of the inner world: a complex system of spiritual exercises aimed at purifying the innermost recesses of the mind to make it suitable for entering into the furthest reaches of spiritual contemplation. Evagrius is particularly important for our purposes for two reasons: his psychological investigations were extremely detailed; and he bequeathed to future generations a technical vocabulary of spiritual direction that influenced generations of Christian monastics.

For Evagrius, the spiritual path consisted of progressive stages of development.[19] Before being able to scale the heights of prayer and contemplation, one had to master the *praktikos* (practical) stage. This stage involves prescriptions concerning things such as fasting, sexual abstinence, and nonpossession of material objects—all of which demand intense confrontation with, and extensive knowledge of, one's internalized representations of self and world. Experiential knowledge of one's mind and the psychological laws that govern its dynamics, rather than merely academic knowledge of these, is thus a prerequisite for mastering the practical stage of contemplation.

The exercises of the practical stage, when followed faithfully for long periods of time, gradually cleared away the deep-seated biases and obscurations that normally clutter the interior layers of the mind, and ultimately distort how we perceive ourselves and others. Evagrius and other Christian monastics had noted how such subtle distortive influences occlude our abilities to attune with the spiritual dimensions of existence.

Evagrius became a masterful observer of thoughts (called *logismoi*). It is the constant bubbling up and arising of these *logismoi*—amalgams of sense images, feelings, and memories—that Evagrius saw as perturbing the deep states of tranquility that were required to advance in contemplative prayer. *Logismoi* were understood to derive largely from sense experiences, memories, and fantasies, but once a monk was lured into entering into an internal conversation with *logismoi* they would veil accurate perceptions of reality.

Evagrius recognized that some thoughts are voluntary and purposeful. Such thoughts can be fairly neutral, like the ones we have when completing routine tasks. Other thoughts are much stickier, shot through with pangs of feeling, and these are much more likely to disturb mental equanimity. Today, we know that these are thoughts that intermingle with the charge of feeling tones originating from deep-down emotional brain circuits like the amygdala, which registers emotional charge, or the dopamine-based neural circuits that anticipate future rewards. In scientific jargon, one would say that this class of thoughts is very high in saliency, tyrannically commanding all available resources of attention.[20]

For example, long-buried memories of perceived insults may suddenly resurface, rapidly triggering rage and thoughts of revenge. Evagrius writes: "Those memories, colored by passion, that we find in ourselves come from former experiences we underwent while subject to some passion. Whatever experience we now undergo while under the influence of passion will in the future persist in us in the form of passionate memories."[21]

Evagrius's observations agree with the findings of modern science that the mindstream is shaped by associative learning: prior experiences accompanied by positive or negative reinforcement are likely to clump together and recur within one's mindstream.[22] They erupt into consciousness with a vigorous force that readily hijacks attention, imparting more and more energy to the evolving thought sequence plunging us deep into the rabbit hole of a biased "ego tunnel" experience of reality.

One of the first practical learnings that Evagrius passed on to monks was how to observe emerging trains of thought without allowing attention to be completely consumed by them. He extorts the novice: "Let him keep careful watch over his thoughts. Let him observe their intensity, their periods of decline and follow them as they rise and fall. Let him note well the complexity of his thoughts, their periodicity, the demons which cause them, with the order of their succession and the nature of their associations."[23]

To be able to monitor thoughts in this way required learning how to maintain interior vigilance (known as watchfulness or *nepsis*).[24] *Nepsis* is a state of awake alertness, a receptive mode of attention that witnesses mental phenomena, without being carried away by them; rather like how a clear mirror simply reflects whatever is put in front of it. The attendant noninvolvement with mental images and concepts allows the mind's arisings to be synthesized within a larger background state of awareness.

Nepsis is accompanied by a second mental factor, known as rebuttal. In rebuttal, the monk moves from being passively victimized by spontaneously arising currents within the mindstream to assuming agency over the contents of one's mind through an exercise of free will.[25] Evagrius knew that there was little or no voluntary control over which thoughts might surface into awareness, but he taught that there *is* control over which thoughts are entertained and allowed to further develop by continuing to invest attention in them. Rebuttal then means to disallow certain thoughts by voluntarily depriving them of attention, choosing a state of interior silence.

The brain scientist Jill Bolte Taylor has made some rather interesting observations that directly echo Evagrian teachings (her story will be covered in more detail in Chapter 11).[26] After suffering a stroke in her left cerebral hemisphere, Jill initially found herself more alive to the present moment and frequently flooded by feelings of euphoria. This was because the left hemisphere seems to specialize in creating elaborate conceptual narratives, deftly concocting stories that try to interpret experiences by layering preexisting priors on top of the raw immediacy of lived experience. With the left hemisphere offline, reality could reveal itself shorn of the usual filters that scientists refer to as *top-down biases*. Once her left hemisphere began to recover, Jill felt many of these old and familiar negative patterns of thought gradually returning. Her discovery was that she now had a degree of choice about whether she would allow herself to hook back into and reengage with such negative thought loops. The first step toward realizing her newfound capacity was in her spontaneous discovery of a practice resembling monastic *nepsis*. "I have found that the first step to getting out of these reverberating loops of negative thought or emotion," she wrote, "is to recognize when I am hooked into those loops." After noting that many of her college students complained that consciously monitoring the content of the brain's incessant chatter required too much effort, she came to the following conclusion: "Learning to listen to your brain from the position of a nonjudgmental witness may take some practice and patience, but once you master this awareness, you become free to step beyond the worrisome trauma and drama of your story-teller."[27] From a neuroscientific perspective, becoming conscious of habitual grooves of thought softens their solidity and makes them eventually more susceptible to modification-a process that, at first, is contingent on repeated observations.

As Kessler writes in his book on mental suffering, though we all share a proclivity to succumb to unhealthy forms of capture, this is "offset by our ability to redirect the wanderings of our attention." This ability for mental redirection constitutes a "countervailing force that pulls many sufferers out of the rabbit hole," even though it is an "uphill battle," as it requires not passively ceding control and allowing emerging negative thoughts and chains of thoughts to shape the flow of the mindstream.

The active maintenance of *nepsis* and rebuttal requires consistent practice. The mental path of least resistance, where one slips into autopilot, is intentionally and actively challenged. In the terminology of modern cognitive-behavioral therapy, this purely reactive style of being is called mindless emoting, where the individual is overwhelmed. When we emote mindlessly, agency is transferred to associatively determined thought aggregates appearing within consciousness. These unhelpful thoughts can then be empowered to direct actions. For example, a thought pops up that a particular person is irritating, and we then allow this thought to continue expanding and intensifying into further hostile inner chatter even up to the point of reacting in aggressive fashion against the offending party.

Nowadays Evagrius' mention of demons may sound strange to modern ears, but one way to understand this term is that it is referring precisely to the ability of certain complexes of thoughts to automatically capture attention and quickly become autonomous entities that take over our inner life, leading to obsession and addiction.[28] It is the state of passive capture, the feeling of being taken over in spite of ourselves, that was labeled as demonic. Evagrius, in agreement with other Desert Fathers and Mothers, wrote that to begin with, many of our thoughts are autonomously occurring what-if types of simulations. The danger lies in the fact that they can very quickly mobilize attention that then quickly escalates into a self-perpetuating avalanche. Someone asked to give a public talk for example will naturally have thoughts centered around the potential for negative evaluation. These thoughts, however, can very rapidly escalate into worst case scenario simulations that paralyze the speaker and render them speechless.

For Evagrius and the other Desert Fathers and Mothers, most of the *logismoi* traveled in packs (or trains of thoughts), and they were, on the whole, negatively tinged. The presence of the *logismoi* was the metaphorical smudge on the lens through which individuals experience the world, causing all kinds of misperceptions. For instance, this might lead you to register a simple greeting from a coworker as a hostile remark, that then triggers an entire sequence of angry thoughts that raise your blood pressure and generally make you miserable. The thoughts seem to possess self-evident truth, even when it is obvious to an outside observer that they are highly subjective and selective interpretations of the true nature of events.

For many of us, these *logismoi* correspond to the ongoing critical monologue that some of us can probably hear as an inner commentary. A large portion of that chatter tends to be overly judgmental of self and others, and frequently defensive in character, mistrustful of the basic goodness of life. From his detailed scrutiny of the *logismoi*, Evagrius developed a taxonomy of the most commonly recurring complexes of thoughts. These thoughts were universal and were the most pernicious obstacles to scaling spiritual heights. He identified eight of them: thoughts about food (gluttony), sex (lust), avaricious thoughts concerning the acquisition of material possessions (love of money), sadness, anger, *acedia* (sloth/depressive listlessness), vainglory, and pride. In some versions of Evagrius's taxonomy, fear makes up a separate category.[29] When such thought-complexes are repeatedly allowed to hijack attention, they can become so powerful that they sediment into what the monastics called passions, or what would in modern terminology correspond to addictions. From a scientific perspective, these are domains of thoughts that interact most strongly with neural systems associated with primal emotions.

Together these thought complexes form a pyramid. A novice monk had to first tame obsessions having to do with food, sex, and money. Note that many of these themes naturally attract attention, as they relate to primordial emotional circuits in the brain, organized around biological survival and social dominance. Sexually enticing stimuli and fantasies, for instance, capture attention readily because they are directly associated with powerful evolutionary drives. The appeal of these drives to the human psyche, and their addictive potential, has not diminished since the time of the fourth-century desert solitaries.

Psychological issues involving fear, sadness, compulsive anger, and depression afflicted monks who were more spiritually seasoned, while thoughts of vainglory and pride were considered to be the most subtle and elusive ones. A wise spiritual elder was usually necessary to help a monk see his own hidden vainglorious and proud thoughts, yet despite their subterfuge, these were also considered the most corrosive to authentic spiritual living.

A Contemplative-Based Stage Model of Capture

The desert monastics paid particular attention to describing the evolution of these passionate thoughts from their inception to eventual capture. The simplest version consists of a four-stage psychological model. It is understood that these stages unravel with extreme speed.

In the earliest phase, provocation/suggestion, an impulse or inclination somewhere between a vague mental image and an abstract concept arises into consciousness. Here, writes Ivan Kontzevitch, "Under the influence of external impressions, or in connection with the psychological working of memory or imagination according to the laws of association . . . provocation enters . . . the sphere of consciousness. This first moment takes place independently of . . . free will . . . in accordance with the laws of psychological inevitability—'spontaneity.'"[30]

After this, the arising thoughts begin to interact with feelings, leading either to sympathy and approach toward the thought, or antipathy and withdrawal from it. Effectively, this stage decides the thought's future. This stage is referred to as that of conjunction/dialogue. At this point, a "sympathetic inclination attracts attention, allowing the suggested thought to grow and turn into an image of fantasy pervading the entire sphere of consciousness and ousting all other impressions and thoughts."[26] It is a kind of deepening conversation with the initial thought seed. Thoughts that are successful in drawing attention grow very rapidly.

This organically leads to the next stage of consent/joining, where a person allows the thought full control over this rising current in the mindstream. At that point, the thought and trains of thought are treated as if they correspond to reality, and one's behavior is now driven by the thoughts. In the case of an addiction-related thought, this would correspond to forming an intention to enjoy a cigarette as soon as the opportunity presents itself, and to begin taking active steps toward actualizing the thought.

The final stage, which follows the passionate thought's fulfillment, is captivity. At this point, thoughts have us, rather than us having thoughts, and the likelihood increases that the thought will spontaneously recur in future mind wandering episodes. In *Way of the Ascetics,* Tito Colliander summarizes this entire sequence in memorable fashion: "The impulse knocks like a salesman at the door. If one lets him in, he begins his sales talk about his wares, and it is hard to get rid of him even if one observes that his wares are not good. Thus follows consent and finally the purchase, often against one's own will."[31]

From the perspective of the Christian sages, passively obeying such momentary impulses makes us their slaves. They would say that the innumerable addictions, anxieties, compulsive rages, and other behavioral patterns that are unhealthy both for us and the people around us beguile us into servitude. The battle with thoughts, especially passionate ones, was understood by these spiritual giants to be the crucible for the formation of genuinely liberated persons.[32]

Getting Stuck Is Easy, Getting Unstuck Is Hard

Dynamic systems theory is a branch of science that can shed more light on the mechanics of capture. As a field, dynamic systems studies complex physical systems consisting of numerous, interacting parts, that collectively specify internal states that change over time. It also examines how such systems respond to external influences. The brain as a self-organizing entity consisting of millions of continuously interacting neurons is a dynamic system *par excellence*. As a result, dynamic systems theory, although technically a branch of mathematics, has greatly informed neuroscience and psychology.

By integrating neurobiology and dynamic systems theory, we can further flesh out how capture might work at fundamental levels. Dynamic systems like the brain often maintain a subtle balancing act, being poised on a delicate edge between stability (resistance to change) and variability.[33] When neurons self organize into networks, they exhibit a preference to fall into relatively stable patterns known as *attractors*. Once neural activity flows toward a particular attractor, it becomes difficult for it to budge from this state, so that stability wins out over variability.[34]

When certain highly salient thoughts (like Evagrius' passions) first arise, they rapidly siphon off attention, amplifying those specific neural networks, and forming a deeper and deeper attractor basin. The end result is that the flow of neural activity becomes as if directed along a canal. Once settled into such an attractor, neural activity becomes increasingly self-perpetuating like a runaway forest fire. Thoughts, feelings, and memories begin to vanish into the gaping maw of a growing basin of attraction, as they become strongly joined to each other. These aspects of the mental life tend to self-organize into patterns that resemble constellations.

Stan Grof, a psychiatrist who became famous for his pioneering work in psychedelic-assisted psychotherapy, termed these constellations of memories, sensations, thoughts, and beliefs organized around an emotionally salient central theme (such as low self-worth) condensed experience systems (COEX).[35] Grof treated COEX systems as spanning multiple layers of the psyche, from conscious to the subconscious, and he found that it was often necessary to unravel their tangles to produce lasting relief in his clinical patients.

Clinical depression is a classic example of a disorder caused by dysfunctional attractors.[36] A depressed person is fully in the grip of destructive self-perpetuating processes. An entire complex of negative thoughts about one's own self, the world, and the future get tangled into a tight knot known to cognitive therapists as the depression triad.[37] Attention is biased to attend only to negative information that acts to confirm self-defeating beliefs. Further, a phenomenon known as state-dependent memory explains the fact that depressed individuals are much more likely to recall other emotionally painful episodes. Maladaptive beliefs become so entrenched in clinical depression, that challenging them starts to require greater and greater conscious effort. Negative thoughts simmer in a stew of distressing emotions like sadness, shame, and anger that in turn increasingly congeal together.

The expanding attractor affects states of the body, no less than those of the mind. As a result, a depressed person adopts habitual stances associated with defeat, like slouched shoulders, chronically tense muscles, a sad facial expression that appears frozen in place, a slow gait, and so on. In severe cases, the entire motor system becomes affected, leading to laborious movements and slowed speech.

Clinical depression is maintained by three interconnected processes.[38] First, there is an unproductive processing loop, in which attention gets easily drawn toward negative material. People with depression have formed a habit of continually elaborating on and dwelling on such negative content, rather than quickly breaking off an internal dialogue with negative thoughts, as Evagrius would have counseled monks under his direction. An expanding depressive attractor is the inevitable outcome, leading to stronger degrees of capture.[39]

Second, depression is prolonged by defective inhibitory processes that would normally dampen the unproductive processing loop.[40] On a neurobiological level, this translates into excessive amygdala activity, alongside reduced activation of prefrontal circuits that are correlated with effortful redirection of attention. Finally, individuals with depression also have dysfunction of positive emotional systems. Not only is their attention automatically attracted to negative stimuli, but they also actively avoid engagement with rewarding, positive material that could buffer them against constantly dwelling on the negative aspects of life.

Once the chronic depressogenic attractor has formed, it grossly distorts an individual's ability to perceive and interact with the world around them in ways that promote healthy outcomes. Clinical depression is so recalcitrant to treatment because new inputs to the system are readily assimilated into preexisting models, as neural activity flows into the already established attractor grooves. By comparison, the destabilization of old patterns and the exploration of new ones, is harder to achieve. Unless counteracting forces unhook a depressed person from unproductive processing loops of thought, the depressive attractor is self-perpetuating.

But it is possible to form another healthier attractor. Dynamic systems theory suggests that the mind requires external perturbations that are strong enough to dislodge neural and psychic activity out of a stagnant pit. Psychedelics like psilocybin and LSD, administered in therapy for severe depression, exemplify a relatively extreme perturbation that appears to be effective. Robin Carhart-Harris, who has researched them extensively, has developed a theoretical model he calls *relaxed beliefs under psychedelics* to explain the clinical efficacy of these substances.[41] He believes that psychedelics succeed in treating depression because they promote relaxation of entrenched core beliefs and allow fresh creative responses that do not rigidly rely on distressing, old, maladaptive models of reality. Spontaneity and variability win out over rigidity and stability.

Carhart-Harris's idea is consistent with the earlier ideas of the German philosopher Martin Heidegger.[42] Heidegger had referred to a receptive, mindful style of thinking (*Gelassenheit*) that helps to inaugurate a renewed perception of reality by releasing one from constantly imposing rigid, preexisting conceptual frameworks onto things. When preexisting belief structures and feelings are released, the world and people around one are able to disclose themselves in a unique and inimitable way. Heidegger used the metaphor of a clearing in the forest (*Lichtung*), in which shafts of sunlight permeate a dense canopy of trees, to evoke the felt sense of such an experience.

Besides novel pharmacologic treatments like psychedelics, multimodal treatments hold promise of improving treatment outcomes. Rather than therapy that only attempts to dismantle self-defeating cognitive beliefs, initiating treatments modalities that involve the body and physiology, behavior, and the processing of strong feelings might collectively provide sufficient force to dislodge an entrenched attractor system. Although contemporary psychology has not paid extensive attention to the spiritual dimension of treatment, hundreds of years of experience suggest that incorporating this deeper aspect of human functioning is required when striving to reach a human being's fullest potential.

Dynamic systems theory also teaches us that when trying to re-channel activity from an already established attractor and into a fledgling one, there occurs a phenomenon known as flickering. During flickering, there is a more or less protracted period of inconstancy, while the system alternates between one and the other form of organization.[43] Clinical psychologists know that, especially during periods of high physical and mental stress or uncertainty, patients are likely to slip back into old, maladaptive attractors. So, while change in the long term is possible, it rarely proceeds along a linear trajectory. Instead, individuals in treatment experience a swinging pendulum, as alternative attractors—one well entrenched, and the other developing—vie for dominance.

Thoughts and Their Seeing Through: Old and New Perspectives

The fourth- and fifth-century Christian monastics of the Mesopotamian deserts developed a sophisticated psychological system concerned with describing pathological *logismoi,* understanding their mechanics, and using an armamentarium of ascetical exercises for healing.[44] Attention, normally subject to passive conditioning and habitual modes of engagement, was believed to be "plastic," or malleable, and capable of being trained and controlled through extended practice.[45] A famous manual for monastics, written by Saint John Climacus, offered practical advice for how to battle disturbing thought-complexes.[46] A tried and true piece of advice handed down to monks was to refrain from entering into analysis, disputation or argument with any line of thought. Whatever arises into mind is simply to be observed in a detached way. When thoughts are subjected to this kind of simple, direct observation, thoughts vanish like fog.

Rather than investing thoughts with belief (for example, because I have a sad thought, this means that I am sad), all attachment is released, and thoughts are let go of, as soon as they arise. Attention is redirected, over and over again, to shift away from thoughts, as well as from memories, feelings, and urges. Contemplative prayer was used as a stable anchor to continuously pull attention upward and prevent it from being absorbed by unhealthy patterns of thought. Particularly when it came to passionate thoughts, such as ones of anger, lust, or sadness, it was considered extremely important to cease internal dialogue with them at the earliest stages.

Constantly repeating the prayer etches a new groove for thoughts to settle into, so that they can be redirected away from their habitual patterns and into the orbit of a new attractor. Over time, the prayer penetrates into the mind's recesses. "Like a drop of ink that falls on blotting paper," a contemporary representative of this tradition explains, "the act of prayer should spread steadily outwards from the conscious and reasoning center of the brain, until it embraces every part of ourselves."[47] Gradually, this process intercepts negative *logismoi* at increasingly earlier stages of their arising.

Through the key practice of constantly observing thoughts, a monk would note those that occurred with particular tenacity. Thoughts linked with negative emotions were considered to be particularly problematic, given the way they command attention and rapidly spawn an entire chain of mental activity. To reduce the likelihood of this happening, the monk would regularly report troublesome thoughts to a senior supportive spiritual guide as soon as possible.

But the therapeutic action went further than disclosure alone. Given the intimate psychological knowledge that a spiritual guide had of the monks, they would offer sharp, individually tailored guidance that helped monks avoid capture by tenacious thought loops. When revealing especially troublesome thoughts, such as ones having to do with physical addictions, the monks were instructed to avoid excessive detail, to minimize attention capture.[48] In modern parlance, detailed elaboration of emotionally charged events risks activating an expansive associative network of related thoughts, feelings, and behaviors that can cause psychological capture.

Contemporary psychotherapists have incorporated concepts derived from contemplative practices into empirically supported interventions. Cognitive and cognitive-behavioral therapies have underscored the importance of identifying, disputing, and modifying negative, dysfunctional, or distorted automatic thoughts and core beliefs (for example, "I am worthless") to relieve human suffering. These interventions are widely utilized based on substantial evidence of their effectiveness.

Yet much in line with the advice of the Desert contemplatives, psychologist Steven Hayes and his colleagues have argued that engaging with, vigorously disputing, and attempting to shape the form or content of negative thoughts can backfire and magnify their magnetic pull and potency. Hayes and his collaborators developed Acceptance and Commitment Therapy, or ACT for short, which incorporates aspects of Buddhism in encouraging mindful and full acceptance of the entirety of experience in the moment—thoughts, emotions, memories, sensations–while committing to pursuing value-driven goals in daily life.[49] Modern researchers have discovered that avoiding and suppressing distressing experiences, rather than confronting or accepting them, often stirs up even greater emotional upheavals.[50] In short, what we resist captures us.

ACT therapists exhort their clients to accept the panoply of negative thoughts, emotions, and self-evaluations that inevitably course through the mind, but not to mistake them for immutable facts. The objective of this admonition is to counter the natural tendency to strongly identify with highly conditioned, negative contents of consciousness. When distanced or "disidentified" from negative thoughts and emotions, people are better positioned to decide how they wish to stand in relation to them from a more liberated, transcendent perspective—whether to heed them, ignore them, or capitalize on them to identify or solve a problem.

Choosing What Captures Us

In the desert lineage of Christendom, the far reaches of contemplation are capable of being reached only after a sustained, frequently lifelong battle with the passions and *logismoi*. Such spiritual heights, reached only by very few, are characterized by *apatheia*, a word that has nothing to do with apathy as it is commonly understood.[51] Rather, as used in this context, *apatheia* denotes a state of authentic emotional freedom, when a person is no longer captivated by passionate and self-centered strivings. *Apatheia* entails a radical restructuring of the way the mind acquires knowledge of the world and other people. No longer beholden to even subtle stirrings of desire and fear, such a person abides in a state of utter mental equanimity and manifests *agape*—selfless, compassionate love toward all other people and, indeed, every living being. Rather than being emotionless, or thoughtless, such a person is freed to encounter reality as a manifestation of Heidegger's *Lichtung,* not as a passing experience but in an enduring fashion.

Perhaps achieving the liberation afforded by *apatheia* may seem too much to hope for. But we can start small. As Kessler has written, capture has a light, as well as a dark, side. The light side teaches us that though we may have no choice but to be captured, we have some measure of control over *what* captures us. Even in cases where genetics and early life experiences have conspired to strongly predispose an individual to be especially sensitive to capture by anxieties, addictions, and grievous things, there always remains a silver lining. This lining is the thread of personal dignity that leaves open the possibility, no matter how unlikely, of choosing to exert effortful control over one's attention. While we spend considerable time on negative attractors, it is possible to consciously work on creating positive attractor sets.

This typically requires enormous amounts of effort that are necessary to inculcate good, health-promoting habits. The philosopher James Smith refers to this process as one of reformation or love-shaping—being very intentional about what we love and want, to shape ourselves into the persons we would like to become.[52] We can become conscious curators of our life stories.

We can be captured by transcendent values having to do with truth, beauty, love, and ultimate goodness. Devotion to such high ideals can be promoted by frequent meditation on these topics and a conscious intention to engage them.[53] A commitment to regular walks in nature, encounter with sublime art, and meditation and prayer can all contribute to establishing positive attractors. When we exert intentionality in curating what draws us, at the deepest levels, we buy back genuine freedom to shape our own lives. As writer David Foster Wallace noted, so many of our maladaptive habits are bad not insomuch as they are "evil or sinful," but rather in "that they are unconscious. They are default settings . . . you just gradually slip into, day after day, getting more and more selective about what you see and how you measure value without ever being fully aware that that's what you're doing."[54]

Current research provides evidence of upward spirals of thoughts and feelings that counteract the downward spirals. For instance, one study found that when patients in partial remission from a depressive episode were assigned to mindfulness-based cognitive therapy, they experienced positive thoughts and moods more frequently over multiple days than did patients assigned to a waitlist control condition.[55] The mindfulness therapy consisted of teaching patients to practice the same dispassionate witnessing of thoughts, along with a suspension of belief in negative thoughts, that the desert monks might have practiced in the fourth century. If we seize back control of how our attention gets used, then we can shift from being hapless victims who are always being had by an influx of thoughts to once again assuming a degree of agency over our mindstreams. Next, we will consider what happens to one's mindstream as it slips further down along the consciousness spectrum, into more interior worlds of imagery and dreams.

7

The Stuff Dreams Are Made Of

Each night we travel between two worlds without leaving the bedroom. With closed eyes and in a state of deepening sensory deprivation, we retreat into ourselves. We begin to drift away from what the ancient Greek philosopher Heraclitus called the *koinos kosmos,* the "common world" of shared, consensus reality we inhabit in our waking hours. Gradually, the state of our neurochemistry and neurophysiology is delicately tuned, marking an entrance into the *idios kosmos*—a private world that we can never directly share, even with those most intimate to us.[1]

The fact that we are conscious while we dream makes clear that wakefulness is distinct from conscious awareness. A person can be asleep and behaviorally unresponsive while still being conscious and aware. "Since dreams are conscious experiences occurring independently of sensory input and motor output, they reveal consciousness in a very basic form—the way consciousness is when isolated from those very systems it normally flawlessly interacts with."[2] So writes the philosopher and cognitive neuroscientist Antti Revonsuo. He and others suggest that dreaming offers the most direct path to glimpsing the nature of consciousness. Dreams serve as primary evidence of how convincing self-in-world virtual-reality simulations can be. And, as we will see later in this chapter, dreams have been used in many traditions, such as by Tibetan Buddhists, to deepen meditation practice.

Neurobiology of Waking, Sleep, and Dreaming

Falling asleep is a tightly orchestrated symphony that reverberates from the molecular level up to the phenomenological. From an objective, or third-person, perspective, the most characteristic signatures are ones routinely collected in any modern sleep lab equipped to perform EEG-polysomnography. This comprehensive diagnostic assessment involves not only an electroencephalogram measurement of the brain's electrical activity but also the use of other sensors to detect changes in heart and respiratory rate, eye movements, gross muscle tension, and any overt behavioral responses.

The onset of sleep is a gradual process (usually lasting minutes) that involves progressing through different levels of sleep. Among its most characteristic features is the progression revealed by EEG recordings, that brain activity steadily shifts from high-frequency and irregular neuronal oscillations to low-frequency and highly synchronous waves. Sleep scientists have organized these continuous transitions into two major phases, the first of which has three stages.

Here is how the progression unfolds.[3] As we start to retreat into ourselves and shift from alert wakefulness to drowsiness, marked physiological changes occur, including slow, rolling eye movements that indicate deepening drowsiness. Their intensity peaks right before entry into Stage 1 sleep, at which point most behavioral indicators of responsiveness fade. Stage 1 represents our initial foray into the domain of sleep. While there is increasing relaxation and quiescence of multiple physiological systems, our threshold for awakening is still low. Were someone to wake us up from Stage 1 sleep, we might even sincerely deny that we were entering into sleep.

Stage 1 progresses into the deeper Stage 2. Here, awakening becomes more difficult. Several specific signatures show up in the EEG. These include a waveform known as a K-Complex, which helps to suppress the brain's arousal by distracting sensory signals that may still be arriving from the outside world. This is one way that the brain organizes itself to protect and maintain a cocooned seclusion from the outside environment.

Stage 2 bottoms out into the slow-wave sleep of Stage 3. This is the deepest sleep, a highly restorative period marked by the preponderance of high-voltage, low-frequency (1 to 4 Hz) delta waves prevalent in the EEG. In Stage 3, awakening becomes least likely, even when loud noises occur. The brain continues to be highly active throughout this deep-sleep stage, but, rather than building images of and responding to the world beyond the skin, it is working in a profoundly introverted mode.[4] The thalamus, which conducts sensory inputs to the rest of the cortex, is largely silenced, inaugurating a natural form of sensory deprivation.

This deepest stage of sleep is interrupted by a brief climb back up to Stage 2, and then the periodic recurrence of a different phase of sleep, named for the distinctive physiological activity that accompanies it: rapid eye movement (REM). The shifts into REM sleep mark such a contrast that the three prior stages are collectively known as the non-rapid eye movement (NREM) phase. REM sleep has also been called "paradoxical sleep," because, physiologically, it shares many characteristics with wakefulness. Thalamic transmissions of sensory signals up to the cortex are nearly indistinguishable in the two states, and monitoring of brain activity reveals in both the same kind of low-amplitude, high-frequency oscillations.

The REM phase's flurry of eye-movement activity corresponds to a sleeper's visual scanning of a dreamscape.[5] Sensory signals still arrive to the brain, but they are incorporated into the sleeper's inner, phenomenal world only sparingly and indirectly. Selected aspects of these signals may work their way into the dreamscape; for example, the sound of rain may get translated into a dream about traveling on a boat. Yet the virtual-reality simulations that get run during REM are not dependent on outside influences. The threshold for waking during REM appears as high as it was during the deepest stage of sleep.

During REM sleep, the body's voluntary muscles are effectively paralyzed, with the exception of the eye muscles. The NREM phase involves a delicately tuned neurochemistry in which the brain is suffused with the neurotransmitters serotonin and norepinephrine, which suppress REM generation. Approximately ninety minutes later, levels of these drop and brain stem nuclei begin pumping massive surges of another neurotransmitter, acetylcholine, to inaugurate the REM sleep period.

As the cycle repeats, as it does four to six times in the span of a night, the experience of sleep becomes a carefully choreographed dance of alternating stages, repeating with systematic regularity. Again and again, we make the descent to deep sleep, climb back up to Stage 2, enter REM, then return to Stage 2, and dip back to deep slow-wave sleep. All told, an average adult spends about 25 percent of the night in REM sleep and the remaining 75 percent in NREM stages. But the cycles do not all feature that same allocation. The longer we have been asleep and the closer we are to morning waking, the greater the proportion of a cycle is devoted to REM sleep. Conversely, in the early hours of a sleep period, NREM deep sleep is most dominant.

The Phenomenology of Inner Worlds

The slow shift down the consciousness spectrum unfolds in stages in much the same fashion as the physiological changes just described. On the phenomenological front, the changes that occur as one floats toward sleep include increasing upsurges of imaginative, fantastical materials, which are primarily images or visionary states and which are characterized by the disappearance of a sense of voluntary control over the contents entering into awareness.[6] In addition, orientation with respect to objective space and time becomes progressively less reliable and eventually vanishes. As the brain and mind become reorganized, and we gear down from waking to sleep, the needle on our reality compass swivels, making a slow rotation from pointing to the world outside to the interior. The stream of consciousness becomes much more like a rushing river, one populated with entrancing images, into which we are steadily absorbed. The ability to distinguish internally generated imagery from perceptions of the shared world of *koinos kosmos* gradually slips through our fingers, like metaphorical sand, probably somewhere during the trek from Stages 1 and 2 to deeper levels.

Liminal Dreams: Hypnagogia and Hypnopompia

When we go to sleep, we pass the threshold from shared consensus reality to variegated inner worlds twice: once in the descent from wakefulness to sleep, and then again in the ascent from sleep to waking. The first passage is known as *hypnagogia,* the second transition is termed *hypnopompia.* Both exist in that murky, in-between territory of the consciousness spectrum, encompassing the deepening phase of drowsiness that shades off into Stages 1 and 2 of sleep.

The study of hypnagogia (including official coinage of the term) was first undertaken by the French psychologist Alfred Maury in the late nineteenth century.[7] Since then, hypnagogic hallucinations have come to be the paradigmatic example of sleep-onset phenomenology. Luminaries of their time, including Charles Darwin's cousin Francis Galton, made important contributions to their exploration. Early on, Galton undertook an extensive study of hallucinations in otherwise healthy individuals, noting their widespread prevalence. He was surprised, having initially missed their occurrence, that he was able with some training and motivation to observe a veritable "kaleidoscopic change of patterns and forms" which most people habitually ignore in drifting off to sleep. The suggestions offered in In-Sight 7.1 are meant to increase your engagement with and the likelihood of obtaining deeper personal familiarity with the hypnagogic state.

In-Sight 7.1

We invite you to explore the fascinating twilight zone of the hypnagogic state using two methods. The first comes from a classic book on hypnagogia by the psychologist Andreas Mavromatis (*Hypnagogia: The Unique State of Consciousness Between Wakefulness and Sleep,* London: Thyrsos Press, 2010). Simply lie down or sit in a chair comfortably, and with your eyes closed, progressively relax parts of your body and concentrate gently on any mental images that arise. To prolong the hypnagogic state, Mavromatis advises lying in a relaxed state with your arm in a vertical position resting on your elbow. As you begin to drift seamlessly into sleep, your arm will fall and awaken you, allowing you to repeat the method and reenter the hypnagogic state if you wish. The state ends naturally when you succumb to sleep. Prior to trying either method, a strong self-suggestion to enjoy and learn from the experience, be receptive to whatever happens, and remember what transpires can help in getting the most out of the exercise.

For several nights, experiment with both methods or combine them; that is, develop an image to induce hypnagogia and use the

(*continued*)

In-Sight 7.1 (*continued*)

vertical arm method to prolong the state. Before full-blown sleep creeps in, alert yourself and write down your experiences. You might address questions such as these: Did you experience both sights and sounds? Did you have control of the imagery, or did it slip from your mental grasp? Did you notice stages of hypnagogia, which Mavromantis describes, ranging from vivid imagery, to hallucinations, to loose narrative thoughts that ultimately shape dreamscapes? Did your experiences vary greatly from night to night? Finally: Could you remember your experiences clearly and describe them in words?

If you find it challenging to resist descending into sleep using either method, then set an alarm for ten to fifteen minutes after you start trying to induce hypnagogia, and then write about or sketch a picture of what happened.

We experience hypnagogic phenomena most often through sight and sound. Visions can range from formless shapes to regular, geometric patterns to faces, figures, animals, and objects, written text, and all the way to panoramic scenes.[8] A frequent experience involves revisiting activities from the day. For example, psychologist Donald Hebb recounts how "A day in the woods or a day-long car trip . . . sometimes has an extraordinarily vivid aftereffect. As I go to bed and shut my eyes . . . a path through the brush or a winding highway begins to flow past me and continues to do so till sleep intervenes."[9]

Psychologist Andreas Mavromantis developed a triphasic stage model of hypnagogia that begins with a light stage, during which the sleeper begins to relax and might experience pronounced changes of one's sense of the body, including light floating, as well as brief flashes of light and blots of color. Progressing into the middle stages entails slowly increasing vividness and intensifying hallucinatory phenomena, where trains of thought are likely to be translated into sequences of images.

Hypnagogic visions evolve from intricate and scintillating geometric patterns, early on, to the beginnings of personal participation in fully formed scenes by deeper levels. During the early-to-middle stages of hypnagogia, by contrast with full-fledged dreaming, the sleeper is much more likely to be an impersonal witness to a creative light show, while remaining somewhat emotionally detached from, and uninvolved with, rapidly alternating images and impressions. The images themselves, at least to begin with, do not tend to cohere into the kinds of narrative structures typical of dreams. Hypnopompia may be somewhat less common, compared to hypnagogia. Visual impressions, sounds, or even speech, may spill over from the dream state to the waking world, and persist long enough to be noticed.

Dream researchers have tended to view hypnopompic phenomena as marked by a distressing, almost nightmarish, emotional quality.[10] Sleep paralysis is a paradigmatic example. Sleep paralysis combines elements of waking with REM sleep, and is most likely to occur during sleep-wake (or wake-sleep) transitions. The individual is aware of their actual physical surroundings, is capable of opening one's eyes, but unable to move due to continuing loss of muscle tone associated with REM sleep.[11]

Paralyzed, the person can exhibit terrifying hallucinatory experiences that spill over from REM stage imagery into semi-wakefulness. These episodes, which last anywhere from several seconds to minutes, can be extremely realistic and upsetting, involving hypervivid visual, auditory, tactile, and kinesthetic sensations.

Analyses of sleep paralysis reports indicate three commonly occurring factors in hallucinatory experiences that accompany sleep paralysis.[12] The first is that of an intruder, involving a highly convincing sensed presence in the room, often associated with overwhelming dread. The second factor is linked to the sensation of a physical assault, including felt pressure on the chest, and difficulty breathing. The final factor of unusual bodily experiences refers to sensations involving floating, flying, and autoscopic (out-of-body) experiences. Although these common features are highly consistent, the sleep paralysis experience is subsequently interpreted in accordance with a rich overlay of social and cultural layers, that can ascribe these episodes to phenomena such as visitations by malevolent spirits (including the prototype of an "old hag"), incubus/succubus attacks, ghostly possessions, or alien abductions.

What such experiences illustrate is that our sense of reality is configured internally. It can be impossible to subjectively distinguish hallucination from perception.[13] The dream researcher Stephen LaBerge puts it succinctly: "To the functional systems of neuronal activity that construct our experiential world, dreaming of perceiving or doing something is equivalent to actually perceiving or doing it."[14]

As discussed in Chapter 4, the function of self-in-world simulations is to direct and guide behavior. At a very deep level, the fidelity of simulations during dreaming are, most of the time, indistinguishable from waking simulations and sufficiently complex to guide sophisticated behaviors. Definition of the "real" is state- and observer-dependent, as demonstrated by sleep paralysis. A medical condition called REM Sleep Behavior Disorder (RBD) further demonstrates this fact.

A patient suffering from RBD no longer exhibits the loss of muscle tone and mobility that is normally associated with REM sleep. As a result, the sleeper can begin to manifest complex, coordinated, and sometimes whole-body movements during the REM stage that reflect the drama that unfolds in the sleeper's dreaming consciousness.[15]

Dreaming in REM and NREM Sleep

At some point, the phenomena of the sleep-onset period mature into scenarios where we are no longer passive observers of fleeting internal imagery. Instead, we become absorbed into richly layered dream worlds. What is the stuff that virtual dreamscapes are made from? Much like our waking experiences, the stuff of dreams consists of a setting, which includes some sense of self, a place/surrounding for the dream events, which unfold in mental time; populating elements, including dream characters, as well as animate and inanimate objects; and events taking place there, potentially including intentional actions, emotions, thoughts, sensations, and speech.[16] As simulations go, the dreaming consciousness is a highly convincing replica of our waking life.

The perceptual content of dreams is chiefly visual, with auditory components and body sensations following and smell and taste occurring only rarely. The sights and sounds of dreams cohere together into meaningful storylines. Dream scenarios are similar to movie scenes, albeit more fragmentary. The emotional coloring of dreams can be quite strong and are often accompanied by negative emotions.[17]

In the vast majority (approximately 90 percent) of dreams adults report, one of the most stable features is the dreamer as a central character who interacts with other characters and surroundings in the dream.[18] As Revonsuo states:

> The self in the dream is the character who represents the dreamer. This character, the dream self, usually possesses a body image much like the one experienced during wakefulness. The dream self is positioned in the center of the dream world. The dream setting and events are seen and experienced from his or her point of view. The dreamer feels as if he or she were embodied inside the dream self's body, more or less the same way we feel our own embodiment inside our bodies during wakefulness. Thus, the dream self has a bodily existence and location in the dream world. In this respect the dream self is not all that different from the waking self.

The dream self is what endows dreaming simulations with that indubitable sense of *presence.* One difference between our dream and waking senses of selfhood is that dream selves tend to be much more changeable.

Revonsuo gathered the following example from studies of dream reports. The first comes from a twenty-five-year-old female student, who says in one dream excerpt, "It's the Second World War and I am a dark-haired, strongly built, Finnish male soldier." And then later, as part of the same dream sequence, "I could see myself in a mirror. Now I was a blond, strongly-built woman."[19]

A much rarer possibility, accounting for perhaps 10 percent of dreams, is the disappearance of an embodied dream self altogether.[20] In such dreams one might be viewing oneself as just another dream character, seen from an external perspective, or the entire dream might be experienced as if from a "camera's-eye perspective." Even in these unusual cases, however, there is still a center point or coordinate system (for example, the camera's point of view), a frame of spatial reference from and in relation to which the rest of the virtual world radiates.

Dreaming is most consistently and highly correlated with REM sleep. Awakenings at this sleep stage are most likely to be associated with immersive hallucinations, where the dream self finds itself navigating different environments, engaged in all sorts of activities, often with strong emotional tones. While dreaming dominates REM, approximately up to one-quarter of dreams throughout the night occur during NREM sleep.[21] The probability of NREM dreams is especially pronounced as the sleeper approaches eventual awakening. The conventional consensus has long been that these NREM dreams are somewhat emotionally flat, more mundane, shorter in duration, and less bizarre than the fantastical and theater-like productions of the REM stage. More exhaustive and detailed analyses suggest, however, that the supposed qualitative differences between NREM and REM dreams are minimal.

As sleep progresses, there are pronounced deactivations of frontal brain areas. More specifically, the dorso-lateral sectors of the prefrontal cortex (located on the top surface and extending toward the sides) reduce their metabolic activity during the wake-sleep transition, with perhaps the strongest reductions evident during REM sleep.[22] Since the dorso-lateral prefrontal cortex is associated with executive functions, like working memory, the ability to organize ordered sequences of thought is impaired, and with it the capacity for highly selective retrieval of autobiographical memories. These changes could help to explain some of the unusual cognitive features of dreams, including their abrupt, illogical narrative developments and the vagaries of selfhood described earlier. The ability to maintain meta-consciousness—being aware that you are aware—is also strongly impaired in dreams, another likely consequence of frontal brain inactivity.[23]

Why Do We Dream?

Theories about why we dream (what dreams are ultimately *for*) are as numerous as the theorists who have thought deeply about the subject. Despite its immense popularity in the first half of the twentieth century, Sigmund Freud's conceptualization that hallucinatory wish-fulfillment is the chief function of dreaming is largely discounted by modern psychology.

J. Allan Hobson's activation-synthesis theory depicts dreams as a type of delirium or reversible psychosis, explained by the chaotic firing of brain stem nuclei bombarding the cortex, during a time when the cognitively sophisticated prefrontal regions are massively deactivated.[24] The cortex's hodgepodge attempt to interpret random volleys of brain

stem excitation, coming from below, results in the bizarre constructions of our nightly dreams.

In more recent times, Hobson advanced his "proto-consciousness"
theory of dreaming. According to this version, the REM state (which
Hobson sees as the primary engine of dream making) is used by the
brain as an innate virtual-reality module: it simulates biologically
meaningful interactions of self in the world and refines one's predictions about how to sequence adaptive actions in space and time. Emotionally significant memories are also revisited during this time.

Hobson hypothesizes that the dream is an ideal setting for running
behavioral rehearsals, since the rehearsals can be tested without risk:
running away from a dream tiger is not going to hurt you. He views
dreaming as a kind of training wheels for waking consciousness.
During dreams, various brain / mind functions are exercised and integrated, so that "dream consciousness guarantees the binding of sense
of self, motility, sensation and emotion. It is upon this base that waking
consciousness is built."[25] Others have speculated that the primary
function of dreams is to support problem-solving, simulate responding
to physical and social threats, support the brain's capacity to regulate
emotions, or facilitate consolidating recently acquired memories.[26]

Ultimately, any functional theory of dreaming, which would pur-
port to discern some definitive reason why we dream, has to contend
with the sheer variety of types of dreams. Some cultures, for example,
make a distinction between mundane (little dreams) and dreams that
are packed with emotional significance and meaning, ones that put us
in touch with a sacred dimension of life qualitatively different from
everyday concerns and themes (big dreams).[27] Among traditions with
strong contemplative roots, one finds elaborate dream classification
systems. For example, in the *Caraka Samhita,* a foundational treatise
of Indian medicine, seven different types of dreams are mentioned.[28]
Some pertain to the fulfillment of unrequited desires, some are the
product of intense trains of thought and the day's waking residues, some
are related to anticipation of future events, and some are indicative of
the body's biochemical processes. In the Buddhist tradition, ordinary
types of dreams, manifesting from recent traces in the brain / mind and
body, also come in distinct varieties, tracing their origins to the forms
of psycho-physiological tensions with the strongest momentum in
the sleeper.[29]

Dreaming: A Default Setting of the Brain/Mind?

Dream research pioneer G. William Domhoff has critically reviewed some of the leading contemporary theories of dreaming and found all of them to be inadequately supported by empirical findings.[30] Domhoff concludes that dreams emerge as a natural by-product of a specific type of neurocognitive processing, rather than having evolved to fulfill a particular function per se.

Domhoff is far from suggesting that dreams are meaningless. He acknowledges that they are replete with psychological meaning, a stance compatible with how some religious traditions, including Tibetan Buddhism, regard them. Here, dreams are viewed as providing a highly opportune inner laboratory for mind research and development, but, as the Dalai Lama says, "if you ask why we dream, what's the benefit? There's no answer in Buddhism."[31]

If Domhoff is correct, then of what type of underlying process is dreaming a by-product? Stated simply, dreaming is an especially imaginative, immersive, and hypervivid extension of waking thought. Domhoff's hypothesis is buttressed by neuroimaging findings that the neural substrate of dreaming overlaps strongly with the default-mode brain network (DMN).[32] As we mentioned in Chapters 2 and 6, the default-mode network is one of an approximately dozen resting state brain networks that are spontaneously active during quiet rest. The DMN is activated when an individual is not focused on the outside world. With the default-mode in ascendency, we begin to entertain increasing amounts of mind-wandering (daydreaming), frequently transitioning from one train of thought, to another. As the psychologist Daniel Schacter has phrased it, when engaged in this kind of cognitive doodling-around, what we are doing much of the time is "remembering the past to imagine the future."[33]

Given that large portions of the DMN are activated during sleep, Domhoff claims that dreams represent a supercharged version of spontaneous waking thoughts and mind-wandering. They are different in degree, rather than kind. "Dreams are an especially dramatic simulated enactment of the dreamer's conceptions and personal concerns," Domhoff writes.[34] The reason that our dreams are so much more immersive and perceptual-like, during sleep compared to waking daydreams, is explained by a slightly different neuronal configuration of dominant brain networks. Not all of the default-mode brain structures are activated in identical ways during sleep relative to waking. In addition, brain areas involved in internal perceptual imagery are also active in dreaming. While during waking hours, the DMN is inversely correlated with activity in visual brain areas, during REM sleep there is a large scale topographic change in brain network connectivity resulting in increased communication between the DMN and sensory areas, such that internally oriented mentation is perceived *as if* occurring "out there."[35]

Additional evidence concerning the continuity of waking and dreaming comes from REM dream reports themselves. The conventional depiction of REM dreams as highly bizarre (Hobson refers to them as a form of psychotic delirium) has been grossly overstated.[36] Careful attention to dream reports suggest that highly novel and unusual elements in REM dreams are fairly infrequent, though they may be particularly likely to be remembered and then rehearsed during wakefulness. Researchers have concluded that the average REM dream report is a "clear, coherent, and detailed account of a realistic situation involving the dreamer and other people caught up in very ordinary activities and preoccupations, and usually talking about them."[37]

There are numerous potentially fruitful and mutually enriching lines of overlap between Domhoff's account of the dreaming process and explanations drawn from various contemplative schools of thought. For example, in the Vendantin tradition of Sankara, dream images are understood to sprout from seeds of persistent impressions, particularly ones with an emotional tinge, which are deposited into the stream of consciousness throughout the day.[38] Dream images are filtered through the prism of habitual predispositions and latent impressions (*vasanas* and *samskaras*), which we might metaphorically depict as energetic foci that, due to one's prior history of either rewards or punishments, gathers strength and momentum in the sleeper's mindstream. In the Tibetan Vajrayana (Diamond Vehicle) tradition, the churning mental activity conditions waking and dreaming consciousness, in equal measure.[39] We next turn to some of the exciting ways in which a contemplative perspective can drastically expand and enrich the current scientific understandings of dreaming consciousness, as well as the ways in which these divergent strands can be combined in future explorations.

Contemplative Approaches to Dreaming

The radically unique starting point, for the contemplative approaches to dreaming that we will examine here, is that our everyday, waking experiences, far from disclosing the true nature of things, are in themselves a dreamlike state. When we dream, things happen to us, and in the dream we take them as true. A predator is chasing us, and we feel afraid and try to escape. Or we win the lottery in the dream, and we feel ecstatic. Things happen, and we react, becoming the playthings of stimulus and response chains.

The traditions that we will now discuss see these responses as not dissimilar to how we behave and react when we think we are awake. The kinds of visionary dreams that we have upon falling asleep are understood as a "double illusion," or a dream within the larger dream of our everyday life, where we tend to function as if half-asleep most of the time.[40] When we dream, we think that is real. But then we wake up, and we think *that is* real. Many Eastern spiritual traditions believe that, whether dreaming while asleep, or having experiences while awake, in both cases we are stuck in a persistent pattern of fundamentally misperceiving the mind's true nature.

When awake, we go on reacting, mostly passively, to things that happen, while our default-mode brain network churns away like a well-oiled machine. Most of us rarely pause to question the subtle aspects of our minds and their workings. In contrast to the deeper, underlying dimensions of our consciousness, the mental appearances that populate our waking experiences are in many ways as insubstantial as dream tigers.

The Buddhist contemplative claim, backed up by generations of practitioners, is that without prolonged training in stabilizing attention and in observing the flow of experience, we remain at the mercy of the unfolding dream narrative. In bringing a state of open awareness to the hours spent in sleep, dreams can instead serve as a portal into a state of heightened clarity, allowing us to carry over that lucidity across both sleeping and waking hours.[41] Contemporary science has largely neglected such claims. Nevertheless, study of the links between awareness training and dreaming carry the potential to use dreams as an avenue for mental training, with benefits that extend to the waking state.

At the Borderlands: Witnessing the Play of Consciousness

The initial difficulty faced by a would-be psychonaut who desires to explore the hypnagogic and related liminal states is to notice their existence in the first place. A qualitative shift that happens when moving down into the more interior segments of the consciousness spectrum is the gradual stripping away of our familiar mode of doing to reveal much more diaphanous experiences of pure being. Without at least some preliminary training in stabilizing attention, and maintaining concentration, noticing states of pure being is not something most of us are particularly good at. We tend to largely miss their occurrence for lack of any free attention to reflect on the actual experience itself.

It should not come as a surprise that the experiences existing at the borderlands between the *koinos kosmos* and *idios kosmos,* are intimately familiar to serious meditative practitioners. Probably every contemplative path mentions such phenomena. For example, in the Japanese Zen tradition, the term *makyō* explicitly refers to a realm of "demons and devils," encompassing the wide range of hallucinatory phenomena that can manifest during states of profound meditative stillness.[42] The states are referred to as demonic since their great danger is that the meditating subject, in a state of absorbed attentive fascination, can easily be lured into unconscious identification with visionary phenomena. The practitioner then treats them as substantive, independently existing things (that is, the hallucinations are reified). Succumbing to the temptation, one imperceptibly enters into a state of grave and chronic delusion, mistaking lofty transient experiences as indicators of authentic realization of the mind's true nature.

Vajra Essence, a treasured text belonging to the Nyingma school of Tibetan Buddhism, likewise contains extensive descriptions of the signs that will begin to appear to the meditator with maturing practice.[43] These signs, *nyams,* are of numerous types, but they can include temporary experiences of bliss, as well as hallucinatory visions of rainbows or sacred deities. Whatever happens, the text repeatedly warns, try your best to remain uninvolved with these phantom appearances, which are no more than manifestations of the mind's perpetual propensity to function as an imagination engine. At root, no matter how seductive on the surface, the panoply of visionary phenomena are no less cinematic productions than the contents of ordinary dreams of ignorance.

The practice of yoga nidra, which has become more widely known and available in recent decades, for many people today offers a particularly accessible avenue to begin intimately acquainting themselves with the dynamic workings of our "500-million-year-old virtual-reality simulator." Modern versions of yoga nidra can be traced to Swami Sivananda and his disciples, but it is rooted in much more ancient yogic and tantric practices that emerged in India, and that are alluded to in the Upanishads texts, dated to somewhere between 700 and 500 BCE.[44] The name yoga nidra refers to the paradoxical state of awake sleep—a state of consciousness that the practitioner enters into, and maintains, at the precarious edge between waking and sleep. Rather than an elaborate or abstract philosophy, yoga nidra is intensely experiential, as through a set of guided steps, the practitioner learns to embody a state of profound physical and mental relaxation.

Unaccustomed as most of us are to inhabiting such a state for prolonged periods, the natural tendency is to slip into a dull sleep and then to effortlessly identify with the internal imagery that eventually arises as consciousness slides down the spectrum toward the realms of hypnagogia and dream. The uniqueness of yoga nidra is that it fosters a highly receptive and awake form of psycho-physical relaxation, one in which the practitioner becomes a pellucid receptacle for the manifestation of dreamlike fragments, including imagery and the automatic affective tendencies to react to them. Here, one remains fully conscious during the inauguration of distinct states along the consciousness spectrum.[45] The desires to approach, merge with, and grasp some experiences, as well as to withdraw and resist others, are skillfully avoided in equal measure.

As taught by clinical psychologist and yogi Richard Miller, yoga nidra guides the practitioner through an exploration of the several sheaths (*koshas*) that ordinarily veil the deeper dimensions of consciousness.[46] At the coarsest level of the first kosha, this means gaining awareness of physical sensations, and from there the exploration continues through successive layers of emotional and feeling textures, through learned patterns of thoughts, beliefs, and images, and finally, at extremely subtle levels, to the persistent feeling of being a separate individual to whom, and for whom, phenomena appear.

Having traversed these veils, the yogi eventually arrives at what Miller terms the Great Undoing—a deep realization of the natural or spontaneous state (*sahaj*).

Essentially ineffable, *sahaj* is a pristine field of pure experiencing potential that is no longer subject to either arising or passing away, capable of neither enhancement nor diminishment, as all dreamlike appearances, and their emotional qualities, are. The ultimate aim of yoga nidra is to become a *purusha*—one who unwaveringly embodies and abides as this deepest dimension of conscious being. At this level lies liberation from the state of passive enslavement to wants, desires, aversion, and hatred, as one flowers into a clear and receptive channel in harmony with the creative flux and play of life itself.

Maintaining the uninvolved stance of a passive witness to the captivating play of consciousness may be comparable to the skill of a master martial artist or a tightrope walker. It is a skill that ordinarily requires many hours of training to overcome the powerful pull of prior conditioning. For this reason, contemplative dream practices are often undertaken only after exhaustive preliminary training of attention, as well as in the context of an ongoing purification of the practitioner's ethical and emotional life.[47] A dedicated involvement with such paths takes a lifetime of work and substantial effort. By and large, contemporary scientific theorizing does not concern itself with, or seriously entertain the possibility of transcending the default brain-mind settings.

The Tibetan Vajrayana Perspective

The most systematic teachings explicitly dealing with the dream state originate with a masterful Indian yogi, Padmasambhava.[48] In the eighth century, Padmasambhava brought his teachings into Tibet, where they were combined with elements of the indigenous shamanic practices. Refined over the years, the sleep and dream practices have become a staple within the Vajrayana tradition of Buddhism. The intention to undertake such explorations is never for the sake of mere novelty, but always reoriented back to a deep intention to achieve realization for the benefit of all beings. Traditionally, training in the sleep and dream practices is also believed to prepare the practitioner for the changes in consciousness that will happen at the moment of dying.

The Vajrayana perspective envisages dreaming as the ideal research laboratory where we can gain a practical, empirically based understanding concerning the appearance/emptiness aspect of all phenomena. Perceptions in the waking state, by contrast, are already too sedimented into their habitual patterns. In dreams, the entire sphere of consciousness is more malleable.[49]

In the Vajrayana understanding, every possible content of consciousness, across both sleep and waking, is indelibly marked by appearance/emptiness. A rainbow is the quintessential example of appearance/emptiness. The rainbow is not a solid, self-existing entity. When the electromagnetic energy from the sun hits rain droplets in the atmosphere, the composite wavelengths of the spectrum are reflected to the eye at different angles. In interacting with the retina and our nervous system, bright peacock colors emerge in consciousness. Yet we cannot say that the rainbow as such exists in some definite place—it is neither "out there" nor to be found inside our skulls, but rather something manifested when suitable conditions are aligned. It is impossible to ever grasp the solidity or thing-ness of a rainbow. To some degree, the rainbow exists (we experience it, it is a definite appearance). But when we drill down to the ultimate level of reality it disintegrates into an emptiness.

What is true of the rainbow, in the Vajrayana context, is true for all the other conscious experiences you may have. They appear, they seem to be in front of you, they are undeniably present, but they are simultaneously empty of any substantive, solid, independent existence in and of themselves. According to this analysis, even a hefty oak desk, although it appears to enjoy a certain degree of solidity in a relative sense, is nowhere to be found in an absolute sense. That is, what or where exactly is the "desk"? If we drill down, it is a collection of particles, which are mostly dancing waves of probability and empty space, at an ultimate level, and there's no "deskness" to be definitively found at any point. Appearances in dreams, and appearances during waking hours both materialize from a shared matrix of luminosity that underlies them, a deeper dimension of consciousness. Accumulated layers of cognitive confusion (solidified sense of inner and outer, self and other distinctions) and affective obscurations (the glue of emotional potentials, wanting to push away or pull toward the simulated world) shroud the inherent emptiness of phenomena, so that we are left compulsively grasping at the appearance aspect, which becomes something like congealed "light forms" in consciousness, eclipsing their essential emptiness.[50]

Neither should the word emptiness, as used here, be confused for some kind of vacuity, or nothingness. Quite to the contrary, Vajrayana teachings specify that this is a special kind of "emptiness." Metaphors of space are sometimes used, but it is a luminous space, which is mysterious, and intrinsically aware, teeming with unrestrained potential to appear as anything at all.[51]

Padmasambhava's dream and sleep yogic practices, to be undertaken in order, consist of daytime practices, dream yoga proper, and luminosity yoga. Luminosity yoga is the most advanced state that utilizes deep, dreamless sleep—in scientific terminology, slow-wave sleep—and it will not be mentioned here. (We will, however, return to it in Chapter 12.)

Daytime Practices

The *daytime practices* serve as preliminaries to more profound engagement with the techniques of dream yoga. They are a prerequisite step on the contemplative way to realizing the potential of working with nocturnal dreams. Even prior to engaging with them, a practitioner will have attained very high levels of calm, abiding meditation, in which attention is stabilized to extreme one-pointedness. Here, we sketch only a brief outline of what the daytime practices entail, which always require the mentorship and supervision of an advanced teacher for their actual implementation.[52]

The initial practice cycle begins with taking frequent pauses, during the day, to conduct experiential probes, reflecting on the dreamlike characteristics of waking consciousness. For example, the practitioner may recall a detailed memory from the previous day and inquire from a purely subjective perspective whether there is any difference between this memory and last night's dream. Or one reflects on how future moments are no more real now than a dream vision or a distant memory and that the present moment is ever-elusive. Repeated engagement with such exercises provides a deeping experiential exposure to the insubstantial, dreamlike qualities of everything appearing to consciousness. It also begins to accustom the individual to not take surface experience at face value, but to conduct so-called "state checks" (that is, to intentionally probe whether one is dreaming while awake, or vice versa).

Next, one might begin to work with mirrors, a method recommended by Padmasambhava.[53] By closely inspecting objects in the mirror, including one's own face, one further reinforces the conviction that they are illusory appearances.[54] All attempts to finally grasp the reflections in the mirror as definite solids fails, and a realization slowly dawns on the practitioner that all of the mind's images are similarly evanescent, temporary appearances in the mind's "mirror."

Then we might experiment with sounds to facilitate an understanding of their ultimate insubstantiality. One powerful practice, which can be done in valleys or caves, is to shout and listen to the echoes. In a state of concentration and mindful awareness, one conducts a detailed subjective comparison between the sounds of one's voice, and the distant echoes, and comes to realize that both of these aural phenomena are equally ephemeral. In addition to sights and sounds, similar experiments are conducted with the many thoughts that emerge and disappear in the spotlight of consciousness. Dissecting and probing into them, one finds that thoughts, which initially seemed to be existent things, dissolve like particles of fog under closer scrutiny.

The final stages of daytime practices involve elaborate and extremely detailed mental visualizations, often of different deities. The practitioner learns to see them with their mind's eye in all their vividness, as if they were really present. These illusory bodies are understood to be phenomenologically similar to the bodies seen in dreams: intensely vivid, and yet nonexistent, like rainbows. After learning how to manipulate and project such visualizations, their forms are intentionally dissolved into the practitioner's own body, facilitating a complete commingling of inner and outer.

Dream Yoga: The Art of Lucid Dreaming

The nighttime practice of dream yoga involves reiterating one's intention to wake up in the dream, right before falling asleep. The practitioner next assumes a sleeping lion posture while lying down to sleep, which involves lying down on one's right side, as Buddha Shakyamuni did at the time of his passing away.

The sleeper then begins with a detailed set of visualizations, picturing specific objects (either one's primary meditation teacher, a deity, specific syllables, or spheres of light) which are located at various centers of the body. Characteristic anatomical locations at which these different visualizations are maintained could be at the crown of the head, the throat, or heart level. A wide variety of techniques were developed over the years to suit the individual style of a practitioner. People who have light sleep or suffer from insomnia are given different instructions than those who have deep and very dull sleep. Of most importance is an unwavering attention during the falling-asleep process, remaining fixed on one's intention to have a lucid dream.

Although it might require numerous attempts, eventually the practitioner is able to achieve a state of wakeful-like lucidity in the midst of dreaming. Initially, the periods of lucidity may alternate with non-lucid sleep. With increasing proficiency, the practitioner next moves on to practice the transformation of dream appearances. A common practice, at least initially, is to either multiply or subtract isolated components within the dream, like transforming a dog into an entire pack of dogs. The dreamer may also transform characters in the dream into sacred deities, or zoom in and out of different dream objects, or change the setting of an ordinary room into an elaborate meditation cave. A very pliable and subtle control of attention is called for here, since in the beginning, the sheer excitement of the possibilities tends to startle the dreamer into waking from sleep. The purpose of this phase is to gain direct, firsthand experience that everything that appears in the dream is simply a product or manifestation of one's own consciousness. Nothing that appears in the dream has any inherent existence of its own; everything is like a rainbow. Perception is much more similar to an imaginative construction than we ordinarily realize.

Since the content of dreams is held to emanate from accumulated tendencies, it is virtually inevitable that dreams sometimes take on negative, distressing emotional qualities. As discussed above, studies suggest that, on the whole, dreams are biased toward the presence of negative emotions. For advanced practitioners, this offers an ideal setting to realize the transformative potential of dream yoga. If the dream begins to segue into a nightmarish scenario, the practitioner is advised to embrace it. "Whenever anything of a threatening or traumatic nature occurs in a dream such as drowning in water or being burned by fire," advises the Tibetan lama Tsongkapa, "recognize the dream as a dream and . . . make yourself jump or fall into the water or fire in the dream."[55] By doing so, he teaches, the dream yogi learns that these aversive events, too, are appearance/emptiness, and they become fuel for the transformative fire of yogic practice, allowing for "seeing through" their illusory-like nature.

Lucid Dreaming in Contemporary Psychology

Contemporary psychology and neuroscience approach lucid dreaming largely as a circumscribed (and often very niche) research topic, rather than as a phenomenon that can lead us into the deeper dimensions of consciousness. As many as half of Americans report that they have had lucid dreams in which they are aware they were dreaming, about a fifth report they dream lucidly once a month, and a smaller fraction—about five percent—experience lucid dreams on at least a weekly basis.[56]

Modern psychologists have taught people to dream lucidly, first by encouraging them to notice periodically during the day that they are awake. Repeating this deliberate act of "testing reality" equips them to differentiate sleep and dream states, and then to know, if while sleeping they experience something bizarre or highly improbable, that they are in fact asleep and must be dreaming.[57] Researchers have further encouraged would-be lucid dreamers to set the intention just prior to sleep to have a lucid dream and even to visualize and rehearse a dream in which they will be aware they are dreaming. Another way to increase the odds of dreaming lucidly is to interrupt sleep after five or six hours with an alarm, for example, and to remind oneself to be aware of dreaming (perhaps by saying aloud, "when I fall back asleep, I will know that I'm dreaming"). All these techniques can be combined to cause lucid dreaming.

We now have objective evidence that lucid dreamers are undoubtedly within the REM state while having their experiences. Independent scientific teams in the United States, Germany, France, and the Netherlands have demonstrated the validity of so-called interactive dreaming.[58] The scientific teams used recordings of eye movements and facial muscle contractions to successfully enable two-way communication between the experimenters and lucid dreamers during sleep. Using previously agreed-upon communication rules, lucid dreamers were capable of giving responses to simple math problems, answering yes/no types of questions, and discriminating sensory stimuli presented by researchers, all the while not leaving REM sleep.

As these authors stated, these findings are analogous to "finding a way to talk with an astronaut who is on another world, but in this case the world is entirely fabricated on the basis of memories stored in the brain." Lines of communication between the *koinos kosmos* and *idios kosmos* stand opened. In the future, such 'world spanning' communication channels can be used to gain a much more precise understanding of dreaming consciousness without having to awaken research participants. The possibility of using these tools with accomplished dream yogis and contemplatives as research subjects can open completely new lines of mutual inquiry.

Neuroscience studies have demonstrated characteristic changes in brain physiology that accompany the lucid dream. As compared to an ordinary (nonlucid) REM dream, the strength of neuronal oscillations at around 40 Hz (gamma range) appears higher during the lucid dream, at levels intermediate between sleep and waking.[59] Neuronal oscillations of this frequency are putatively involved in the perceptual binding of experience into the unity of the conscious self-in-world that we experience subjectively. In addition, the synchronization of brain waves at multiple frequencies is much higher in lucid compared to nonlucid dreams. During a lucid dream, brain physiology shifts to a state that's intermediate between REM sleep and waking. This pattern of findings suggests that the lucid dream involves a hybrid form of consciousness, which incorporates components of the waking state with the internally generated imagery of dreaming. Specifically, the increased activation of precisely those frontal brain areas that we previously said are typically inactivated during classic REM sleep, is what likely accounts for the emergence of self-reflective capabilities of meta-consciousness that characterize lucid dreaming.

A scientifically informed understanding of lucid dreams opens fresh paths for discovering new ways to promote the likelihood of lucidity during sleep. The same research group whose work was just mentioned used electrical stimulation of frontal brain areas, precisely at the 40 Hz (gamma) frequency that they discovered to be a brain-state indicator of lucid dreams. Stimulating the brain with low intensity electrical currents at the gamma frequency during REM sleep increased subjective reports of dream lucidity. About two-thirds of individuals who received this form of brain stimulation experienced lucid dreams, compared to none of the study subjects who received a placebo form of brain stimulation.[60]

Another way to further boost the likelihood of lucid dreaming is through manipulating neurochemistry. For instance, a study discovered that administering galantamine, a molecule that increases the presence of the neurotransmitter acetylcholine, increases the frequency of lucid dreams.[61] In this case, individuals who were instructed in specific cognitive techniques for inducing lucid dreams were strategically awakened throughout the night and administered different doses of galantamine or a placebo pill before returning to sleep. Compared to the placebo, galantamine was up to three times more likely to lead to a successful lucid dream induction. The combination of a prepared mental set and pharmaceutically assisted boost in acetylcholine function may now be one of the most effective strategies for inducing lucid dreams in individuals lacking contemplative training.

There are likely to be numerous points of mutual contact between ancient contemplative techniques and the modern armamentarium of tools for enhancing dream lucidity. The spiritual traditions can benefit by being able to incorporate some of the most empirically promising techniques. In their turn, modern scientific programs can also be enriched by a deeper appreciation of the enormous potential for experiential learning and integrative psychological functioning that lucid dreaming can offer, beyond existing as a little more than an exotic research area.

Having considered how dreams provide a unique vantage point into virtual-reality models in their more pure form and the continuity of our phenomenal worlds during the day and night, we are now ready to consider the next question: Is it possible to discern in waking experiences something about the imaginative and dreamlike roots of our ordinary perceptions? Discovering the answer will take us down into foundational issues concerning the origins of our experiences.

8

Psychedelics, Sensory Isolation Tanks, and Dreamy Mental States

In 1958, the psychiatrists John C. Lilly and Jay T. Shurley presented an unusual paper at a conference held at Harvard Medical School. The symposium's theme was sensory deprivation, an area of scientific research just then beginning to hit its stride. The authors set out to test a fascinating question: When all sources of sensory stimulation are reduced to their absolute minimum, does consciousness collapse into sleep, or does it become attuned to inner worlds? Subjects were immersed in a sound- and light-proofed tank of gently flowing water, kept at body temperature. A breathing mask ensured continuous oxygen delivery. The tank was partially below ground and specially constructed to eliminate physical vibrations. Insulated from external sensory input, with the body afloat in neutral buoyancy, what would one experience?

After a few minutes, the background awareness of the body drops away, and it feels like one is flying through an enormous, dark space. What results is not so much a gap in consciousness as a flip in the direction of awareness, a turning inward of the searchlight that normally illuminates the exterior world. Subjects in the Lilly and Shurley experiments relived fragments of earlier memories, mentally time-traveled into imaginary futures, and experienced sensations of other presences, lights, voices, dog barks, fragments of music, and dream-like characters.[1] Lilly and Shurley concluded that, "when given freedom from external exchanges and transactions, the isolated-constrained ego (or self or personality) has sources of *new information* from within. Such sources can be experienced as if they are outside with greater or lesser degrees of awareness as to where or what these sources are (projected imagery, projected sounds, doubling of body parts, emotional states of euphoria or anxiety)."[2]

In Chapter 7, we reviewed examples of nightly dreams as hallucinatory experiences that can be partially understood as the invasion into the sleeping brain of waking-like consciousness. We now begin to examine the converse phenomenon: situations in which waking consciousness begins to feel like dreaming. The reason these cases are especially informative is that they can reveal to us the processes that enable the virtual-reality simulation of a self-in-the world, regardless of whether those models play out during sleep or waking. As will become clear, a neuroscientifically guided model of the mind suggests that we dream our lived reality into existence. To fully appreciate this counterintuitive argument and its far-reaching implications, we will first review evidence that our brains simultaneously straddle the "real" world and an imaginative one. Taking psychedelic substances or undergoing prolonged sensory deprivation makes it possible to dredge up "waking dreams" that are ordinarily below the threshold of awareness. Entire realities are conjured, made entirely from and in imagination.

A Phantastical Pharmacopeia

In the late nineteenth century, a pioneering German toxicologist created a classification system for psychoactive drugs, substances that exert noticeable effects on the human mind. Lewis Lewin's taxonomy encompassed commonly known stimulants like coffee and tea, euphorics, like the leaves of the cocoa plant, and sleep-inducing substances like chloral.[3] Yet arguably the most mysterious category was the one that Lewin poetically labeled *Phantastica,* with substances in this class possessing the capacity to supercharge visual imagination and fantasy. Lewin focused on peyote, a spineless cactus containing numerous psychoactive alkaloids—most notably, mescaline.

Today, the drugs in Lewin's broad category of phantastica are referred to by their less imaginative name, *hallucinogens.* Probably the best known among them are the psychedelics (like mescaline), which act on a specific neurotransmitter receptor for serotonin and are unrivaled in their ability to mimic dreaming.[4] Marked qualitative changes in consciousness occur under the influence of hallucinogens. These involve sensory hallucinations and misperceptions (with visual ones being most prominent), pronounced changes in one's sense of self and relationship to the body, and altered emotional and thinking processes. A classic analysis of psychedelics summarizes their effect on human psychology: "In its imagery, emotional tone, and vagaries of thought and self-awareness, the drug trip, especially with eyes closed, resembles no other state so much as a dream."[5]

The curious effects of psychedelics on the human mind captured the attention of some of the foremost psychiatrists of the early twentieth century. Many of them gained personal familiarity, conducting self-experiments with substances like mescaline and publishing their findings in premier scientific periodicals. "Everyone who has taken mescaline," wrote John R. Smythies, one of the towering figures of early neuroscience, "will make it plain that it is necessary to experience these phenomena oneself in order to fully understand them."[6] These pioneering papers are replete with rich phenomenological descriptions of an almost unfathomable cornucopia of fantastical perceptions. Frequently, the visions evolved from "vague patches of color" to "more complex sensa . . . mosaics, networks, floating arabesques, interlaced spirals, wonderful tapestries" amalgamating into "great butterflies . . . fabulous animals, soaring architecture . . . cities . . . fully formed scenes."

Curiously, mescaline left the observer's "observational integrity intact and . . . critical judgment unclouded" such that one felt and appeared fully awake, even though the "sense-fields . . . contain sensa which have no counterpart in the physical world." As Smythies noted, the visionary phenomena he experienced with mescaline occupied the same inner theater, as did his ordinary perceptions in waking life, suggesting the difficulty in drawing a clearly marked boundary where so-called hallucinations end, and the "real world" begins. These findings captivated many serious scholars who grasped that such substances opened a connecting link between the worlds of dreaming and waking. These early investigators recognized an unrivaled opportunity for experimentally addressing perennial philosophical problems about how the mind creates experience.

The alterations in consciousness induced by psychedelics resemble a hybrid state, falling somewhere between the ordinary waking mode and the one we shift to when we enter rapid eye movement (REM) sleep. Similarities between REM dreaming and psychedelic hallucinations were already remarked on in the world's first ever LSD trip, the one that would set expectations for future generations of amateur psychonauts. With World War II at its height, a young chemist, Albert Hofmann, was ensconced in the offices of Sandoz Laboratories in neutral Switzerland, examining the medicinal potential of ergot fungus. Following a hunch that a particular lysergic acid derivative (LSD-25) might be worth further investigation, the cautious Hofmann accidentally absorbed a miniscule amount of the substance through his fingertips. Hofmann's own account follows:

Last Friday, April 16, 1943, I was forced to interrupt my work in the laboratory in the middle of the afternoon and proceed home, being affected by a remarkable restlessness, combined with a slight dizziness. At home I lay down and sank into a not unpleasant intoxicated-like condition, characterized by an extremely stimulated imagination. In a dreamlike state, with eyes closed . . . I perceived an uninterrupted stream of fantastic pictures, extraordinary shapes with intense, kaleidoscopic play of colors.[7]

Anecdotal accounts that closely mirror Hofmann's are plentiful, but to quantify the patterns in subjective experiences scientists have sifted through thousands of written experience reports using what is called latent semantic analysis.[8] This method is based on the idea that words with similar meaning should appear with similar frequency in written text. Using latent semantic analysis, substances like LSD and Lewin's favorite, peyote, have been discovered to bear the closest similarities to subjective reports of lucid dreaming. By comparison, the flowering plants of the *Datura* genus, colloquially known as jimsonweed, are most similar to nonlucid dreams, in which the dreamer becomes completely identified with their dream.

One of the most detailed and comprehensive psychological analyses of psychedelia comes from Benny Shanon.[9] A cognitive psychologist by training, Shanon systematically documented hundreds of his own sessions with the Amazonian brew known as ayahuasca, a drink brimming with psychoactive alkaloids. He supplemented his observations with those of numerous indigenous and nonindigenous informants. One of Shanon's contributions was to elaborate on a framework for describing the visionary phenomena commonly encountered under the brew's influence, building on the early mescaline experiments. Shanon replicated a commonly observed effect that visions range from relatively simple forms, such as colored clouds and formless blobs, all the way to highly realistic visions of majestic landscapes, animals, and historical figures. Shanon discovered that this waking dream can also superimpose itself on top of the external world, as if holographic hallucinatory projections were shimmering across objectively "real" surfaces. The fact that the psychedelic trip readily combines influences from the external environment with imagery sourced from interior sources of fantasy and memory is what distinguishes it most from nightly dreams. That the person under the influence of a classic (serotonergic) hallucinogen is able to maintain lucid awareness throughout the intoxication period is another hallmark feature, further indicating that the state of consciousness induced by psychedelics melds features of waking and dreaming awareness.

In a typical nonlucid dream, by contrast, the dreamer simply accepts the bizarreness of whatever surreal elements bubble up, even if the very same material would be considered extremely unusual, incongruous or illogical in one's daily life. During a REM dream one would think nothing of seeing Salvador Dali-esque flying tigers or melting clocks, and taking them in stride. During a psychedelic state of consciousness, as Shanon observed, he was able to maintain a more or less continuous commentary on his vivid, imaginary experiences and to be constantly amazed by them. This ability of psychedelics to reveal the intermingling of the interior play of consciousness with sense perceptions derived from the external world offers a strong clue as to how our experiences are constructed.

Strobes and Dream Machines

Some two hundred years ago, Jan Purkinje, a Czech physiologist among the most well-known scientists of his generation, noticed that when he spread out his fingers and waved them in front of a bright light, he began to see "light and shade figures" that had no basis in

Examples of Purkinje's light and shade figures elicited by a flickering light source and optical forms elicited by mechanical eye pressure. © 2023 Hodari Nundu.

the physical world.[10] Rather than dismissing this as a trivial observation, Purkinje made detailed drawings of the figures and recognized that the phenomenon revealed something critical about the foundations of visual experience.

Purkinje inferred that the most elementary stages of visual perception consisted of a number of these basic "quanta" or "atoms" of visual consciousness. The rhythmic stimulation provided by waving one's hands in front of a light source provided sufficient energy to trigger the perception of Purkinje's forms. Suddenly, one could literally "see stars" even though they were not actually present in the exterior world, because their forms were being activated from within the brain itself.

Shortly thereafter, other scientists noticed that these phantom patterns could also be induced by rapid motion alongside sunlit railings or mechanical pressure on the eye. Extremely vivid illusory colors appeared when one stared at rotating disks. Nowadays we could say that these were early discoveries attesting to the fact that it is possible to "wiretap" into the visual brain and force it to populate our sensory field with images that have no corresponding existence in the physical world, much like what happens when we dream.

In the years shortly after the end of World War II, electronic stroboscopes were becoming readily available, allowing vision scientists to revisit and build upon the discoveries of their predecessors.[11] This new technology permitted precise control over the light source, creating the ability to dial in specific frequencies at which flickering lights stimulated the visual system. The outcome was an explosion of new findings concerning the ability of flicker to evoke hallucinations. Among the most frequent reports were those involving distinct geometric patterns such as filigrees, whirlpools, mandalas, and perceived motion such as pulsating mosaics and unearthly colors of breathtaking brilliance.

In patients with brain disorders, but also in otherwise healthy and neurologically intact persons, flicker could also elicit much more complex imagery, even up to "complete scenes, as in dreams, involving more than one sense," populated by "small animals . . . human figures." At times, the imagery could be directly traced to elements stored in memory, resulting in visual reconstructions of past events. The visual reconstructions mixed equal parts of memory and imagination, to create novel arrangements that then dominated consciousness. One subject reported the following: "I see myself, last night, trying to push my car, which broke down . . . and I can see, very clearly, the drops of water from the rain trickling down the windscreen . . . but it's funny because, though this really happened last night, my car is green and different."[12] All of this occurred while subjects were fully awake and looking at a completely featureless source of flickering white light.

John R. Smythies, the same scientist who investigated the mind-altering properties of mescaline, also mapped out the effects of flicker on visionary phenomena. He found that when subjects closed their eyes, and then had both eyes stimulated by the strobe, this elicited simple geometric shapes.[13] When only one eye was lit by the stroboscope, the other eye tended to encounter a bubbling cauldron of amoebalike shapes that would sometimes arrange themselves into full blown scenes, such as visions of landscapes or fish in an aquarium.[14]

Smythies also discovered that hallucinogen doses, which were too low to elicit obvious perceptual changes, could catapult people into intense visionary states when in the presence of strobing light.[15] Their perceptual world would be completely transfigured by brilliant colors, rarely or never seen in ordinary waking life, pulsing geometries, and dreamy scenic panoramas like seeing a woman walking down New York's Fifth Avenue with a poodle.[16]

The ability of flicker to evoke these unusual and startling effects appears to be particularly pronounced at frequencies matching those at which neurons in the visual brain naturally oscillate.[17] In other words, hallucinatory phenomena are especially pronounced when the rate of flicker approximates the rate at which groups of visual neurons are spontaneously humming before the strobe light is activated. This is a strong clue that part of what explains the fantastic visions is that the flickering light taps into the spontaneous neuronal chatter of those brain regions.[18] The background chatter of the visual brain ordinarily does not make its way into the bright spotlight of waking consciousness, but it can be made to do so when amplified using rhythmic flicker.

Aldous Huxley had attested to the visionary potential of stroboscopic light in the 1950s.[19] By the 1960s, with much of this knowledge being publicly available, there occurred an unusual intersection between visual neuroscientists and artists. The eccentric Beatnik writer,

William S. Burroughs, played a key role in this unlikely meeting point between science and counterculture.

Burroughs and the Beatniks were no strangers to exploring non-ordinary states of consciousness and regularly dipped into these waters for their creative inspiration. Intrigued by the similarities between the visions elicited by hallucinogenic drugs and flicker, Burroughs and his friend and fellow artist, Brion Gysin, seized on the opportunity to manufacture ways of hallucinating without drugs. Gysin was inspired by what Burroughs told him about the visionary potential of strobo-scopic lights and even more by the things he saw when peering into flickering light displays. He went on to construct a device he called a Dream Machine, which consisted of a perforated cardboard cylinder, rotating on a turntable lit by a bulb. When the rotation of the turn-table approached the natural frequency of alpha-band rhythms in the visual brain, sitting next to the Dream Machine predictably elicited an entrancing inner theater, all seen with closed eyes.[20]

Gysin intended for his device to enter into mass production and eventually replace home television screens, allowing people to access "interior vision" and the brain's intrinsic creative capacities. While the mainstream potential of Gysin's invention failed to materialize, a dig-ital version of the Dream Machine is now available as an app (https://dreamachine.co/).

Experiments in the Dark

In the 1950s, the psychologist Donald O. Hebb helped to initiate an entire line of research concerning the effects of sensory deprivation on the human nervous system. Hebb theorized about cell assemblies, groups of neurons that he believed are formed through synchronous firing (think of fireflies flickering in unison on a summer evening) and that serve as the building blocks of complex mental life. Hebb believed that appropriate transitions in the flow of cell assembly activation re-quired a continuous influx of stimulation from the external environ-ment. He theorized that artificial situations, where sensory exposure was drastically reduced far below usual levels, would produce dramatic alterations in the state of consciousness.

Hebb turned to the students in his laboratory at McGill Univer-sity in Canada to test his theories. In an ironic historical twist, it has since been revealed that Hebb's work was covertly funded by the CIA, which hoped to discover ways of making individuals susceptible to novel beliefs—an arrangement that has to some degree tarnished his legacy.[21]

The McGill studies consisted of putting people into prolonged isolation, lasting two to six days, in a cubicle that was sound insulated by headphones and ventilation noise.[22] Participants wore translucent goggles that prevented patterned vision and gloves/cardboard cuffs to minimize tactile stimulation. They spent most of their days confined to a bed but were permitted meal and toilet breaks. Needless to say, the research team found it exceedingly difficult to recruit subjects, and those who were recruited, hated the experiment so much that they frequently elected to terminate participation early. The research subjects spent most of their initial time asleep as a way of passing the time. When not sleeping, they were increasingly irritable and hungry for any source of sensory stimulation.

Among those who persisted, one of the most curious effects were vivid hallucinations, which the subjects compared to dreaming while awake. The hallucinations were predominantly visual and usually started with abstract shapes such as spots, circles, and other geometric patterns, much like described earlier in the context of psychedelics and flicker.

Over time these simple visions would morph into complex arrangements, up to dreamlike scenes filled with landscapes, architecture, and animal and human figures. These scenes were like quick snapshots of a movie, but without an accompanying narrative. The appearance of these visions in consciousness was completely involuntary. Some of the imagery tended toward the cartoonish, such as a "procession of squirrels with sacks over their shoulders," "prehistoric animals walking about in a jungle," "eyeglasses marching down a street," "rows of little yellow men with black caps on and their mouths open."[23]

Secluded in a bare room with nothing but the hum of the air conditioner fan, most subjects reported hearing things or feeling pronounced distortions in their body sense. Another common experience was that of being watched and in the presence of some other person or entity. The feeling of being in the presence of the "other" is a fairly common consequence of sensory isolation.[24] In fact, it has been observed enough to have earned a name: the *third man factor* (or syndrome).[25] The effect can range from experiencing a vague sense of a nearby presence to believing one is hearing, seeing, or even having tactile contact with said presence. The presence can be aloof, or it may interact with the person experiencing the effect. The former possibility is more likely in artificial research settings, while the latter is much more common in life threatening situations, where the presence can be a source of great comfort and guidance. Under the influence of ketamine, a powerful dissociative anesthetic, John Lilly experienced contact with numerous entities (some benevolent, some malicious) while floating in the isolation tank for hours on end.[26]

The researchers in Hebb's laboratory wanted to rule out the possibility that the students were just misreporting little dreams ("dreamlets") from their frequent naps. To do this, they attached electrodes to monitor the brain's electrical activity. The patterns of brain activity were coregistered with reports of hallucinations and used to confirm that these imaginative experiences were happening while the physiological record clearly indicated that they were awake.[27]

The basic pattern of these findings has since been replicated under less austere conditions. In one study, sighted individuals were blindfolded for five consecutive days. Two days into this sudden visual deprivation, more than two-thirds of the subjects were reporting florid hallucinations that similarly graduated from initial amorphous shapes and geometric outlines to (sometimes frightening, and often bizarre) visions of faces and objects.[28]

The studies coming out of Hebb's McGill laboratory caught the attention of John Lilly and Jay T. Shurley, who took the experimental setup to another level by creating the most stringent forms of sensory exclusion possible in water suspension. A unique aspect of the isolation tank is that it simulates a quasi-null-gravity environment by virtue of the strong water buoyancy created by thousand pounds of magnesium sulphate (Epsom salt). The tank keeps sensory stimulation to a minimum through ensuring that it is light, sound, and vibration insulated. The feeling of weightlessness further reduces the numerous computations that the nervous system is regularly performing to counteract the effects of gravity on our bodies.

Lilly's research was seen as central to space travel, a topic that was beginning to receive increased funding with mounting Cold War tension. Lilly himself was a frequent subject in his own studies, spending hours at a time in his float tanks, sometimes under the influence of powerful hallucinogens. (Lilly served as an inspiration for the fictional character Dr. Edward Jessup in the 1980s sci-fi cult classic *Altered States*.) Lilly's crucial insight was that the removal of external signals created an ideal environment for observing the brain's spontaneous internal dynamics.

Upon initially entering the tank, Lilly had experiences consisting mostly of the residual mental chatter and the buildup of a sort of craving for outside stimulation. Breaking through this barrier, Lilly found that a new sequence was initiated. He describes it rather dramatically: "the black curtain in front of the eyes gradually opens out into a three-dimensional, dark, empty space in front of the body." This "dark, empty space" could be filled by "small, strangely shaped objects with self-luminous borders" and a "tunnel whose inside 'space' seemed to be emitting a blue light."[29]

Arabesque, geometric patterns (by now, these should be sounding intimately familiar to you) popped up here too, as well as scenes from memory and imaginative reconstructions.[30] Lilly combined hundreds of subjective reports written by numerous people who experimented with his isolation tank.[31]

What Lilly was discovering was that floating in the darkness, in profound physical isolation, he had exactly the right set of environmental conditions to work with what he referred to as a "projection display." The projection display is the three-dimensional phenomenal bubble whose layout we first discussed back in Chapter 4. Proving that he was at the forefront of adopting the latest scientific breakthroughs of his day, Lilly (partly brilliant, partly mad) theorized that the isolation tank provided an opportune setting for visualizing and modifying the successive layers of "programs" he believed to be stored inside of the "human biocomputer," allowing individuals an opportunity to confront and change their prior conditioning.[32]

Although he was apparently not aware of it, Lilly's experiments were treading a terrain whose cartography had already been mapped, and in many ways mastered, by Eastern shamans and contemplatives in the wilds of Tibet. For hundreds of years, a protracted, seven-week retreat in the darkness of secluded caves provided the setting for one of the most dangerous and advanced mind practices of Bönpo, the indigenous religion of Tibet that existed prior to the arrival of Buddhism.

This visionary practice, known as *Tögal,* has always been shrouded in secrecy and undertaken only after years of the most demanding preparations. Advanced Buddhist and Bönpo yogis continue to progressively deepen and stabilize their meditative realizations in solitude through practicing the postures, gazes, energy, breath control, and *Tögal* yoga even today, in both East and West.[33]

In *Tögal,* the practitioner utilizes the protracted dark retreat as a way of cutting through habitual emotional patterns and achieving stability in the spontaneous manifestation of one's own energy, as it is vibrantly expressed in the appearance of "sounds, lights, and rays."[34] The "sounds, lights, and rays" are internally generated phantasmagoria, and the greatest danger of this practice is that the practitioner might develop a delusional belief in their reality, precipitating a psychotic break. The intent of *Tögal* is quite the opposite: "to realize that the vision of ordinary normal life is equally illusory and insubstantial," as explained by Lopon Tenzin Namdak, one of the most well respected contemporary Bönpo teachers.[35]

Tögal practitioners are disinclined to reveal the details of their experiences. Yet today we have a surprising amount of information concerning their discoveries. Having received the appropriate training, an American couple, Robert and Rachel Olds, spent nine years in retreat in the mountainous regions of Northern California, drawing on their professional backgrounds in the arts to depict the rich splendor of *Tögal* visions.[36] Even a cursory look through their colorful sketches reveals structural similarities between these *Tögal* visions and the shapes that participants in Hebb's and Lilly's studies reported.

An unprecedented testimonial to this yogic practice has been discovered at Lukhang, the secretive island temple once occupied by the enigmatic figure of the 5th Dalai Lama. A mural on one of the temple walls depicts some of the common *Tögal* visions, with geometric designs that would have been readily recognizable to many of the research subjects mentioned earlier.[37]

Demonstrating an impeccable penchant for systematizing their phenomenological findings, Tibetan contemplatives had elaborated a taxonomy of *Tögal* visions as progressing through four developmental stages, known as the Four Visions.[38] A stagelike model is certainly attractive, purely from a desire to systematize phenomenology. In reality, *Tögal* is more of a fluid and organic unfoldment than a neatly organized, linear progression. With that caveat in place, we can try to discern some of the major landmarks along the way.

The first stage involves small iridescent spheres (traditionally called *thigles*, but what a modern scientist might refer to as phosphenes). *Thigles* join together into chains or networks at the second stage. At this point one could begin to catch glimpses of isolated body parts of deities, as well as other human or animal entities. The seedlike *thigles* coalesce their bodies into full-fledged forms at the third stage. These forms are traditionally interpreted within their respective ethnocultural context as a variety of deities with peaceful and wrathful properties, serving as visionary vehicles for specific psychological energies. Entire generations of yogic practitioners have attested to this gradual evolution of visionary states from simple to more differentiated stages.

The fourth and final stage is known as the Exhaustion of Reality. During this stage, the teeming and protean multiplicity of visions dissolves back into a luminous, unified source, which is nothing less than the empty, space-like nature of the source awareness. It is precisely in this phase when the practitioner stands to gain a fundamental realization into the mind's innermost nature.

Tögal visions imply that across different cultures, people have been recording a core set of phenomena that appear to be drawn from a kind of universal library of possible visual experiences. The visual experiences can be evoked either by occurrences in the external world, or alternately, they can be triggered independently of any corresponding stimuli arriving from the outside world.

Staring into the Big Nothing

Another clue concerning the curious phenomenon of waking dreams comes from research initiated by the Gestalt school of psychologists, an eclectic band of German thinkers who in the 1930s attempted to integrate the study of the mind with the latest developments in physics. One member of the Gestalt school, Wolfgang Metzner, had come up with a peculiar experimental setup. He seated observers in front of a smooth, whitewashed wall illuminated by a projection lantern. Curved wings extending from this flat surface helped to maintain a homogenous optical field. He called this preparation a *Ganzfeld* (whole field).

A more cost-effective way of creating a Ganzfeld is using a halved ping-pong ball to cover the eyes while sitting in front of a nearby light source. When one stares into a Ganzfeld one's vision dissolves into a mist or sea of light. The usual experience, in which one's visual sense cuts up the world into a clear foreground set against a backdrop, completely falls away.[39] The background becomes the foreground and then melts into insubstantiality.

Visual scientist James Gibson, reflecting on his studies with the Ganzfeld phenomenon, offers this account: "What my observers and I saw under these conditions could better be described as 'nothing' in the sense of 'no thing.' . . . There was no surface and no object at any distance. . . . What the observer saw, as I would now put it, was an empty medium."[40] The idea of an empty medium that underlies all of our experience and serves as a universal backdrop (normally unnoticed) is a topic we will return to in subsequent chapters.

As the amount of time spent looking into the Ganzfeld increases, viewers may eventually have the sense of confused boundaries between inside and outside and experience this featureless and vacuous void as coming from inside their heads.[41] The void can also become a sort of projection screen for seeing the superimposition of various spots or geometric shapes.

The cavernous light and space installations of the American artist James Turrell have brought the Ganzfeld phenomenon into popular consciousness. Turrell has built room-sized Ganzfeld installations to create a unique no-point-of-view perspective, enabling the experience of "looking at the looking" itself. His so-called "perceptual cells" are dome-shaped capsules into which the viewer slides, only to be immersed in edgeless expanses of colored light.[42] Alterations in a viewer's sense of self-extension in space and time are pronounced. Some have found the effects so dislocating that they have crawled out of Turrell's installations. Experiential reports culled from visitors to the perceptual cells mention some not being able to tell if their eyes were open or closed. They report, for example, "becoming one great blob of red, soaking in baths of color," feeling "no body and no limits," with colors experienced as "inside my head, rather than around me," so that "you lose yourself totally," not being able to distinguish "where the color ends & I begin . . . more like feeling than seeing," a "feeling of dreaming and being lifted out of myself."[43]

Italian researchers have recently conducted several studies that immerse subjects into whole body egg-shaped chambers, flooded with homogenous lighting. Upon entry into these egg domes, participants report numerous dreamlike visual sensations, including fairly elaborate imagery of suns, trees and forests, even spaceship-looking objects.[44] For many subjects, the experience of time was also profoundly altered, including the seeming disappearance of time altogether, while visceral sensations of the body (interoception) were heightened.[45]

Something even more strange occurs when uniform sensory stimulation is simultaneously applied to other senses, for example by piping a constant white noise into the ears while people stare into a Ganzfeld. After several tens of minutes of exposure to this environment, people sometimes begin to experience dreamlike, hallucinatory perceptions.[46]

The report below comes from an experimental psychologist undergoing a prolonged immersion in a multimodal Ganzfeld setup:

> For quite a long time, there was nothing except a green-greyish fog. It was really boring, I thought, 'ah, what a non-sense experiment!' Then, for an indefinite period of time, I was 'off,' like completely absent-minded. Then, all of sudden, I saw a hand holding a piece of chalk and writing on a black-board something like a mathematical formula. The vision was very clear, but it stayed only for a few seconds and disappeared again. The image did not fill up the entire visual field, it was just like a 'window' into that foggy stuff."[47]

In-Sight 8.1 provides readers with basic instructions for creating their own Ganzfeld experiment.

In-Sight 8.1

You can make your own Ganzfeld setup with just a little bit of effort. You will need to gather a few things: a ping-pong ball (choose one without a logo), a precision knife or similar cutting tool, clear

---

medical tape, a lamp, headphones, and an audio recording of white noise. To optimize comfort of the eye cups, draw a wavy line that bisects the "equator" of the ping-pong ball (where the two halves have been glued) four times. Next, cut along the drawn lines to produce two shaped halves that will fit over the eyes and smooth the edges after cutting. Once the eye cups fit comfortably, secure them with the medical tape. For the light source we recommend a red incandescent light bulb. You can vary the distance (and therefore brightness) from the light source and experiment for different effects. Finally, don the headphones and play your white noise of choice—perhaps the constant, relatively unchanging sound of a waterfall.

What happens when you relax into this experience? Are you able to enter a dreamy state? See if you can notice any changes in your style of thinking, perhaps becoming aware of whimsical trains of thought that flitter up to consciousness. Do you experience visual shapes like transparent spheres, geometric patterns, honeycomb-like lattices? How is your emotional state affected? The Ganzfeld can become a blank screen for projecting the internal dynamics of your brain activity onto the clear phenomenal canvas before you.

As we saw in Chapter 7, in the gradual drift from waking to drowsiness and to various stages of sleep, the qualities of consciousness morph from relatively focused, abstract, and language-mediated forms of thought to more imagistic forms that often seem to enjoy a degree of autonomy. These images intrude into awareness seemingly out of nowhere.

One might ask: Are the fantastic fragments of dreamlike imagery spillovers of hypnagogic, sleep-onset or, even light-sleep phenomena? The evidence from concurrent recordings of the brain's electrical activity excludes that possibility. Rather, the findings unequivocally demonstrate that participants remained firmly rooted in waking consciousness during the intrusion of dreamlike sequences into awareness.[48] Participants are often firmly aware that they are simultaneously awake and present in a real-world setting, unlike an average dream experience in which there is unreflective fusion with the hallucinated world.

Specific signatures of functional brain activity also emerge with prolonged immersion in the multimodal Ganzfeld that differ from simply resting wakefully, with one's eyes closed. This latter observation points to a unique state termed *hypnagoid*—that is, neither strictly hypnagogic (transitional between waking and sleep) nor ordinary waking and encompasses an extended stretch of the consciousness spectrum. In this state-space, the waking mind is especially receptive to visionary phenomena conjured from internal, rather than external, sources.[49] Characteristic of the hypnagoid family of conscious state is the combination of high (waking) levels of alertness together with diffuse attention.

Hundreds of years before scientific psychologists ever considered creating perceptual Ganzfelds, various spiritual paths created their own versions of them. Doing so facilitated hypnagoid-like states of consciousness to serve as the beginning of a deepening inquiry into the foundational substrates of the human mind. The combination of high levels of waking arousal, with diffuse attention, and very sparse sensory content, seems to increase access to deeper layers of the mind that might be routinely obscured by fascination with specific contents of consciousness that routinely absorb attention. Beyond the dark retreats mentioned earlier, yogis were also practicing with uniform sensory fields like vast expanses of the sky. An instruction from an ancient Kashmiri tradition provides one example of such a practice: "In summer when you see the entire sky endlessly clear, enter such clarity."[50] Or again, the Tibetan mystic and scholar Longchen Rabjam: "Through watching the sky of the outer world, which is taken as a symbol of awareness, the awareness thereby symbolized will arise. The secret sky will subsequently manifest, that is, the realization of primordial wisdom."[51] Here, primordial wisdom refers to the subtle backdrop of consciousness, which is normally concealed from awareness.

To get an experiential feel for these types of contemplative practices, try In-Sight 8.2. Ideally, you should try this exercise on a cloudless day, looking out at the endless blue sky, or while gazing out onto a large expanse of water.

In-Sight 8.2

Assume a physically comfortable seated position. For a few minutes, breathe naturally and calmly, gently noticing the in-and-out breaths. The more that you can induce a state of physical and mental relaxation the better. If any sense of effort arises, invite it to collapse in on itself and dissolve into an open, receptive mode.[52]

When you are ready, open your eyes and gently gaze into the open expanse of space. Your eye muscles should be relaxed—you are not straining to see anything in particular, but instead just allowing your eyes to calmly rest, and easing your unfocused gaze out into the multidimensional space before you. While maintaining relaxation, feel that your breaths are moving in and out into the luminous sky space, gently filling the spacious and all-embracing expanse. As your gaze absorbs the empty space in front of you, allow the free mingling of "inside" and "outside" spaces, and feel how they freely dissolve, one into the other. Allow any thoughts that arise simply to merge back into this state of continuous dissolving, like waves into an all-pervasive sea of present moment awareness. As you continue to breathe, allow the inside and outside to mix and mingle, until all distinctions between the space out there and the space inside your head have melted. Enjoy the openness of this space for as long as you can. This exercise may help you gain an initial experiential realization of the clear space of awareness, the background context that provides a container for your thoughts, feelings, and sensations.

Dreaming While Awake in the Surgeon's Operating Room

The annals of neurology contain surreal accounts of waking dreams. In 1957, Wilder Penfield, one of the most famous neurosurgeons of all time, read a paper before the National Academy of Sciences that described a reliable way to elicit dreamlike experiences from awake epilepsy patients on his operating table.[53] He discovered that electrical stimulation of the temporal lobe consistently elicited these strange hallucinatory episodes. In the first such instance, as the tip of the electrode activated the target, his female patient suddenly said she felt that she was reliving a previous experience, while being fully aware that she was lying on the operating table. Over the next several years Penfield made many similar observations in other patients. The overall pattern was always the same: patients experienced hallucinatory audiovisual scenes, as real as other waking experiences, all while aware of being on the surgical operating table. Many of the experiences seemed to correspond to memory traces. The experiences evoked by electrical stimulation of the brain were not just dim recollections, but entailed vivid reexperiencing.[54]

Given that Penfield never seems to have taken the extra step of attempting to corroborate the precise details of the autobiographical memories that his patients were reporting, and that many of them clearly and repeatedly compared their experience to a "dream," it is more likely that what was being dredged up to the surface was a hybrid of the real and imaginary. One of his patients offers the following, haunting account: "Something is coming to me from somewhere. A dream . . . I keep seeing things. I see people in this world and in that too . . . I keep seeing things—I keep dreaming things."[55]

The dreamlike quality of direct brain stimulation is also supported by a more recent report in the literature, this time of a middle-aged man with a slow growing lesion owing to a diffuse, low-grade glioma (brain tumor).[56] During awake surgery (the brain has no local pain receptors), the doctors discovered that they could transiently and reversibly "disconnect" him from the external reality, while preserving consciousness. When their electrode touched the white matter fibers in the left posterior cingulate, the patient suddenly stopped a simple naming task the doctors had him performing and became unresponsive for a period of about four seconds.

When doctors subsequently asked him what happened during this unresponsive interval, he reported, "I was as in a dream, there was a sun," indicating that while he was transiently disconnected from the objective world, he was immersed in an inner dream realm. The surgeons then found that they could repeatedly activate this dream generator by reapplying currents to the same area. Again, the patient would be transported into a parallel, inner world, this time reporting that he found himself somewhere on a beach. This example differs from Penfield's observations insofar as the experience of being in two places at once was missing. Rather, the patient was being "teleported" by his neurosurgeons between the inner and outer realities. The Penfield examples and this one both suggest that the right amount of electricity, applied to the right place in the brain, can bring to consciousness a complex mix of dreamlike experiences that blur the seemingly stark boundaries between dreaming and waking consciousness.

The electrical currents that unleash this inner movie or virtual-reality simulation need not come from the surgeon's electrode. The neurologist Hughlings Jackson coined the term *dreamy state* to describe a peculiar state of altered consciousness he observed in patients suffering from temporal lobe epilepsy, in which electrical storms of activity arising from an intrinsic source of irritation spread across the cortex.[57] Dreamy states are most commonly observed before the onset of an acute seizure. One of their defining characteristics is a kind of reminiscence, sometimes building up to a strong feeling of *deja vu*—the persistent sense that what is happening now has already happened at some point in the past. Of course, it seems entirely possible that the feeling of *déjà vu* itself may arise precisely from interference between the waking and dream worlds, suggesting a lived memory but actually resulting from a hyperlinked, echolike, present-moment impression.

A tentative conclusion, at this point, is that something about certain very extreme and unusual situations—ranging from psychedelic ingestion to sensory deprivation to stroboscopic lights, unvarying sensory environments, electrical brain stimulation, or abnormal brain function and structure—dupes the brain into mistaking its streams of self-generated imagery as coming from an external source. If it is true that these edge cases are really revealing something fundamental about how our experiences get put together, then it ought to be possible to gain glimpses of the brain in the process of hallucinating its reality even under less extreme scenarios.

David Foulkes, a dream researcher, thought to test this very possibility. He had people simply sit in a reclining chair, in a quiet room, until they felt relaxed but awake (a state confirmed by monitoring their brain wave activity). He discovered that as much as 24 percent of the imagery reported by people while awake was just as bizarre, surreal, and dreamlike as the imagery occurring in REM sleep.[58] It seems that our presumably "ordinary" consciousness is far stranger than we suspect.

Another study tested the possibility that a slight shift in attention could foreground the brain's continuous spinning of dreamlike narratives, which typically goes on completely outside of awareness.[59] Participants seated in a quiet, bare room were instructed to remain sensitive to the "immediacy" or "raw feel" of their perceptions, thoughts, and emotions, rather than become preoccupied with thinking *about* their experiences or labeling them with words, a technique that corresponds very closely to meditative practices developed by generations of yogins and yoginis. After as little as ten minutes, the outcome was a dreamy subjective state of unreality that was, in many ways, similar to what is reported under the influence of psychedelics.

It would be easy to add to the already extensive pile of evidence gathered here.[60] Collectively, these disparate observations converge to highlight the fluid nature of subjective life: they show that, in our dynamically constructed models of self-in-world, the boundaries between subjective imagination and objective reality are blurry and shifting. What is left for us now is to make sense of these disparate facts rather than treating them as the neuroscientific equivalent of a carnival sideshow. As Chapter 9 will explore, these various strands of evidence point to some rather startling conclusions about the nature of nervous systems and how evolution has equipped them to dream lived realities into existence.

9

Dreaming Reality

At the age of twenty-seven, Georges Seurat finally completed a painting he had been laboring over for two years. Since that date in 1886, *A Sunday Afternoon on the Island of La Grande Jatte,* a canvas ten feet wide, has become the most famous work in Seurat's colorful oeuvre and one of the preeminent examples of pointillist technique. *Sunday Afternoon* presents a cross-section of nineteenth-century Parisians relaxing on the banks of the River Seine. Upon closer inspection, these faceless people frozen in their timeless poses, as well as the backdrop of the island are only amalgamations of innumerable luminous dots cohering into a palette of spectral hues. In Seurat's universe, reality is composed of a dynamic agglomeration of constituent particles: tiny, vibrating spheres of light and energy. You can imagine swarms of such particles, in ceaseless motion, ever arranging and rearranging into medleys of patterns that fill consciousness.

In this chapter, we introduce an overarching framework that can help us to understand how immersive worlds emerge in consciousness from a rich variety of possible, dreamed realities. Since our conscious world, both during waking and dreaming, is so heavily dominated by the visual sense, much of this chapter will focus on the roots and origins of visual consciousness. The same principles covered here, however, also apply to sound and other sensory phenomena. Like Seurat's paintings where the visual mosaic is a prismatic display of lower-level ingredients, we will now undertake an exploration of the basic building blocks that make up the seeming solids of worlds experienced in consciousness.

"Brain Hacks" and the Origins of Visual Experience

In 1974, Jurij Moskvitin, a Danish mathematician, philosopher, and pianist, published an insightful essay informed by his own phenomenological analyses of hypnagogic visions.[1] This prescient work took quite seriously the question of what an exploration of non-ordinary states of consciousness might tell us about the rudimentary origins of mind—a perspective out of keeping with mainstream academic trends. Perhaps this is why it has largely languished in obscurity, having long since been out of print.

A seed of the idea that drove Moskvitin's wide-ranging essay came to him unbidden one spring day as he was enjoying a state of diffuse attention—alert but very relaxed, somewhere between waking and sleep. In this state, while squinting at the sun, he became aware of smoky and patchy patterns flitting across his visual field. Eventually, Moskvitin trained himself to maintain a stable focus of attention while making his observations, so that he could watch as these protean forms transformed into arrangements of various geometric patterns that "upon close and intense observation became the elements of waking dreams, forming persons, landscapes, strange mathematical forms."[2]

Moskvitin's next bit of ingenuity was to hypothesize that hallucinations and ordinary waking perceptions shared a common substrate. In both cases our visual consciousness is constructed from the same basic material, which could be triggered entirely from within or by impulses arriving from the outside world. The initial "smoke-like forms," "sparks," and the various patterns that Moskvitin witnessed arranging themselves into more organized and higher-order patterns were the primitive ingredients, or "selective forms," as he called them, from which the prismatic mosaic of visual experience is built. The perceptions of which we become aware are sampled from a larger internal library of possible visual forms.

Stated another way, our phenomenal world is composed only of the sensorial impressions of which our organs of perception are capable. A sequence of external events and impulses triggers within us a corresponding sequence of internal impulse-reactions: "Whatever we experience of a world external to ourselves is nothing but the release by external impulses of something that is in ourselves. We do not perceive an object external to ourselves but something in ourselves which has been called up by impulses from that object and which presumably corresponds to it—colours, sounds, smells, and sensations are in us, and by projecting them onto the unknowable world around us we build it up like a mosaic from elements in ourselves."[3] A still more provocative way to put this is that we "dream" continuously with the difference that some of these dreams (the dreamlike simulation that we call our everyday, waking life) correspond more closely to the world beyond our skin than ones encountered during actual dreams.

The propensity of a properly connected neuronal network to spontaneously auto-generate images (hallucinations), Moskvitin felt, was a natural function, while the capacity to perceive an organized, external environment was, at least partially, acquired through a learning process. Might it be possible, he wondered, that our waking perception "somehow derives from the spontaneous capacity for having hallucinations?"[4] Of late, Hobson's proto-consciousness theory hypothesizes, in a similar vein, that the brain is genetically equipped with a virtual-reality generator; his theory proposes that the functional significance of REM sleep dreaming is in supporting the development and maintenance of waking conscious experiences.[5]

Having chanced upon a clever technique to "hack" his visual brain, Moskvitin was able to witness and document the genesis of experience, catching the steps that would ordinarily lie outside of awareness. He was all too aware that getting this "behind-the-scenes" glimpse into the constructive processes that generate perceptions is highly unusual. The fact that the consensually real world materializes only as a kind of selection and arrangement of internal elements is usually concealed from consciousness, since evolution has optimized organisms to survive and procreate precisely through creating highly convincing virtual-reality interfaces that afford seamless interactions. Fully inhabiting a virtual-reality simulation is more important to successfully escaping a predator than having deep insight into how the fabric of experience is woven. Precisely because they can bypass routine ways of experiencing, non-ordinary states open a window of opportunity for studying the ingenious machinations that nature uses to enchant us.[6] Departures from the "ordinary" configurations create opportunities for accessing the deeper and more causal reaches of the mind.

Discovering a Neuro-Mental Dictionary

Moskvitin was far from the only person who found ways of using innovative brain hacks to lift the hood on the construction of visual experience. When multiple explorers, working independently from each other in time and space, converge on a related set of observations, scientists learn to take notice. There is an obvious correspondence between Moskvitin's "selective forms" and Jan Purkinje's documentation of "light and shade figures" evoked by a flickering light source and mechanical pressure applied to the eyelids.[7] The neurologist Heinrich Klüver extensively cataloged universal features that were consistently present in the visions occasioned by psychedelics like mescaline, and he undertook a meticulous effort to group them into a kind of basic vocabulary of forms, which we now know as Klüver's form constants.[8] And centuries before any of them, yogis and yoginis were practicing rigorous psycho-physiological techniques of *Tögal* and describing how visions become built from elementary seeds of light (*thigles*).[9] All of these convergent strands of evidence are concerned with the same fundamental building blocks of visual phenomenology.

Pooling these different observations, we can now describe an inventory of visual phenomenological building blocks that we will refer to as a "neuro-mental dictionary."[10] This dictionary is not a static entity. It offers a large menu of possibilities for constructing immersive worlds and appears to be organized hierarchically, spanning many different levels of complexity.[11]

At the first, and most primitive level of this dictionary, we find poorly differentiated components. These can include phenomena like light flashes, spots and spheres, phosphenes, streaklike patterns, or visual snow consisting of things that resemble clumps of cloud matter. This initial level corresponds to entoptic phenomena—so called because their origin can be traced to the anatomy of the eye, like debris floating in the jelly-like liquid that fills the cavity between the front and back of the eye. Other entoptic examples are caused by light illuminating the rich networks of vasculature that nourish the retina.

Many of Purkinje's drawings are entoptic in origin, being representations of the eye's structure.

At the next hierarchical rung of this neuro-mental dictionary we encounter Klüver's form constants. Klüver noticed that the plethora of mescaline visions shared a smaller number of geometric universals. He categorized them into several different varieties, including tunnels and funnels; spirals; gratings, lattices, and checkerboards; and cobwebs.

Tunnels

Spirals

Honeycombs
Gratings

Cobwebs

Examples of each of Heinrich Klüver's form constants. Lisa Diez / Wikimedia Commons / CC BY-SA 4.0 DEED.

Combinations of these classes can also occur to create novel arrangements. Color and movement (rotation or pulsation) add a further dimension to the Klüver forms. The shape and structure of Klüver's form constants begin to incorporate intricacies of brain anatomy and function, including the connections between the retina and visual cortex.[12]

At the top of the hierarchy we find fairly complex imagery, including objects, landscapes, body parts, animals, human figures, and so forth. These are the "mental solids" that populate much of conscious experience. Complex elements, like figures, can sometimes be experienced in non-ordinary states of consciousness as assembled on or climbing on top of a scaffolding of more basic geometric designs. Much of the complex imagery, like the internal models rehearsed in nightly dreams, are drawn from memory traces of previous encounters with additional imaginative license.

Even at the stage of complex (iconic) visual features, there are certain universal constants. The range of combinations is extremely large, but not infinite. Our inner neuro-mental dictionary also contains obvious subgroupings of exemplars. In short, the complex images are some (albeit fairly large) sampling of latent possibilities that recur with some nonrandom probability. One study of more than five hundred hallucinations occasioned by LSD found clear regularities in the content, such that consistent categories of complex imagery were discovered.[13] These included religiously themed symbols and images (crosses, mandala-like patterns) and animal and human forms, both mythical and mundane.

So far we have restricted our discussion to a static snapshot of possible visual phenomenological realities. There is also regularity when it comes to the dynamics of how these various elements build on each other and evolve through time, with a simple-to-complex sequence that seems characteristic of many different states of consciousness.

This simple-to-complex pattern has been discovered and rediscovered independently, many times over, by numerous investigators.[14] This general sequence holds true regardless of the means eliciting visionary phenomena (sensory deprivation, mind-altering substances, hypnagogia, stroboscopic lights, and others).[15] The existence of a definite sequence hints at the possibility that the spontaneous self-organization of neuronal ensembles in the visual brain performs an exploratory walk along the length and breadth of a hierarchically organized neuro-mental dictionary.

Research at UCLA, by Ronald Siegel, a pioneer in the study of hallucinatory phenomena, confirmed this stagelike development in studies of chemically induced visions, which is sometimes formalized as a three-stage process.[16] A first stage involves both amorphous (for example, splotches of light which could resemble bits of cloud matter) and geometrically regular patterns, including zigzag lines, as well as the various Klüver form constants. At the next stage, many of these geometric regularities can be integrated, like Lego blocks, and interpreted by the brain as slightly higher-level, meaningful percepts such as crawling insects, or undulating snakes.

Usually, the initial eruption of a more complex vision, such as a natural landscape, is preceded by lattice-tunnel forms, with the subjective impression of passing through a kind of spatial vortex. Tunnel passages, as intermediate stages, leading to more full-fledged and complex hallucinatory phenomena, figure prominently in contemporary accounts of hypnagogia and psychedelic-elicited hallucinations and as a very common trope in reports of near-death experiences. And exactly such tunnel and vortex shapes would be predicted to occur based on the structural-anatomical details of how the eye is wired up with the visual cortex, and how information flows within separate areas of the visual cortex itself.[17] So this recurring subjective experience of passing through tunnels reflects a neurophenomenological fact: the people experiencing it are, in some sense, peering into the interior functional anatomy of their brains.

Siegel discovered that it was only at much later stages of chemically induced altered states that the complex visions moved to the center stage. At this final stage, a person can witness full-fledged visionary phenomena that are structurally identical to dreams. At this point, "recognizable complex scenes, people and objects" appear, "often projected against a background of geometric form."[18]

The following account, elicited from a guided session involving the administration of peyote (mescaline), offers a prototypical sequence of visionary phenomena:

The first impressions began to appear in about an hour. With the eyes closed I saw multicolored lines and streaks, faint at first but growing stronger, in a sort of moving kaleidoscope. . . . These images involving play of light, color and movement, were replaced by a static series of patterns of rich hues and abstract design, somewhat like tapestries. They followed at regular short intervals, changing automatically like a slide projector. . . .

As they gradually became more perceptible, scenes involving human forms and architecture began to emerge accompanied by play of light and color. . . . Most of the scenes were oriental— brilliantly illuminated landscapes, strange towers, pagodas, and temples, furnishing the background to exquisite lovely dancers. . . . The colors seemed to glow with an inner light. It seemed a glimpse of something timeless and primordial, a sort of breakthrough into the realm of the absolute.[19]

The Brain's Dark Energy

Now we know something about the phenomenological qualities of the visual neuro-mental dictionary. But since that dictionary is neuro-mental, we need to also consider the neurological side of this story. To do so, we reintroduce key elements of how brains work, first described in Chapter 2, which point to the fact that the brain is mostly engrossed in a ceaseless, ongoing monologue with itself.

We discussed how the majority of the brain's activity is spontaneous (or, as some say, intrinsic), rather than being determined by influences from the external environment. We stated in Chapter 2 that "we only become aware of those externally occurring signals that mesh with or get entangled within the brain's self-created webs of spatiotemporal activity." In some states, the brain chooses to ignore the contribution of signals from the exterior, preferring to shield its own ongoing self-in-world model simulations from outside interference, as it does during REM sleep.

One neuroscientist, Marcus Raichle, has referred to the large share of intrinsic and ongoing brain activity as the brain's "dark energy."[20] For much of the history of neuroscience, it has been difficult to conceive of a function for spontaneous brain activity. So the brain's dark energy has been shouldered aside as little more than random background noise. Today, we know that the noise hypothesis is woefully inadequate.

Unlocking the secrets of the brain's dark energy has been greatly helped by optics-based technologies for imaging functioning neurons, which has revealed that, far from being haphazard, the spontaneous activity of the cortex exhibits a highly organized structure.[21] Moreover, the likelihood of a single neuron firing at any given time depends on the activity of the local neighborhood within which it is embedded.[22] When the brain is presented with a particular stimulus, like a visual image, there is substantial variability in its response from one presentation to the next. Most of this variability can be explained by the ongoing churn of spontaneous brain activity.[23] This is hardly something one would expect if the background chatter was mere noise.

Perhaps most intriguing is what neuroscientists learned when they directly compared brain activity patterns in the complete absence of visual stimuli to ones recorded when the brain was being bombarded by visual images. In one such experiment, researchers presented simple black-and-white gratings, with different orientations to anesthetized cats.[24] Some of the gratings were vertical, some horizontal, and many others were slanted at various orientations between those two axes. After multiple presentations of each such stimulus, they captured the stereotyped patterns of response in the visual brain of the cats for each of the different orientations. In other words, each orientation of the black-and-white grating evoked a preferred signature of neuronal ensemble activity.

To capture spontaneous patterns of brain activity, the scientists closed the cats' eyes and dimmed all lights. They discovered that the patterns of neural activity recorded in the absence of all visual images were nearly identical when compared to those recorded when the cats were viewing black-and-white gratings. Deprived of input, the visual brain did not sink into quiescence. Instead, it kept spontaneously cycling through the configuration of neuronal firings associated with the various grating orientations, briefly settling on one pattern, then transitioning to the next, like a bee moving from one flower to another.

As one visual neuroscientist summarized these surprising findings, without any incoming information from the outside world, under anesthesia, with shut eyes, and shrouded in darkness, it was "as if the cortex were to be spontaneously 'hallucinating' a physical stimulus with a particular orientation," playing out an entire fantastical sequence of different grating orientations.[25]

Neither were the findings of this study isolated. Similar data have been gathered from mammals other than cats, and in waking, just as in states of anesthesia.[26] Today it is widely recognized that every part of our human brain, from single neurons to the entire cortex, shows signs of spontaneously self-organized neural activity in ways that closely resemble the patterns evident when responding to the world.[27]

A large portion of this ocean of spontaneous neural activity sketches a vast realm of possible responses to the external world. When the brain responds to a sensory input (say, an image), then a particular intrinsic configuration, out of many possible ones, is selected for some short period of time, before it decays, to be replaced by the next sequence. In other words, streams of sensory data coming from the world are in fact choosing from a palette of intrinsic brain activity patterns that form a large storehouse of possibilities, entries in a neuro-mental dictionary. The world around you is not passively imprinted onto the brain, in the way that a seal is impressed onto candle wax.

Even the physical wiring of connections in the brain bears out the conclusion that perceptions are sampled from a catalogue of internal model configurations. In their classic review neurophysiologists Rodolfo Llinás and Denis Paré presented overwhelming evidence that only a tiny minority of neural connections are actually transferring sensory information from the external world.[28] The majority are recurrent loops that allow large clusters of neurons to sustain rich repertoires of oscillatory activity that can operate either in the presence or complete absence of sensory data. Left undisturbed, the brain continuously spins through a ceaseless iteration of hypothetical scenarios, trying to reduce its uncertainty about the outside world.

After carefully poring over considerations of brain anatomy and function, Llinás and Paré concluded that "consciousness is fundamentally a closed-loop property" and that the only difference between waking and dreaming is that, in the former state, external stimuli are more effective in shaping the internally generated neural chatter. This means that ordinary perception is a dreamlike state that gets pruned by sensory data.[29] When the inputs from the world around us become weak (as, for example, in sleep, or during sensory deprivation), then the hum of the brain's dark energy is the primary source of activity with minimal stabilization from outside signals.

Possible Worlds

Our brains are put in charge of solving an immensely complicated problem: buoyantly floating within the skull's enclosure, they have to somehow build convincing and useful simulations of the world existing on the other side of the skull's wall. This challenge must be solved without direct access to the outside world, but only by referring to the internal configurations of its own components.[30] In modern jargon, this is referred to as a self-evidencing system: it points to things outside of its boundaries solely through modifying its own interior parts.[31]

Fortunately for the brain and nervous system, it happens to be attached to a body. So in its quest to derive a model of what is happening outside of the skull, it can propel the body into the world and then receive the consequences of those actions through sense data. You move a limb, or program an eye movement, and then receive the updated data that inform you about the consequences of those movements. Action and perception are linked in ongoing loops wherein each helps to shape the other, so that the organism scaffolds its way to the most useful simulation of being a self-in-the-world.[32]

One of the most famous scientists of the nineteenth century, Hermann von Helmholtz, realized that the only possible way that the brain could meet the daunting challenges set for it would be to constantly make conjectures about what was causing its internal states.[33] By now, it has become commonplace to say that the brain is a prediction engine. It constantly generates guesses (or better, hypotheses) about the workings of the world (which, of course, also includes itself), then acts on the basis of those predictions, and then uses the resulting outcomes to further improve its model of itself in the world.

While talk of the brain's building internal models may sound abstract, it is not. Your lived experience is a phenomenal experience of such a model, as is your sense of being a separate *you*. Brains building models is objective lingo, from a third-person perspective, but as it is experienced from the inside-out perspective, that model looks like what you see through your eyes. Modern neuroscience claims that your lived reality is an incredibly intricate and sophisticated model, built of many, many layers, nested one within another.[34]

The free-energy principle, championed by Karl Friston, happens to be the most mathematically sophisticated version of this predictive theory of the brain; it sees it as actively involved in the construction of a self-in-the world model.[35] For our purposes, we need to understand only the broadest and simplest of summaries. And since our focus so far has been largely on the visual building blocks of our phenomenal worlds, we will stay with examples of the visual sense, but with the understanding that similar principles apply to just about everything that the brain does.

The basic sketch goes something like this: information coming into the eye is conveyed along a long chain that is made up of numerous "stations." One of the first major stations is a neural structure known as the thalamus. The thalamus sends projections to a part of the visual cortex called V1. From V1 there are connections going up through a hierarchy of connected cortical areas: V2, V3, V4, and so on. Each area can be understood to instantiate a particular layered model of the surrounding rich, visual world. Lower areas like V1 are closer to the periphery (or surrounding world), and areas situated further up the hierarchy (V2 and beyond) are progressively further removed from the initial inputs from the world.

Traditionally, neurons in these various areas were believed to duly build up increasingly complex representations of the information coming in through the eyes. From V1-level information such as primitive lines, edges, and contours, neurons at higher levels would advance to objects and faces, as information passively percolated up through the hierarchy. The Fristonian free-energy theory specifies something very different and also deeply strange. According to it, the world is not being passively absorbed and then undergoing piecemeal assembly by the brain; rather, brains are using the information coming from the world primarily to constrain and narrow prolific inner fantasies and conjectures.

Each neural station generates a prediction about the activity occurring at the level below it and sends a signal down to that level to keep it informed of this guess.[36] For example, in the visual system, V1 generates predictions that get sent down, via a descending connection, to the thalamus. Since predictions come from one level up in the hierarchy, they are said to originate as top-down signals. Returning to an earlier analogy, you might picture predictions as samples from what we have called the neuro-mental dictionary and then conveyed down along a conveyor belt. V1's predictions look a lot like the geometric shapes and patterns described by Klüver. At the next level, V2 generates predictions that get sent down to V1, and so it goes on up the hierarchy. The end result is a linked and multilayered model of the visual world, which is constantly being refined to accommodate and explain more and more of the unknowns.

At any given station, whatever inputs do not match the prediction coming from the layer above it—the discrepant bits—continue to get conducted up to the next station. This prediction error or residual—the surprising, unpredicted part—is precisely the free energy that the brain tries to minimize. Free-energy minimization is a dynamic process that takes place through constantly updating and refining prior predictive models, to generate better future predictions.

Unlike the predictions, which are said to be imposed top-down, prediction errors ascend in bottom-up fashion. Facets of the world that were poorly explained by an initial model are successively refined, to explain away as much of the uncertainty as possible. If at first a predictive model does a poor job of predicting what it expects to receive, then it revises its predictions. These continuously refined predictions are precisely the world you see around you. At any given moment, your conscious experience is the best current model that your brain has access to at that point to explain a highly uncertain and complex world beyond it. In this way, the brain's "hallucinations" get pared down to scale, to provide a good enough fit to the evidence that your senses provide.

One of the most direct ways that you can experience the dynamic character of predictive updating is through an experimental phenomenon called binocular rivalry. To invoke it, scientists use a clever mirror set up to present different inputs to a subject's left and right eyes. One eye might see an image of a face, the other an image of a house. Since houses typically do not have human faces, there is no prior predictive model by which that particular combination would be expected.

In isolation, however, houses and faces are both perfectly feasible predictions of visual input. The way your brain handles the unusual situation of rivalrous inputs imposed by the mirrors is to periodically flit from one predictive model to another.[37] As a result, the subjective experience in binocular rivalry is one of alternating between having a face or a house dominate conscious awareness. At first you might see a face for several seconds, but this interpretation is only stable for some limited time. This is because the entire time that you are seeing a face there is a steadily accumulating buildup of prediction errors, as there are still signals consistent with a house coming into one of the eyes, which are not well explained. To resolve this growing discrepancy, your brain assumes that the prediction of a house is now the most likely interpretation and so, involuntarily, the face suddenly disappears and is replaced in your consciousness by a house.

There is substantial evidence that the spontaneous churn of brain activity corresponds to maintaining and updating an internally generated self-in-world model.[38] The spontaneously occurring brain signals embody continually changing predictions that are subsequently refined by streams of prediction errors.[39] At least one part of the brain's "dark energy" is related to maintaining a statistical model of self and world. This large set of priors or predictions are constantly changing, as the brain's activities are sculpted by its repeated contacts with the world. The world instructs the brain as to how it ought to subtly modify the way it "talks to itself."

At the early levels of the visual system, like V1, the kinds of predictions that the brain talks to itself about are simple ones, concerned with lines, edges, and other fairly primitive geometric patterns (like Klüver's form constants). At subsequent stations, like V2, V3, V4 and so on, the predictions get more complex, consisting of features like colors, motion, faces, bodies, natural landscapes, and so forth. Steps going up the predictive hierarchy, which we discussed in reference to the visual system, are associated with areas of the brain that are further removed from any single sense modality.[40] Models that exist at these levels pool and converge inputs from multiple senses. When this happens, the hierarchy ends up containing even more abstract predictions, including sophisticated narratives of selfhood, which are believed to be anchored to activity in a brain network that is furthest removed from the particulars of sensory data. These abstract and complex levels of the predictive hierarchy are rich in a property called *counterfactual depth,* because they can entertain possibilities and speculations that are highly fantastical and not specific to any particulars in the actual sensory data streaming from the world.

You might be wondering where the brain's predictive models come from in the first place. Some have been acquired through evolution, and they are "pre-installed" at birth. This is because, at least in a broad statistical sense, the structure of the world encountered by our ancestors over the millennia is not all that different from the world we are embedded in. For example, the visual system is born with specific predictions, also called priors, concerning what shape human faces and bodies ought to have. The visual brain also expects to encounter landscapes with sources of illumination that originate from the skies above. Other predictive models are unique to each of our life histories. They are acquired and ceaselessly sculpted to provide a unique match to the precise sequence of experiences that you, as an individual, encounter as you live life.

When we fall asleep and dream, sensory inflows get exceedingly quiet, so that what is left over is a largely hallucinatory world. We romp through expansive predictive models of what the brain has acquired during waking hours.[41] Then, as we wake up, the majority of these fantastical worlds become suppressed to allow us to navigate around within a very carefully "controlled hallucination."[42] Suppression of the brain's background conjecture-making, and those models that provide weak fits to the outside world, are critical to keeping a person from being inundated by counter-factual fantasies—it is what keeps one grounded in the immediacy of the moment.

During waking then, the brain is no longer spontaneously careening through its storehouse of predictive models, but rather its search path is being selectively guided by signals from the external world. Hand in hand, the world is metaphorically walking the brain to the most likely interpretation of reality. The world can look up very particular entries in the neuro-mental dictionary, rather than rifling through it, seemingly at random. "Dreaming can be viewed as the special case of perception without the constraints of external sensory input," wrote the psychologist Stephen LaBerge, and "conversely, perception can be viewed as the special case of dreaming constrained by sensory input."[43]

It is crucial to understand that there is absolutely no solipsism implied at any stage: no implication that the world "out there" is somehow illusory and nothing but a story that the brain tells to itself.

In fact, it is precisely from the *interaction* of a spontaneously creative brain and the real external world, that our perceptions get stabilized into an adaptive virtual-reality type of interface, allowing us to manage the exigencies of life. This partnership or handshake between brain and world is what allows one to cross a busy intersection without getting annihilated in the process.

Waking Dreams Revisited

At last, we have most of the background covered that we need to start making sense of the seemingly bizarre examples of "waking dreams" cataloged in the last chapter. To fully appreciate the mechanics that underlie the eruption of waking dreams into consciousness, it is necessary to add one more technical term, known as *precision*. *Precision* refers to the amount of confidence that we place on the predictions (the priors that the brain wants to impose on the world) or to the importance of the prediction errors (the mismatch between our priors and the true state of the world). A couple of concrete examples may suffice to get a better understanding of the technical meaning of the term precision.[44]

Walking through your house, at night, without a source of light, you are likely to rely primarily on the stored predictive model you have of the general layout (built and refined over countless previous interactions). For example, you are careful not to stumble into the coffee table, the approximate position of which you know, even if you cannot see it in the dark. Since the sensory data are very noisy and of a poor quality, the precision of your predictions (priors) in this case is quite high, while the precision of prediction error is low. When the precision of sensory data is low, the prior internal model enjoys substantial confidence, and it is less likely to budge in the face of information from the world. The brain's top-down predictive models become the primary determinant of what is "real" in the moment.

By contrast, consider driving through a completely unfamiliar neighborhood, perhaps in a foreign city. As you navigate through the streets, you are wise to place greater emphasis—precision—on prediction errors. The exact and concrete sensory details of the locale in which you are driving impress themselves with especial vividness and saliency when you are seeing them for the first time, perhaps trying to remember a few landmarks that can later serve as useful points of reference. Because you have no stable priors to rely on, the prediction errors play a preponderant role in making the "real" spring into sharp relief. Now we can begin to sketch out how the unusual situations considered in Chapter 8 can be explained using this short neurobiological primer on the predictive brain.

Anomalous Sensory Environments

Consider the example of floating in the complete darkness of an isolation tank, or prolonged exposure to the featureless field of a Ganzfeld with constant, unvarying audio stimulation. In these situations, when faced with sensory data that provide very little in the way of novel information, either because the sensory stimulation is so low (as in sensory deprivation) or constant and featureless (as in the Ganzfeld), the ascending streams of prediction errors become muted. This happens while the neuronal signal that represents descending feedback coming down into sensory areas from stored predictive models starts to wax stronger.

When it is being continuously confronted with low precision sensory data, the brain naturally begins to give free rein to fantastic romps through prior models of the self and world. The brain's inner monologue is no longer being strongly constrained by objective reality, so the sheer weight of top-down signals overpowers the weak, bottom-up ones. The buoyant virtual-reality simulator inside the skull becomes decoupled from any return signals that would otherwise indicate a poor fit and impose guardrails around the brain's endogenous explorations. In the absence of such reality-testing opportunities, hallucinations gradually get amplified enough to overcome the ordinary suppression, as they start to grow stronger and then blossom in seemingly bizarre varieties that spill over into consciousness.

The psychologist Louis West offered a beguiling metaphor for this entire process. Imagine a person, at dusk, looking out from a living room window.[45] As darkness descends, the outside world begins to fade, while the fireplace inside the room (analogous to the continuing brain arousal that sustains a state of waking consciousness) illuminates the interior. Pieces of furniture in the room's interior (analogous to the brain's existing predictive models) are now projected onto the window, creating the illusion that everything inside is apparently outside. By analogy, a similar process seems to explain the production of waking dreams during sensory deprivation and prolonged immersion in a Ganzfeld environment.

As we have already mentioned, as subjective visionary phenomena begin to erupt into consciousness, there is a sequence that progresses from initially vague, formless clumps of impressions to various simple shapes and geometric patterns, and finally all the way up to more complex, lifelike scenes. It may now be easier to appreciate that such a simple-to-complex sequence is very consistent with the hierarchical organization of multi-layered predictive models. These widely observed patterns indicate the existence of steadily propagating waves of message-passing within extensive nested hierarchies of predictive models. As the precision of predictions is dialed up from the weak precision of prediction errors, successively "later" and more complex samples from the neuro-mental dictionary get activated.

In terms of the more abstract, deeper layers of this dictionary, one often encounters multiple echoes of memories and memory fragments. These shards of memories often appear rearranged and stitched together in creative new varieties. Elements from the past become a prism for perceiving the present and a template through which to imagine the future, while all three—past, present, and future—are, equally, constructions.[46] Like a good tennis player scanning the court and anticipating an opponent's next move, the brain never tires of running through endless varieties of possible scenarios.[47] The actual content of these waking dreams can be indicative of emotional memories that have never been fully processed, and which still command substantial amounts of attention, as the brain keeps reverting to these priors to prepare for an uncertain future.

Psychedelia

The wide spectrum of fantastical dream worlds and alterations in the sense of self evoked by psychedelics emerge from alterations in the brain's predictive processes. In some ways, the effect of these mind-altering substances represents a mirror version of the scenario created by profound sensory deprivation or an immersion in a Ganzfeld environment. Whereas the latter are examples in which the low precision of sensory data shifts the scales in favor of prior predictive models, the former appears to entail a massive relaxation of prior predictions along with an attendant upsurge of bottom-up prediction errors.[48]

When predictions coming from the top down are low in precision, this effectively means that they have very weak felt confidence. As a result, the importance given to fresh streams of incoming sensory data is dialed up. The evidence from brain imaging studies increasingly points to the fact that after psychedelic administration, neural traffic is increased from relay stations in the thalamus up to the top surfaces of the cortex, while the signals coming down from the cortex and impinging on the thalamus are reduced in strength.[49]

Aldous Huxley's experience with mescaline taught him how to relish the beauty of common, everyday objects around him.[50] A walk in the garden, listening to music, the view from hills overlooking the city—all of these became occasions to live the crisp freshness of the world, such that even a regular chair was suddenly transformed into the most portentous of objects. For the title of his essay on mescaline, "Doors of Perception," Huxley borrowed from famous lines penned by the visionary poet William Blake: "If the doors of perception were cleansed every thing would appear to man as it is: Infinite. For man has closed himself up, till he sees all things thro' narrow chinks of his cavern."

Blake's "narrow chinks" are precisely that thick overlay of one's prior predictive models, which can sometimes be so heavy as to suffocate the actual vibrancy of the world as it is. His is an inimitable way to express what it is like, subjectively, to encounter the world as scintillating mystery, once it has been released from the thick veneer of predictions that we ordinarily impose on it through our priors. So often, for example, we never actually experience a tree in the park, but more so see our "idea" or a cognitive "schema" of a tree. In other words, we relate less to the reality behind our predictive models and more to our stored models of the world solidified through previous interactions.

Of course, the ability of psychedelics to reduce felt confidence in the brain's prior predictions affects more than just one's sensory systems. The "chink-opening" effect of these substances spreads to also reduce one's confidence in prior models of selfhood, to the point that even the deeply held belief that there is a stable reference-coordinate system for who "I myself am" (the belief, that is, that "there is a solid, independently existing *I,* a center from which everything is experienced") becomes deeply challenged.[51] One leading psychedelic researcher, Robin Carhart-Harris, hypothesizes that these substances act primarily to reduce felt confidence in past priors within these more complex, deeper levels of the predictive hierarchy.[52]

One dramatic expression of this is the so-called *drug-induced ego-dissolution* phenomenon, in which neural activity is altered within select brain areas to disrupt the narrative story of being a separate, boundaried self.[53] With such dramatic loosening of higher hierarchical levels, bottom-up prediction errors begin to "find freer register in conscious experience, by reaching and impressing on higher levels of the hierarchy."[54] Carhart-Harris believes that this surge in the salience of bottom-up neuronal signaling can help to account for the variety of alterations in visionary phenomena and alterations in sense of time, among other effects. Transformative changes in social-relational responding can also be facilitated under psychedelics. We frequently experience the people around us not as the incredibly multifaceted and interesting persons that they truly are, but rather through stored categories, built up from prior interactions and stored across numerous layers of complexity in inner self-world models of reality.[55]

Under the influence of a psychedelic, the brain can oscillate between explorations of different segments of the neuro-mental dictionary as it meanders through multiple layers of predictions that enjoy transient periods of dominance before being swamped by rising tides of resurgent prediction errors. The wide-ranging consequences of psychedelic ingestion are too complex to be summarized as invariably involving *only* a loosening of prior predictions. The rich visual phantasmagoria, described so evocatively in Chapter 8, can under some circumstances be better explained as a dip into existing layers of predictive models. When one's eyes are closed, it is possible to become more deeply immersed into these waking dreams, as there is reduced competition from sense data carrying prediction errors.[56]

Strange interpretations, not often entertained during ordinary waking perception, can also gain traction in the acute phases of psychedelic influence. Using the same type of binocular rivalry experiment that was described earlier, psilocybin increased the likelihood that viewers would fuse dissonant inputs to the two eyes.[57] While ordinarily a viewer would sequentially alternate between seeing one type of input and then another for some duration of time, under the influence of the psychedelic viewers experienced hybrids stitched together from typically discordant elements. Surreal, dreamlike interpretations become much more tractable than they otherwise might be.

Brain Stimulation and Stroboscopic Lights

The neuro-mental dictionary of visual forms can be directly activated by an electrode. A study that combined both EEG and fMRI neuro-imaging, for instance, discovered that periods of acute visual hallucination during flicker were characterized by disconnections between the thalamus (relaying signals from the exterior world) and the visual cortex. Meanwhile, brain communication among cortical areas increased during flicker-induced hallucinations—what people were seeing was increasingly determined by the brain talking to itself, exploring its own visionary potentials, while ignoring signals from the outside environment.[58]

The entries within the visual brain's vast neuro-mental dictionary have a rather anatomically orderly spatial layout. As a result, the hallucinatory percepts elicited by direct electrical stimulation of the brain change depending on location of the electrode's tip. Stimulating the early or lower regions of the visual brain, like V1, elicits reports of simple spheres of light or the Klüver form constants.[59] When the electrode is moved further up the hierarchy, patients can begin to see faces or the colors of rainbows or houses, superimposed over everyday objects. In one study, for instance, a simple box could suddenly appear to "grow" a face when the right brain area was stimulated.[60]

Flickering lights provide a much less intrusive way to activate some of the same portions of the predictive visual hierarchy that are activated by direct electrical stimulation. Exposure to stroboscopic lights is capable of igniting a broad spectrum of hallucinatory visual percepts. The nature of visual experiences is shaped by the frequency of the flickering lights, such that a strobe with speeds somewhere between 10 and 20 cycles per second can evoke a bevy of spiraling geometric patterns, while speeds below 10 cycles per second are biased in favor of radial shapes.[61]

The geometry of these subjective visual sensations can be predicted by elegant mathematical models that take into account the transmission of traveling waves of excitation that move from the eye to the early relay stages of the visual cortex.[62] In fact, the shape of the reported visual hallucinations corresponds very closely to the actual structure and layout of physical wiring in the visual cortex. In other words, in seeing the elaborate crosses, mandalas, arabesques and whirlpools, one is seeing the functional anatomy of the visual brain.

When stroboscopic lights reach very slow speeds, around 6 to 3 cycles per second, there is a much greater likelihood of activating more complex visions—faces, buildings, scenes—corresponding to higher levels within the hierarchy of possible visions.[63] The eruption of such full-fledged scenes into consciousness requires a gradual evolution or buildup time, taking perhaps tens of minutes. Presumably, this period reflects the duration over which excitation accumulates within the higher visual regions from successive volleys of rhythmically pulsing light. At some point the buildup is sufficient to overcome any suppressive influences that usually interfere with becoming conscious of these images.

The Birth of the World (Leaping over the Skull)

We will close this chapter by considering the esoteric contemplative practice of *Tögal*, mentioned on several occasions already. Translated into English, the name of this practice is "leaping over the skull," and this connotation is precisely what can point to the twilight domains lying beyond where contemporary neuroscience stops.[64]

Situated within the Dzogchen and Bön traditions of Tibet, *Tögal* is undertaken as an advanced set of psychosomatic practices, following rigorous preparation in training the stability and focus of attention. In fact, to undertake *Tögal*, before the completion of such preliminaries and without guidance by a senior teacher would be vigorously condemned, as a way of safeguarding the overall mental and spiritual well-being of the practitioner.

Clearly, there are vastly different presuppositions in the overall framework and worldview of this ancient contemplative practice and modern science. And yet it is also true that they share the following basic insight: everyday subjective reality is constructed as a dreamlike experience through an intrinsically creative process.

Tögal is often practiced in the context of a prolonged (usually, seven weeks long) dark retreat. Another common setting involves prolonged gazing into a cloudless sky, or indirectly looking at scintillating rays of sunlight (like Moskvitin), or through crystalline materials that scatter a concentrated light source. The practitioner adopts and maintains specific, rather intricate, body postures, while also carefully regulating respiration and eye movements. In some cases, physical pressure may be applied to the practitioner's body (for example, by gently pressing on the eyeball or the carotid artery in the neck), with the goal of "releasing" interior visionary phenomena into a seemingly external space. Another example is the rushing of blood, a type of "inner sound," which becomes audible when plugging up the ear canals.

In the traditional understanding of this practice, lived reality is taught to be the dynamic exuberance, play, and rearrangement of such self-generated "sounds, lights, and rays." A classic metaphor that these writings offer is that of an earthen vessel (corresponding to the human body), whose interior cavity is lit by a candle.[65] Like a magic lantern slide what we then see as outside is a projection from within the interior. For *Tögal* practitioners, even more so than for contemporary neuroscientists, the range of visionary phenomena encountered during contemplation are treated as direct gateways into exploring the mind's true nature and its proclivity to be captured by the contents of experience.

As the yogi or yogini progresses in *Tögal* practices, an entire gamut of visionary phenomena are released into a nebulous "out there" space. Although the visions are never intentionally willed, the melding of perceptions undergoes an organic maturing process, like the ripening fruit. The visions evolve from very simple, elementary particles, and then gradually grow in stability and complexity, as the primitive dots, lines, zigzag patterns, and so on, combine together in novel ways to generate more sophisticated arrays.[66] The practitioner is instructed to maintain a passive, witnessing stance that refrains from interfering with the visions, however subtle the inclination to do so might be. Any degree of wanting to hold on to some visions, or to look away from others, or even to become curious about them, needs to be deflected by a true master of this practice.

To the extent that the practitioner is able to achieve this detachment, to still the ordinary state of agitated attention and to understand that the almost endless variety of visions is a self-originated cinematic show, their authentic spiritual vision is then understood to undergo progressive purification. Their ability to see "reality" is stripped, layer by layer, down to a pristine, unadulterated source. The purification, in this context, refers to a shedding of habitual overlays of predictive priors, including conceptual filters and affective tendencies of fear and desire. As this process deepens, the practitioner has a firsthand realization of just how spontaneous the mind's image-generating potentials actually are, and the extent to which this clothes an ordinary experience of lived reality.

Undoubtedly, the key to the spiritual potency of Tögal practice lies in the following understanding: the Tögal visions, although evanescent as soap bubbles, are, ultimately speaking, no different than the visions of everyday waking reality. Over and over again, practitioners are provided with detailed instructions for making direct phenomenological comparisons of the spontaneity of Tögal visions to those deriving from ordinary experiences. "The reason why we do [Tögal] practice is in order to realize that the vision of ordinary normal life is equally illusory and insubstantial. We think that . . . the world as we see it as human beings, is solid and real and concrete. But this vision is all a projection." Later in the same talk, he adds that "we can bring the knowledge gained from [Tögal] practice to bear on the vision of normal life. We compare them and we discover that all of our so-called normal life is an illusion. If we cannot make this comparison, then perhaps it is better just to watch TV."[67]

And yet, despite the persuasiveness of these instructions, one might still have a compelling feeling that somehow everyday visions and the ones encountered by a contemplative engaged in Tögal really do differ, in some subtle, but nonetheless very consequential sense. What might be the source of this very strong and persistent sense that ordinary perception is more "solid," closer to an independently existing objective reality?

Traditional Tibetan teachings insist that it is only through a kind of misperception akin to an optical delusion that the visions of ordinary, waking consciousness are more firmly stabilized or fixed. The apparent solidity of perceptions is understood to be the work of our conceptual structures that help to reify and stabilize what we see around us, thereby occluding our ordinary experience through a dynamically occurring veiling process. From a modern neuroscientific perspective, we could attempt to express that insight by saying that our default tendency is to overweigh the precision of priors so much that bottom-up signals are unable to make contact. Think again of what we said earlier—how often in seeing a tree do we stay fixated on our idea of a tree, rather than being present to the fresh spring of continuously arising sense data.

From a Buddhist perspective, the ordinary consciousness is clouded by the grime of many such accumulated karmic traces, which can be understood as a readiness to respond to experience on the basis of previous interactions. The stickiness of the overlay on raw experience becomes even stronger as the affective tinge becomes more pronounced. The emotions of desire, fear, envy and so forth, function to solidify the given concreteness of lived experience, resulting in the complete entrenchment of biologically adaptive self-in-world models.

The evolving stages of Tögal practically can allow a practitioner to undo the apparent solidity of conscious experience. The contemplative texts of the Dzogchen and Bön tradition say that the culmination of Tögal is the phase known as the "Exhaustion of Reality." This phase, more than any of the three preceding ones, is ineffable. The best approximation is found in the classic texts which describe how all of the stuff of conscious experience, in its seemingly endless and zoo-like proliferation, finally dissolves and becomes reabsorbed back into the mind's source, a source that presumably contains and holds every imaginable permutation of conscious experience, but which we normally fail to explicitly notice as an ever-present component within every state of consciousness.

In the Dzogchen understanding, at the core of all that is or can ever be is this mystery, which they believe to be an unoriginate, indescribable, and inexhaustible "base," shorn of all polarities, including inside versus outside, and pleasant versus unpleasant.[68] According to *these ancient teachings, it resists any conceptualization whatsoever; it* is taught to be completely insubstantial, resisting any positive statements. It also eludes characterization as being either personal or transpersonal. Namkhai Norbu Rinpoche, a gifted teacher from this tradition, writes that the term *the Base* is "used to denote the fundamental ground of existence, both at the universal level and at the level of the individual, the two being essentially the same; to realize one is to realize the other. If you realize yourself, you realize the nature of the universe."[69]

As we will consider at greater length in Chapter 13, this notion of the Base refers to a more objective characterization of ultimate or first-order reality—what lies beyond the dreamed version of reality that is believed to be a more or less subjective approximation to the world beyond the skull. This helps to make us understand why *Tögal* means for the contemplative to take a "leap beyond the skull" and enter into a more direct access to the Base of reality. Indeed:

> all of the world's beings, objects, and appearances are said to rise up from the "ground" (gzhi) [Base] of reality, which in its primordial state is a field of pure possibility, beyond differentiation. Awareness serves as the dynamic, knowing dimension of this ground and acts as a kind of luminous vibrancy that "lights up" (snang) from the ground, creating appearances through its "dynamic energy" (rtsal). In this view, all appearances are simply the "play" (rol pa) or the "radiation" (gdangs) of awareness, with some appearances (such as visionary ones) being awareness appearing in its unclouded intensity, while others (like ordinary objects) are only its dimmed derivations.[70]

As we will see further in Chapter 13, the Dzogchen view, as well as that of many other nondual traditions, holds that this Base is the wellspring of certain fundamental qualities, which are capable of being experienced as pure awareness, infinite compassion, and several other qualitative features like interconnectedness and unity, that are the perennial source of many mystics' insights down through the millennia. Why would a Dzogchen practitioner strive to attain this realization? Because it is held to liberate a person from inveterate self-centeredness and allow a natural form of so-called *nondual consciousness* to emerge which is spontaneously more loving, compassionate, and in direct touch with the reality of life. It remains to be seen whether neuroscience can even begin to accommodate such intuitions and contemplative insights. It is suggestive, at least, that recent psychological models hold that excessively rigid models of self-in-world are behind a great deal of mental suffering and that their alleviation promotes health and resilience (we will return to this topic in Chapter 11).[71]

Through initial exposure to many of the concepts introduced here, one gains a preliminary intellectual understanding of some key principles of how our conscious experiences materialize and then stabilize. One learns, for example, that this dynamical process involves imagination as much as it does perception. But the form of knowing afforded to us by contemplative paths—the kind we call knowing-through-identity—presents a quite different challenge. Gaining this knowledge is a more gradual and arduous process, and requires exposure to much more than the information and concepts that can be presented in words. As we will see, it requires active participation on our parts.

10

The Imaginative Brain and the Origins of Myth

At this point, it behooves us to pause and take stock of what ground we have already covered, and anticipate where the book's journey will take us next. Wherever you find yourself right now, put this book down and take a momentary pause. Look around you and reflect on the totality of whatever it is that you are experiencing, the variety of sensory impressions that are shuttling through your mind. Go deeply into the very core of your experience for a few minutes. See how detailed you can get in dissecting the particular contents filling your consciousness.

This book's fundamental premise is that what you encounter around you is something akin to a very convincing, fully immersive virtual-reality simulation. Of course, you do not have a conscious experience of any kind of virtual-reality system, since it seems to be simply life "as it is"—a transparently clear window that opens out on an independently existing world "out there." The "out there" is perceived from a particular perspective. Your sensory organs *seem to be* passive recording devices, analogous to cameras and microphones. On the surface, it appears that this is all there is to the story.

You could, however, as we have seen, remove that seemingly independent world, bit by bit, by quieting sensory inflows, and you would still be immersed in an equally rich and multisensory experience of being a self-in-the-world. This is exactly what happens whenever you dream. When the neurobiological and phenomenological data is actually examined, the commonsense world starts to appear pretty far from ordinary. Reality is far stranger from what it may seem to be on the surface.

From everyday waking consciousness, we have been burrowing into more subtle manifestations of the mind—into a reality often as "subjective" as "objective," rich in a spectrum of feeling tones and dreamlike visionary phenomena. Our argument has been that these inner, imagination-rich recesses are actually more foundational to the makeup of your experience than that "daylight" consciousness that seems so ordinary. In the murkier reaches of the consciousness spectrum, the mind's intrinsic potential for creative imagination becomes supercharged, as constraints by the outside world loosen. The mind becomes less reliant on thinking framed in concepts, less dominated by logically structured narratives, more dominated by flights of imagination.

Our mundane, everyday waking perception is closer to the realms of dream, imagination, and hallucination than we commonly appreciate, and considerably more so than we are often comfortable admitting. Personal experiences with the liminal segments of the consciousness spectrum—including hypnagogia, lucid dreaming, and psychedelic-induced states—are powerful demonstrations of how much of our experience is constructed and shaped by subjective factors.

There are common underlying principles in how experiences get constructed, whether you are awake and crossing a busy intersection or lucid dreaming about flying through visionary landscapes. Waking consciousness works with the basic stuff that dreams are made of. At the depths of the consciousness spectrum we are drawing on the raw, neuro-mental dictionary of dreamlike forms that also combine into the forms we commonly experience as the consensually valid "real" worlds of waking. This impressive feat of creativity is how we end up with the vividly convincing sense of having a transparent window onto reality.

Throughout the preceding chapters, we have been shimmying down the consciousness spectrum and penetrating its hidden root networks. In this trip down the spectrum, it becomes increasingly likely to encounter the building blocks (the entries and symbols in the previously mentioned "neuro-mental dictionary") as they exist in their most pristine forms. At these deeper layers we can begin to uncover the raw materials of our lived experience. In the more interior expanses of mental life, states of pure being reign over the states of consciousness that are organized to promote active doing. Ironically, here is also where we become most disconnected from academic fields of neuroscience and psychology and find ourselves much more at home with topics plumbed by a minority group of artists, poets, depth psychiatrists, and contemplative masters.

This chapter represents an intermission in our extended story as we pause to consider how direct encounters with the building blocks of our lived experience can be charged with an overwhelming sense of awe and meaning. We take a speculative leap to suggest that these encounters "of the liminal kind," submerged far below the upper, daylit regions of the consciousness spectrum, have played a pivotal role in some of humanity's oldest cultural artifacts and ritual practices. These cultural artifacts have been developed and transmitted through generations precisely to facilitate such awe-filled contacts and draw on the revitalizing energies of mythology.[1] At these depths of the human psyche, we find the wellsprings of creativity and meaning. This chapter will be the last stop that still occupies itself with experience of "stuff" and beyond this, we begin to venture from subtle to the *very* subtle manifestations of mental life: the bizarre Antarticas of the mind where consciousness becomes radically emptied of all content altogether.

Cave Art and Cultural Artifacts

From the southwestern coast of France, and extending along the autonomous region of Cantabria, in northern Spain, there sprawls an intricate network of limestone caves, containing some of the world's most mysterious archeological puzzles. Journeying into the darkness of these caves, some of the oldest art painted by anatomically modern humans is engraved on stone walls, floors, and ceilings. The Lascaux and Chauvet cave paintings are probably among the most famous examples of this art.[2]

The caves themselves are natural marvels, carved out by the pounding surf of underground rivers some millions of years ago. At some point during the period known as the Upper Paleolithic, between twenty-five thousand and fifteen thousand years ago, humans with nervous systems like our twenty-first century ones began to visit these caves. Despite the caves' being inhospitable for long-term habitation, the humans left behind cryptic signs that continue to resound through the millennia.

Cave access appears to have been decisively cut off by tumultuous geographical changes near the end of the last Ice Age, somewhere between eleven thousand and ten thousand years ago. The artistic legacy of these prehistoric humans continued to hibernate in obscurity until the early decades of the twentieth century. Since their resurfacing into the light of public awareness, cave art has continued to fascinate both specialists and laity alike.

Perhaps the most enigmatic instance of Upper Paleolithic art is known as the "signs," made up of a wide variety of abstract geometric patterns, such as grids and lattices, parallel lines and chevron patterns, dots, filigrees, and cruciform shapes. These geometric markings outnumber the more readily identifiable animal and human figures by a ratio of about 2 to 1.

Archaeologists have puzzled over the possible meanings of these geometric signs, which often overlay and are interspersed with more complex figural elements. Many anthropologists have wondered whether the regular, abstract patterns were humanity's first forays into the world of symbols. The best evidence to date fails to support the hypothesis that these patterns were pictograms suggestive of a proto-language. Nevertheless, it is unlikely that the signs were mere scribblings. In light of their wide prevalence, and geographic universality, they seemed to have held great importance for early humans.

The Canadian paleoanthropologist Genevieve von Petzinger used computational tools to establish a database of geometric signs collected from cave art found at over a hundred different cave sites in France. Her work has identified 32 distinct geometric signs that recur regularly throughout the European Paleolithic caves, including symbols like the circle, lines, cross-hatches, zig-zags, spirals, ovals, quadrangles, and ovals.[3]

The geometric signs, moreover, are not confined only to the Upper Paleolithic caves in Europe. Older instances of the signs have since been unearthed. One example is the Blombos cave in South Africa, where objects like bone awls, spear points, and blocks of red ochre bear similar geometric patterns. These artifacts are believed to be approximately 70,000 years old, predating the abstract and figurative drawings discovered in Europe by some 30,000 years or more.

Microscopic and chemical analyses of the pigment layers discovered in Blombos reveal that they were intentionally applied.[4] We now know that early humans were leaving imprints on multiple kinds of media at sites scattered all over the globe.

There is a remarkable cross-cultural universality in these geometric markings: different instances of the thirty-two signs identified by von Petzinger recur in rock art discovered throughout African, Australo-Asian, and American sites. This is a striking observation given that this assortment of geometric signs is such a small subset of the virtually infinite set of possible arrangements of lines and markings that could have been made.

In previous chapters we have already made frequent allusions to pulsating, luminous, geometric percepts that can be spontaneously generated by the nervous system under a wide variety of eliciting conditions. Whether in the case of hypnagogia, strobe-induced pattern formation, sensory deprivation, psychedelic ingestion, or other types of non-ordinary states of consciousness, we are repeatedly confronted with the same primordial geometries.[5] It is reasonable to hypothesize that the abstract, nonfigurative patterns preserved in rock art from around the world are among the most ancient of the documented examples of Klüver's "form constants" and Moskvitin's "selective forms." Moskvitin himself had a hunch: "The patterns to which I had gained access," he wrote, "were so similar to certain patterns in art, especially in religious art, or art and ornamentation created by civilizations dominated by mystical initiation and experience, that I could not doubt that my experience was a very old one and widely known."[6] He wondered about such "holy signs" painted on "walls of the cave."[7] Another scholar fascinated by these parallels is the archeologist David Lewis-Williams, a specialist in cave art. He was struck by the numerous similarities in the rock engravings he was studying and the structural features of visual hallucinations reported by participants in contemporary neuropsychological studies on non-ordinary states of consciousness. The correspondences were simply too compelling to be discounted as coincidental.

Lewis-Williams concludes that the impressive universality of the geometric signs can be explained by the universal functional architecture of the human nervous system, believed to be essentially unchanged for the last hundred thousand years or more.[8] His central contention is that the mysterious geometric signs (which he calls "the signs of all times") in Upper Paleolithic cave art mean that prehistoric humans were fixing into rock visionary phenomena encountered in non-ordinary states. One can remain agnostic about how such non-ordinary states of consciousness were induced. There are many possibilities, ranging from the profound (natural) sensory deprivation of dark caves, hyperventilation, dancing, rhythmic drumming, hypnagogic rituals, ingestion of psychoactive plants, extreme fasting and sleep deprivation, or some combination of all these.

The first thing that Lewis-Williams did was to conduct a descriptive inventory of the abstract forms that prior neuropsychological research on visual hallucinations had identified. Extending Klüver's classic work on form constants, Lewis-Williams discerned up to six fundamental categories—each of which was expanded in perception through replication, fragmentation, rotation, and the juxtaposition of more primitive elements.

Lewis-Williams first documented these six categories in examples drawn from contemporary research with participants experiencing different non-ordinary states, largely induced by mind-altering substances, and from Upper Paleolithic cave art. He then went further by drawing examples from South African San rock art and Shoshonean Coso rock art originating from the California Great Basin. The latter two, commonly accepted as instances of shamanistic art, feature content generally thought to derive from visions experienced in non-ordinary states of consciousness. Lewis-Williams confirmed the presence of all six categories of forms across all of the data gathered from these four sources.

Although the six basic categories did not account for every sign encountered in Upper Paleolithic paintings, the fact that all six were nonetheless present, seems to offer support for the basic tenets of Lewis-Williams's theory. While we are not aware of any such comparative phenomenological study, we suspect that many of the categories formalized here would find robust representation in the first three developmental stages of the visionary *Tögal* practice of Tibetan Buddhism.

One explanation for the shared structural features and the fact that the repeatedly encountered geometric signs are a small subset from a theoretically near infinite set, is the constraint of a common neural substrate. The signs point us toward a universal neuro-mental dictionary that provides the raw building blocks for our lived experiences. That is, when we look upon these rock engravings and ancient

ENTOPTIC PHENOMENA		SAN ROCK ART		COSO
		ENGRAVINGS	PAINTINGS	
A	B	C	D	E
I				
II				
III				
IV				
V				
VI				

Visual forms induced by (A) mind-altering substances, and (B) migraines, compared to visual forms encountered in different forms of shamanistic art (C through E). Used with permission of University of Chicago Press from J. D. Lewis-Williams and T. A. Dowson, "The Signs of All Times: Entopic Phenomena in Upper Palaeolithic Art," *Current Anthropology* 29, no. 2 (1988): 201–245, figure 1; permission conveyed through Copyright Clearance Center.

cultural artifacts, we are also gazing into the spontaneous dynamics of a brain that ceaselessly "talks to itself."[9]

In his romp through the clinical world of hallucinations, the neurologist Oliver Sacks sums up the remarkable similarities in geometric patterns that recur across so many different contexts as follows: "We can actually see, through such hallucinations, something of the dynamics of a large population of living nerve cells and, in particular,

Visual forms encountered in Paleolithic mobile engravings (F and G) and
parietal (cave) art (H and I) from different locales. Used with permission of
University of Chicago Press from J. D. Lewis-Williams and T. A. Dowson,
"The Signs of All Times: Entopic Phenomena in Upper Palaeolithic Art,"
Current Anthropology 29, no. 2 (1988): 201–245, figure 1; permission conveyed
through Copyright Clearance Center.

the role of self-organization in allowing complex patterns of activity to emerge. Such activity operates at a basic cellular level, far beneath the level of personal experience. The hallucinatory forms are, in this way, physiological universals of human experience."[10]

In the murky and, often gradual, transition from nonfigurative, abstract elements into the more recognizable, iconic forms (whether of animals, humans, or other well-defined structures), there is a heavy overlay of prior learning and cultural expectations. This explains why universal zig-zags and curvy lines might evolve into snakes, or birds, horses, lions, or mammoths, body parts or exotic architectures, in differing geographic locales, depending on which features of the world were mostly encountered by people residing in those areas. Frequency of contact no doubt exerted an important influence on significance accorded to specific signs in local beliefs and mythologies.

In the case of a well-developed and systematically elaborated contemplative practice like *Tögal*, each stage is accorded a very specific interpretation and provides experiential grounding for deeper realizations into the fundamental nature of the mind. Nonfigurative seeds encountered at the beginning stages (the so-called *thigles*) gradually morph into the spectral bodies of deities, which are carriers of specific psychospiritual energies and potentials. By contrast, when similar, early-stage hallucinatory geometric patterns are evoked in someone participating in a research study on stroboscopic-induced visions, these signs are not likely to be accorded any particular meaning. Rather than agglomerating into the shape of deities, the same starting materials might morph into cartoonish figures or other, more mundane perceptual categories.

Lewis-Williams sees a kind of history of art preserved in the prehistoric cave paintings. The very first forays, he suggests, into what later evolved into representational art were not attempts at two-dimensional representations of anything per se. Rather, he hypothesizes that the early humans were literally tracing geometric patterns and dreamlike forms that they were in fact seeing as being projected onto the walls, ceilings, and floors of the caves. Tracing the patterns that these prehistoric "shamans" saw superimposed upon material substances, for example with a finger in sand or on soft walls of the cave, would have been sufficient to preserve them for posterity.

If the above is true, then when we follow the signs etched in humanity's oldest rock art, we are literally witnessing to the waking dreams of our ancestors, radiating through the darkness of millennia. And the very contours of these primordial patterns, emanating as they do from deep layers of the brain/mind, are the same ones through which our own lived experience is filtered.[11]

Hypnagogic and Oneiric Rituals in the Ancient World

Dark caves, grottos, and crypts as liminal, transitional spaces and sacred sites have always exerted a profound pull for humans. As we move from the obscure mists of prehistory into the historical era, we encounter many of the same, recurring themes. To leave behind the more mundane setting outside a cave's entrance and step into its depths is to immerse oneself in an anomalous, multisensory ambience. Visibility drops precipitously—the inner recesses of caves and grottoes progress toward pitch black—and the sounds of the familiar world fall away. Descending into this unusual environment, also exposed to novel smells and air composition, people experience mild and sometimes severe disorientation, and their bewilderment frequently evokes complex medleys of emotions, combining disquiet and claustrophobia, terror and reverential awe. Modern visitors, so acclimated to the comforts and amenities of contemporary living, may be especially unsettled by the qualities of such otherworldly spaces. Caverns, caves, and grottoes are the paradigmatic examples of the sensory deprivation conditions imposed in restricted environmental stimulation therapy (REST), a set of techniques developed by psychologist Peter Suedfeld.[12] They were the naturally occurring, prehistoric antecedents of the sensory isolation tanks employed in the twentieth and twenty-first centuries.

Anomalous environments have radically different space and time features than the ones that our nervous system expects to encounter in the natural world. In addition to the unusual "settings," for many of the historical practices that we are now going to describe, the people entering caves, pits, and similar sites often did so with a very deep "set." This consisted of having a specific intention, like to undergo a vision quest or to seek answers to deeply meaningful questions. These practices were usually accompanied by preparatory practices consisting of protracted periods of fasting, meditation, sexual abstinence, sleep deprivation, rhythmic music, or other similar techniques.

As we know from the discussion of scientific studies on sensory deprivation, and related REST-like techniques such as the Ganzfeld, austere physical environments can precipitate a complex range of visionary phenomena. After some time the nervous system enters into a state that psychologists call stimulus hunger, not dissimilar to hunger that follows from extended food fasting. Beyond the occurrence of mental imagery, with profound sensory restriction and prolonged immersion in such environments, subjective reports of selflessness, spacelessness, timelessness, and infinite extension become possible.[13] Mountain tops and secluded caves have been the prototypical environment of yogis and contemplative masters from all of the world's great spiritual traditions.[14] Darkness, stillness, solitude, quiet, simplicity, and sensory restriction, in addition to the high altitude are highly conducive for the transformations of ordinary consciousness that involve a disruption with ordinary ways of experiencing.[15]

Special geographical and geological conditions in ancient Greece created opportune settings to seek non-ordinary sources of inspiration.[16] One of the most famous religious sites was the temple of Apollo and the oracular center at Delphi. The central point of Delphi itself was a cavern, in whose innermost recesses was found the oracle, a prophetic priestess known as the Pythia. Written records concerning activities at Delphi date back to the fifth century BCE.

The Pythia was a simple woman, living a life of strict social isolation, who prepared herself by undergoing practices of interior purification, including fasting and sexual abstinence, prior to entering the heart of the temple. Here, the Pythia would seat herself upon a tripod. Those wishing to make inquiries of the oracle would then pose questions to her. The Pythia would receive answers, believed to be inspired by Apollo, by entering a non-ordinary state of consciousness that the ancient Greeks recognized as a divine or prophetic mania.

Although a topic of controversy for many decades, more recent geological discoveries have amassed substantial evidence that a geological fault was located in the center at Delphi through which gaseous vapors entered the temple's inner sanctum, in the precise location of the Pythia's prophesying activities.[17] The identity of the vapor remains uncertain, although the best current hypothesis points to ethylene, a colorless hydrocarbon gas. At high concentrations, this gas produces unconsciousness. At lower doses, however, the gas can evoke trancelike states of euphoria and feelings of expansion beyond the body's confines.[18] It seems highly plausible then that, at least at the Delphic center, the oracle's prophetic mania was a largely chemically induced state of consciousness, evoked by exposure to narcotic gasses.[19] At other oracular centers, which were plentiful throughout ancient Greece, the intense sensory deprivation of grottos was probably the primary contributor.[20]

The role of sensory deprivation in ancient rituals is further attested to by the well-documented practice of *incubation,* which appears to have spread widely in the world between roughly the sixth century BCE and the sixth century CE.[21] Incubation involved going down to a dark place, often a cave beneath a sacred temple or a tomb, falling into a deep sleep there, and receiving inspired knowledge and visions. Knowledge was sought through these means for purposes of healing or more generally acquiring depths of insight not possible in ordinary states of consciousness. No doubt the visions included geometric signs, and much grander visions of exotic beings and fantastical landscapes sourced from a vast inner realm of possible worlds.

Peter Kingsley has documented the practice of incubation by some pre-Socratic philosophers of ancient Greece, the progenitors of much of modern Western civilization.[22] In the poetry of Parmenides, with its descriptions of passages through dark subterranean worlds, Kingsley finds more than evocative metaphors, also noting the poet's fairly sophisticated and accurate descriptions of states of mind induced by incubation. For example, there are frequent allusions to sensory phenomena (such as rushing, buzzing, or whooshing sounds) that often precede initiation into more liminal zones of the consciousness spectrum.

Pagan incubation rituals in the Greek world appear to have been greatly modified and censured with the introduction of Christianity, where there was much greater distrust of the imagination, especially when it remained unpurified from egocentric motives. Among the early Christians, incubation-induced dreams and visions were not so readily trusted. Still, incubation practices did survive to some extent, especially in the eastern half of the Roman Empire.[23]

Kingsley and de Becker are in agreement that those practicing incubation sought to hold themselves for prolonged periods in hypnagogic states of consciousness. Besides just darkness, incubation required the utmost immobility of the body and changes in respiration, involving slowing of the breath. Focused concentration was required to maintain the precarious hypnagogic state and not succumb to nonlucid dreaming or insensate sleep. Kingsley writes that "the silence, the deliberate calm, physically not even moving—weren't simply ends in themselves. They were means used for the sake of reaching something else. . . . The purpose was to free people's attention from distractions, to turn it in another direction so their awareness could start operating in an entirely different way. The stillness had a point to it, and that was to create an opening into a world unlike anything we're used to: a world that can only be entered 'in deep meditation, ecstasies and dreams.'"[24]

Incubation practices opened up a line of communication with the deepest, most interior layers of the human mind.[25] These aspects of the mind are ordinarily submerged far below the surface (external) levels of awareness and ordinarily too subtle to be accessible during ordinary waking experiences. In these strange and awe-inducing hinterlands, at the nexus point between the private, interior *(idios kosmos)* and the consensually shared external world *(koinos kosmos)*, there is more likelihood of coming into direct contact with the very basis of meaning-making and the foundations of consciousness.

The Mythological and Fairy Tale–Producing Brain

In this section we venture into more speculative territory considering the possible neurological bases of mythological creation and the phenomenological structure of fairy-tale narratives.[26] Much of this draws from an empirical observation that the human brain/mind has an intrinsic potential to switch to a style of self-organization that is dominated by visionary phenomena that condense and convey cognitive content and meaning.[27]

In 1953, the psychoanalytical anthropologist Géza Róheim published his life's work, *The Gates of the Dream*.[28] The gates to which the title refers are the two distinct points of intersection Róheim identified between the inner self and the outside world. The first and most familiar to us is the gate through which residues of waking experiences pass to influence our nightly dreams. The other, less commonly recognized, is the gate through which inner domains of dream, imagination, and visions travel to shape and structure waking experiences.

Róheim appreciated that the consensual world we inhabit during waking hours (the "reality" of naive realism) is *causally* influenced by the usually concealed interior. Traditional tribal cultures too, had a much deeper understanding of these facts, which helped to account for their animistic views of the cosmos, in which nature itself was charged with potent psychological and cosmic energies.[29]

Róheim combined detailed psychoanalytical interpretation of dream phenomenology with a truly encyclopedic wealth of anthropological material on cross-cultural myths. He found striking correspondences between universally recurring dream images and sensations, such as the feelings of falling, sinking, floating, alterations in embodiment (feelings of body doubles, bodily dismemberment or engulfment), as well as images of natural and supernatural forces, and motifs from the mythological stories of different cultures. Róheim traced the origin of many of the world's oldest creation myths, for example, to the sleeper's ability to enter into immersive dream worlds, filled with majestic scenery, strange characters, and strong emotional undercurrents. Róheim's contention was that, not only creation myths, but many other myths trace their origins to dreams and related visionary phenomena that were interpreted by traditional cultures as no less real than was the consensual world itself. A strong desire to share these highly meaningful private experiences with other group members, according to his theory, then inspired the act of retelling and elaborating them as oral stories and their pictorial depictions on cave walls and rock art. Phenomenological analyses of dreams reports, conducted several decades after Róheim's passing, have found supporting evidence that the narrative structure of extraordinary dreams shares similarities with myths and fairy tales.[30]

One of the most striking anthropological case studies, broadly consistent with Róheim's theorizing, is to be found among the creation myths of Australian aborigines. The diversity of worldviews held by the hundreds of different aboriginal populations of Australia, prior to European colonization, broadly shared a theory of cosmogenesis (the creation of the universe) attributing the origin of all visible phenomena, from natural landscapes to the vegetable, animal and human life, to the dreaming of ancestral spirits.

The Dreaming is the English term that anthropologists gave to this act of oneiric cosmogenesis, but this translation is not able to encompass the all-inclusive nature of the process: all of culture, for these indigenous Australian populations, stemmed from the Dreaming that completely interpenetrates lived reality and forms a continuous fabric connecting the past, present, and future, merging the spiritual realm with the physical land and people.[31]

Mythic tales about the Dreaming were put into song and dance, given life by the evocative sounds of the didgeridoo, a traditional wind instrument. Others were represented through symbolic art, which could be drawn on rock, stitched onto pieces of clothing, worn as body paint, or etched into sand. Many of the symbols resemble the now very familiar set of geometric signs. All such symbols were hypercharged with emotional significance and meaning, which may account for efforts to reproduce them in so many different cultural artifacts.

The association between creation myths and non-ordinary states of consciousness is found in numerous cultures. For example, the anthropologist Gerardo Reichel-Dolmatoff has studied the Tukano Indians of the northwest Amazon and their ritual use of the potent ayahuasca (*yage*) brew.[32] The Tukano cosmos consists of several different planes of existence. For a person who has been duly prepared, through spiritual ascesis, ritualistic ingestion of *yage* inaugurates a distinct, transcendent form of consciousness that allows experiential access (a knowing-through-identity) to these planes of existence. The Tukano peoples recognized that the luminous visions opened through the *yage* brew matured in distinct stages, from dancing dots and kaleidoscopic patterns to dreamlike scenes and then baths of color, light, and more subtle contemplations. Living geometries of light, shining from some ineffable interior source, were at play, ever arranging and rearranging through progressive sequences. These stages roughly correspond to those discovered in contemporary neuropsychological studies, as well as the ones recognized among yogis engaging in *Tögal* practices.

The Tukano Indians subsequently depicted elements of *yage* induced visions on more durable materials, such as on walls and clothing materials. Once again we are confronted with familiar geometric patterns, consisting of zigzag lines, polygons, spirals, and other entries from what we have called a universal neuro-mental dictionary. In the Peruvian rainforest, the cultural artifacts—pottery, clothing, and body paint—of Shipibo-Conibo indigenous peoples are decorated with similar designs.

Traveling further north, in central America, the Huichol people, whose religious beliefs are closely linked with the ritual ingestion of a different plant psychedelic (peyote), used textiles to weave a similar constellation of geometric designs on their shirts, pants, cloth belts and bags.[33] Many of these designs are associated with specific meanings, being identified with universally encountered objects like the sun, stars, rainbows, the moon, snakes, deer, eyes, and so on.

In summary, when we examine the cross-cultural record, we find that our universal set of signs is drenched with significance. Indeed, these archaic patterns seem to constitute irreducible building blocks of meaning, ones that are then elaborated into complex narratives, accounting for the origins of the entire cosmos according to many traditional mythological narratives. If our mind has this built-in potential to weave materials that resemble mythological stories and fairy tales, then this imaginative process ought to be evident across several different contexts.

An important contribution to deep-seated proclivity of the human mind to generate mythological representations can be discovered by reading the first generation of studies on psychedelic-assisted psychotherapy that were conducted in the 1950's and 1960's. Psychiatrists from this period sometimes made use of psychedelics to circumvent and partially dissolve a patient's defenses and complexes. Although the scientific rigor of these studies was often lacking by today's standards, what makes this literature unique is that a patient would undergo numerous depth-psychological sessions with a therapist, both with and without these pharmaceutical compounds. As a result, the clinical case studies offer extremely rich observations of the mind's operations across many states of consciousness.

A classic work from this era outlines the prototypical progression of psychological materials that were discovered across therapy sessions.[34] Of most interest here is that during the advanced phases of therapy, patients were more likely to access the so-called symbolic and integrative levels of consciousness where meaning was expressed through images and feelings.

Characteristic of these deeper stages of psychological exploration was a bubbling up of fairy tale and mythological motifs in imaginative material. A patient's desires, fears, and persistent psychological fixations were increasingly more likely to be expressed by materials that resembled these dramatic forms of storytelling. Often, this material would incorporate themes or characters from ancient mythological stories without the patient seemingly having much if any personal background knowledge of the myths. A patient could experience transforming into a mythological being, for example, permitting the possibility of emotional catharsis through reliving and reprocessing chronic psychological blocks. This hints at the possibility of a universal endowment that the human brain/mind has for spontaneously producing and expressing meaning and profound degrees of knowing through a mythologically inspired mode of expression.

The Origins of Meaning and the Numinosum

In his scientific research on dreams, integrating the literature from biology, psychology, and anthropology, Harry T. Hunt identified a distinct class of dreams that he called *archetypal*.[35] These are quite unlike the more common dreams, which usually involve the recycling of memory residues from the preceding day. Archetypal dreams are filled with more exotic content and overwhelming feelings, sometimes conjuring up elaborate mandala-like patterns and scintillating geometric forms, and sometimes portraying contact with mythological beings. Hunt's research suggests that archetypal dreams might be more common in long-term meditators, especially those who, through advanced training, have gained increased abilities to maintain lucidity. The long-term meditators Hunt studied were also more prone to characterize these experiences as intuitive visions, as opposed to just fantastic dreams.

Archetypal dreams are often suffused with a profound emotional charge. The feeling tone is that of the numinous. The *numinosum* is a unique kind of mental state with a very distinct qualitative feel unlike any of the more familiar emotions. The term was originally coined by the early twentieth-century theologian and philosopher Rudolf Otto to refer to an ineffable mixture of awe, dread, fascination, and reverent sense of transcendent mystery.[36] Carl Jung, who subsequently did much to develop the concept of the numinosum, described it as a state that often simply overtakes a person, seeming to emanate involuntarily from some deeper reaches of the psyche that are felt to be "other."[37] The numinous is characterized by an overwhelming sense that something of enormous significance and portent is being revealed.

The depth psychiatrist Lionel Corbett has compiled triggers that can facilitate contact with the numinous.[38] Some common triggers include experiences in nature (especially solo explorations in wilderness), dreams (of the archetypal variety), profound sensory deprivation and deep stages of meditation, evocative music and art, and ingestion of mind-altering substances. Jung and Corbett both believed that the numinous contains within itself the seeds of a regenerative, healing quality.[39]

An experience of the numinous seems much more likely to occur in the subtle and more interior layers of consciousness. These zones of the consciousness spectrum appear to be enveloped by the numinous as if by a fog. Harry Hunt's research revealed that the other-worldly flora and fauna of archetypal dreams—the living, pulsating geometries and their mythic transformations—are resistant to being broken down into smaller units of semantic meaning. In a more ordinary dream, for example, encountering a dream character like a long-forgotten childhood bully (or friend) might elicit an entire network of cognitive associations upon waking. When it came to the content of archetypal dreams, however, a psychological-cognitive analysis usually stopped being able to proceed any further. It had reached its utmost limit. A luminous geometric form, for instance, was felt to convey meaning directly, without any intermediary steps.[40] It was like an element in a periodic table of conscious experience that could no longer be decomposed to lower constituents of meaning. "It is as if this experience of felt meaning and portent were its own end," wrote Hunt.[41] The experience of this type is frequently attended by a sense of "hyperreality." In other words, what is being disclosed in consciousness is felt somehow to be even more real than the usual forms of waking perception.

This observation should make sense, if it is indeed true that in moving through the more interior strata of the mind we are simultaneously approaching the bubbling spring from which mental life issues forth. Individual characters of a universal neuro-mental dictionary are like elementary hieroglyphics our brain/mind uses to organize its sense of reality, in close collaboration with the physical world surrounding us. In liminal states of consciousness, they are much more likely to be experienced in a raw and spontaneous manner, independent of the external world. These archaic, sensorial forms hold intrinsic meaning and value in and of themselves. This helps to explain humanity's perennial and ubiquitous desire to create representations of these neuro-mental building blocks, using and reusing them as recurring motifs in cultural artifacts, from cave art to clothing textiles, body decorations, and sand paintings.

Encounters of the Liminal Kind

Throughout this book, we have distinguished between doing and being modes of consciousness.[42] Most people in modern societies spend the majority (or nearly all) of time in the former. With a focus on doing, consciousness is task-dependent—organized around getting things done—which means navigating through an uncertain world and successfully interacting with other people to accomplish a shifting set of goals. This corresponds to the very familiar zone of "daylight consciousness," in which we occupy a consensual reality heavily shaped by our roles, responsibilities, and social scripts. The good order of society is dependent on each of us meeting the requirements of our respective roles within the family, in our places of employment, as voting, tax-paying citizens, and so on.

Another, more central, dimension of our humanness is not, however, encapsulated by our performative capacities. And unless this depth dimension of consciousness is acknowledged and included, we are likely to feel a gnawing dissatisfaction in our lives. The celebrated author Saul Bellow spoke to this theme at the ceremony where he was awarded the Nobel Prize in Literature, in 1976. In "contractual daylight" consciousness, Bellow said, we see ourselves "in the ways with which we are so desperately familiar." But beyond this, "there is another life coming from an insistent sense of what we are which denies these daylight formulations and the false life." This other side of life, according to Bellow, reveals itself, for most of us, in flashes of what he called "true impressions." Our real condition "reveals, and then conceals itself. When it goes away it leaves us again in doubt. But we never seem to lose our connection with the depths from which these glimpses come. The sense of our real powers, powers we seem to derive from the universe itself, also comes and goes. We are reluctant to talk about this because there is nothing we can prove, because our language is inadequate and because few people are willing to risk talking about it. They would have to say, 'There is a spirit' and that is taboo. So almost everyone keeps quiet about it, although almost everyone is aware of it."[43]

Following Bellow's lead, we turn to the aspects of our life where consciousness is characterized simply by being rather than by doing. To be dominated by doing is to be in intimate interaction with things outside of us, in a tight interlock of give and take with the outside world. To be, simply to be, is to sink into the moment and gradually learn to identify as a silent witness rather than as one who "does" anything in particular. Contemplative masters are experts at navigating through and permanently abiding in states of being, while many of us visit these slices of our consciousness state-space only very briefly, in glimpses, and at unexpected moments.

The liminal states of consciousness are an in-between land, a portal between the extroverted lands of daylight consciousness and the enigmatic, introverted lands that shimmer in the depths of the mind. From entering into prehistoric caves, to the oracular centers of ancient Greece, to incubation centers, humanity has apparently felt an enduring call to voluntarily venture into these liminal spaces. It appears that some of humanity's most powerful mythologies and belief structures have been recovered like lost sea treasures by those venturing into these deep spaces of consciousness. Humans have always been inherently fascinated with, and have gone to great lengths, to seek such "encounters of the liminal kind." Down there, primordial geometries of light shine out among visionary landscapes, and powerful surges of the numinous enveloped generations of psychonauts.

Some scientists have suggested that there exists a universal biological drive to transiently perturb and alter ordinary, daylight forms of consciousness.[44] The opportunity to reversibly and transiently decouple the brain's endogenous dynamics from their enslavement to the outside world potentially confers benefits.[45] Having opportunities to allow for a decoupling of the brain/mind's spontaneous self-organizing tendencies is what permits for a periodic exercising of imaginative potential, explorations that move the brain beyond the usual forms of self-organization. We probably do this also when we daydream or dream while sleeping.

In his popular book on the science of psychedelics, writer Michael Pollan used the metaphor of "shaking the snow globe" to explain why periodic alterations of consciousness can be beneficial.[46] Periodically shaking up configurations of neuronal assemblies, and allowing them to settle into novel connectivity profiles, could introduce a healthy dose of variability that optimizes the brain's ability to get by in a world that is often highly uncertain, unpredictable, and ever-changing. Ultimately, this might be one way to get out of old, worn-out grooves of thinking, feeling, and behaving, and to reengage the world around us from a fresh perspective.[47] Indeed, the evidence does suggest that psychedelics can facilitate laboratory-based measures of divergent thinking—being able to come up with a greater variety of novel solutions to problems.[48] Periods of voluntary sensory deprivation and seclusion have also been shown to boost creativity and facilitate novel insights.[49]

Is the spontaneous brain activity that results from using psychedelics more complex, in terms of information content (or entropy), than that of ordinary consciousness under daylight conditions? An impressive amount of evidence compiled by Robin Carhart-Harris suggests that, for a wide variety of psychedelics, the answer is yes.[50] When research subjects take LSD, for example, the neuronal repertoire expands to offer a larger menu of possibilities for the brain to cycle through in this non-ordinary state of consciousness.[51] Recent findings suggest, too, that some of the same changes in neurophysiological activity can be induced by non-pharmacological methods of inducing alterations of consciousness, such as exposure to rhythmic flicker.[52] Such an enhanced repertoire of neuronal network states likely maps onto an expanded phenomenological space of possibilities.

The prospect of gaining access to normally neglected sources of knowledge may constitute another reason that humans have been drawn to develop cultural and ritual practices for exploring liminal terrains. One of this book's major themes is that the brain/mind is organized as a very deep, living hierarchical system consisting of models upon models. Each tier in this nested system layers in its own different type of knowing. For example, the ordinary "doing" modes of consciousness are strongly characterized by logical concepts, language, and narrative. This form of knowing is highly conducive to problem-solving and is routinely engaged when a person needs to perform tasks and cooperate with other human beings.

By contrast, the autonomous production of images (the autosymbolic process) seems to ceaselessly operate at more interior, deeper strata of our mental life. These deeper layers seem to tap more readily into a vast reservoir from which humanity has been able to shape mythological stories. The knowledge accessible in non-ordinary states of consciousness seems to center around responding to profound issues that confront us as living human beings, ones having to do with some of the most meaningful things we will face, both as individuals and as cultures, like the search for ultimate meaning, finding ways to cope with individual and collective suffering and facing our death.[53] States of being, unlike those of doing, are resplendent with emanations of images, and feelings. In these zones of the consciousness spectrum, the intuitive and instinctual intelligence of the body and its trillions of cells carries more weight than abstract cognitive information and verbal concepts. The deeper buried layers are evolutionary ancient and also predate the higher-order, language-dominant forms of thinking. It seems quite likely that dreamlike consciousness appeared at much earlier stages of life than waking experiences organized around clear and distinct conceptual categories.[54]

It seems that the deeper, more interior layers of the human mind, which are so often obscured by those models of the self and world that characterize our daily life, know more than we can ever say. Our being can express truths impossible to express in language. In large part, for most adults, these imaginative, as-if simulation processes are disconnected from daylight consciousness, surfacing only in daydreams and during sleep, and then quickly fading back into forgetfulness. Much like the shrouded interiors of Upper Paleolithic caves, these layers of the mind appear to be ceaselessly active, and ready to open up lines of dialogue with our waking consciousness.[55] It is from their depths that we become recipients of those "strange powers" alluded to by Saul Bellow. It would seem that humans have been carving these messages from the depths on walls of stone and elaborating them around flickering campfires since the dawn of our species's origin.

11

Who Is Watching the Show?

On a quiet afternoon, Bernadette Roberts found herself sitting in silent attention and prayer, wrapped in the hush of a monastic chapel. She experienced her mind settling into profound and deep stillness. It was as if a great and ceaseless whirring had finally sputtered to a stop, revealing a pregnant silence underneath. In the past, when she had experienced this silence, it was so powerful that she had been overwhelmed by a primordial fear of annihilation. This rising wave of panic would then break the inner quiet and suck her back into the experience of ordinary life.

This time she felt no fear and the silence refused to budge. Keys were being rattled, the chapel needed to be locked up, Bernadette needed to leave, to go home and cook food for her kids. As she proceeded to do these things, the distinctly novel but also strangely familiar sense of inner silence never left her. Driving home, preparing dinner—all of these activities were happening, but now with a big, silent spot that would not be submerged by busyness. Over the next several days, this experience never left her. "By the ninth day," she wrote, "the silence had so eased up, I felt assured that a little while longer and all would be normal again. But as the days went by, and I was once more able to function as usual, I noticed something was missing and I couldn't put my finger on it. Something, or some part of me, had not returned. Some part of me was still in silence. It was as if some part of my mind had closed down."[1]

Turning to the writings of the medieval Christian mystic Saint John of the Cross to find an explanation for her experience proved unsatisfactory. While walking home one day, with a vast panorama of verdant valleys and hills in front of her, she suddenly had an epiphany: "I turned my gaze inward, and what I saw, stopped me in my tracks. Instead of the usual unlocalized center of myself, there was nothing there; it was empty; and at the moment of seeing this there was a flood of quiet joy and I knew, finally what was missing— it was my 'self.'"[2]

Up to this point, we have been pulling back the curtain to reveal how the vivid phantasmagoria of our conscious lives is created—how we get pulled into an extremely vivid sense of the world's "out-there-ness." We have seen that this happens when we are awake, and that the mind's image-making faculties continue to generate interior self-in-world models as we descend into sleep or occupy the borderlands at the thresholds of waking and sleep. Feelings and thoughts further shape and construct the experience of this lived reality.

We now turn to consider a topic that lies at the very center of all of our experiences of reality: selfhood. Sights, sounds, smells, tastes, the resonant chords of pleasure and the entire palette of the affective life—all of these phenomena belong to some specifiable system, a self. The inner phenomenal theater, in which everything that will ever happen in the course of a human life actually happens, has a characteristic contour or shape. Each of us occupies a center point from which awareness radiates. The world is encountered from a particular perspective, what the philosopher Thomas Metzinger has called an "ego tunnel."[3] In technical jargon, this is known as *perspectival-ness*. We will now go down into the roots of selfhood to consider how it emerges.

Markov Blankets All the Way Down

We begin by introducing the Markov blanket, a statistical model of a boundary between the inside and the outside of a given system, between the self and not-self.[4] The two-layer lipid cell membrane can be modeled as a Markov blanket: it segregates the cell's internal contents from the extracellular fluids that bathe the cell. Another example: Markov blankets make it possible for your immune system to recognize self from non-self, and prevent self-destructive activities. Failures in this essential process result in autoimmune diseases such as multiple sclerosis and rheumatoid arthritis.

A Markov blanket, then, in marking off a boundary between inside and outside, helps to insulate internal processes from a highly complex and uncertain world beyond. Yet it is obvious that total insulation would spell death for any living system. Life is sustained by exchanges across the boundary of Markov blankets. The inside of a system is coupled to the outside via two distinct pathways: an active one in which spontaneous upwellings from the interior core of a system radiate outwards to impact the outside and a passive, sensory route through which the consequences of those emissions are registered. Self and nonself are in a state of perpetual communion.

The only way for the self to know about the nonself is through modifying states of the Markov blanket. The cell knows about the reality beyond its membrane only by the minute electrochemical exchanges that are sensed along that membrane. Markov blankets do not appear in isolation, but organize themselves like Russian dolls, in repeating layers of Markov blankets within other Markov blankets. Consider the cell's inner structures, like the DNA-housing nucleus, which in the eukaryotic cell is bounded by a separate membrane: the nucleus is a compartment, with its own inside and outside, but housed within the larger Markov blanket of the larger cell's membrane. A dizzying hierarchy of nested layers of Markov blankets extends everywhere when you start to look for them. Organelles are housed inside of cells, and cells of the same functional type are organized into tissues, which form organs, which compose organ systems, which come together to form a whole body. It is no exaggeration to claim that reality is like an onion with seemingly endless layers, Markov blankets within Markov blankets within Markov blankets.

In such lowly origins, of a single Markov blanket (like the cell membrane), we may glimpse the most elemental foundations of selfhood, the first appearance of boundaries demarcating differences between inside and outside. Such a primitive boundary is almost certainly not conscious in the way we think of our own selfhood, but it establishes the basic prerequisite for the eventual emergence of subjectivity. If we zoom very far up from this "proto" selfhood, we will eventually arrive at your feeling of being a particular individual.

A Hierarchy of Selves

Our bodies, and the world at large, can be depicted as nested hierarchies of layers upon layers of Markov blankets. Selves, too, are hierarchically arranged. From the standpoint of neuroscience, we draw a distinction between two nested layers of selfhood: a minimal/embodied self and a narrative/autobiographical one.[5]

The minimal/embodied self refers to the nonverbal, nonconceptual, direct, and immediate sense of being someone.[6] The minimal/embodied self, although tricky to define in words, is probably more intimate to your experience than anything else. It is a sense *that* I am, more than a sense of *who* I am. The minimal self was present before you had a name. It is situated right *here,* right *now:* that unshakeable feeling of a presence that looks out from behind your eyes. Can you think of any experience that is *not* here and now for some *one?* It is difficult to even imagine such a dis-incarnate, free-floating conscious experience that occurs in a void, although there may exist an even more incipient state from which the minimal self emerges.[7]

The minimal/embodied sense of selfhood is deeply rooted in being a living body that moves into and senses the world around it. Neuroanatomically, this minimal self is referenced in ancient neural circuits that lie along the brain's midline.[8] These are among the evolutionary oldest neural structures and also among earliest ones to develop in brain maturation. A characteristic signature of these brain regions consists of extremely slow (<0.1 Hz) oscillations that persist through deep sleep and general anesthesia.

The narrative/autobiographical self, by contrast with the minimal one, is extended through time, linking together the past and future. This self has a definite name—the name that you go by. With the narrative self we move into the domain of *who* I am, beyond the direct feeling *that* I am. This self is strongly bound up with language. You can talk about your narrative self at length. The narrative/autobiographical self is reflected in brain regions dedicated to memory, language, and higher-order thought, spreading all the way up and out into the mantle of the cerebral cortex. An upwardly shifted (cortical rather than subcortical) set of midline brain structures, compared to the deep brain ones mentioned earlier, constitute important nodes within the *default mode brain network* (DMN). The DMN is associated with mental time travel and becomes activated when we are thinking about ourselves.

To a very large extent, the narrative/autobiographical self, closely linked to the DMN, exists as an elaborate conceptual network.[9] It is a mega-complex of thoughts and feelings that becomes associated with a particular name. This constructed identity seems to be chronically busy with justifying itself as a substantial thing, anticipating a panoply of future scenarios. This busyness shows up as high levels of DMN activity from an objective, third-person perspective. Subjectively, it is experienced as a whirring cloud of spontaneous thoughts from a first-person perspective, which is rarely silent.

The minimal/embodied self serves as the directly felt bedrock on top of which the narrative/autobiographical self gets layered. Like stacked Russian dolls, these different neurological subsystems elaborate distinct flavors of self. To get an experiential feel for how these two hierarchical levels of selfhood differ qualitatively, try experimenting with In-Sight 10.1.

In-Sight 11.1

To begin with, assume a comfortable seated position and take a few deep breaths. Try to be as fully in the present moment as you can, and oriented to the room you are in.

You will pose the following question, silently: *Who am I really?* When you pose this question, you will close your eyes and sit with feeling how that question resonates in your consciousness. It is as if you are dropping the inquiry and then just feeling how it echoes inside. In answering the question, it is important that you do not reference any of your thoughts, memories, or mental images (like recalling how your face looks in a mirror). This aspect of the In-Sight exercise requires you to stay only with your direct experience and minimize all concepts and narratives. Stay with the sensations only. You can periodically repeat the question, silently. But again, if you catch yourself dredging up memories, thoughts, words, sensory images, or anything along these lines, simply drop that and sit with the direct experience of selfhood.

You can try this for several minutes and then open your eyes. What did you feel? Was your wordless sense of being you localized within your body. For many people, it seems to be anchored within the torso area.

Next, imagine that you are preparing to introduce yourself to someone you have never met. You can actually try to write out a succinct introduction of yourself in four to five sentences. Take notice of what immediately comes to mind when accessing this flavor of selfhood. Did your name immediately pop into mind? What other facts, memories, and stories rose to the surface? Did you mentally visualize your external appearance? Did you feel yourself as anchored within your body when accessing this version of yourself, or did you feel more distant from the body, perhaps pulled up into your head?

Can you feel how this more complex level of selfhood feels different compared to the nonconceptual, here-and-now feeling of selfhood explored initially? Compare and contrast your experience of these two levels of selfhood. By carefully attending to the felt differences, you will understand the distinction between the minimal / embodied and the narrative / autobiographical selves from a more personal perspective.

Without a minimal/embodied self, the strata that are built on top of it simply cannot exist. The reverse is not true. There are numerous cases providing evidence of how the narrative/self can dissolve and melt into the experience of a minimal, nameless presence.[10]

The neuroscientist Jill Bolte Taylor is a famous example of this separation of selves. Taylor suffered a stroke in her left hemisphere. The left hemisphere, being specialized for language functions, is constantly weaving conceptual narratives, deftly filtering new experiences through the prism of preexisting priors and entrenched cognitive frames. With her left hemispheric interpreter going offline, Taylor found the mental chatter growing dimmer and dimmer, allowing her minimal/embodied self to come to the forefront of consciousness. No longer tethered by conceptual narratives of who she was, she found the boundaries between her and the outside world gradually melting into a warm pool of pleasurable sensations. "As the language centers in my left hemisphere grew increasingly silent . . . I became detached from the memories of my life, I was comforted by an expanding sense of grace," Taylor reports.[11] As time went on, her experience deepened, and it was as if she dissolved through several nested layers of Markov blankets, moving from the highly conceptual and constructed sense of being a separate person called "Jill" down to an embodied consciousness close to the very source of life itself. "Life! I am life!," she suddenly realized. "I am a sea of water bound inside this membranous pouch. Here, in this form, I am a conscious mind and this body is the vehicle through which I am ALIVE! I am trillions of cells sharing a common mind. I am here, now, thriving as life. . . . I am cellular life, no—I am molecular life with manual dexterity and a cognitive mind! . . . For the first time, I felt truly at one with my body as a complex construction. . . . when the systems functioned properly, they naturally manifested a consciousness capable of perceiving a normal reality."[12]

Self as an Integral Component of the Brain's Predictive Models

The earlier discussion of Markov blankets gives us some inkling of why some flavor of selfhood, even if it is only minimal, appears to be inevitable in living systems. But can we go further in saying why the brain seems to effortlessly build and maintain a hierarchy of neural systems dedicated to supporting various senses of selfhood? To answer this question, we will start with an apparently trivial example: the fact that you cannot tickle yourself. A specific pattern of light touch, on an especially sensitive part of the body, can elicit peals of laughter, and yet the very same pattern of sensory stimulation, when it is self-initiated, falls flat.[13] Behind this observation lies a deeper observation about the brain's penchant for continuously updating and tracking a self in the act of experiencing.

Distinguishing between self-caused and non-self-caused signals is required for the nervous system to support intelligent behaviors and to stop it from constantly generating hallucinatory perceptions with little relation to the outside world. To understand why this is the case, we need to pause and review some details about the modern understanding of the brain as a sophisticated prediction engine. Consider that theoretical vision of the brain that we discussed at length in Chapter 9:

there we described how the brain is constantly busy constructing models of the world that provide an optimal match to the data that the senses provide at any given moment. The brain's continued existence, as well as that of the whole body, hinges on these models being good enough to permit an organism to survive and thrive. The brain's modeling enterprise is a subtle handshake between its autonomously generated fantastic guesses, which are selectively pruned by being in intimate contact with an objective world. This two-way dance allows the brain to conjure the external world, by placing special emphasis on the surprising, the novel, and the unexpected (what we called "free energy"). The brain begins with faint sketches of the unknown and then relies on the world to provide hints about how "hot" or "cool" it is getting to the actual state of things.

Now think about how this understanding of the nervous system may explain your inability to tickle yourself. To fulfill its mission of coming up with good enough explanations for sensory signals, the nervous system must be exceedingly diligent in its recording keeping. The only information that can be used to build useful models of the outside world comes from observing the sensory effects of acting on it. The nervous system therefore has to make a life-saving distinction between reafferent processes (sensory signals that arise from movements initiated by the organism) and exafferent ones (those sensory signals arising from the world itself).[14] Reafferent signals are suppressed because they are not very surprising—they are entirely expected. The brain anticipates what sensory signals are caused by actions that it initiated from the inside so that it can conserve time and energy resources. When your hand moves to tickle your body, the brain is already accurately anticipating that such an action will lead to specific sensory patterns. These sensory consequences are dampened, and so they evoke no laughter. By contrast, when another person is tickling you, the motor commands in your own brain are not being issued to inform the brain: "motor command for a tickle initiated, expect to feel it." Being more unexpected, the tickling sensations are amplified and you end up laughing. You could become especially ticklish when you are blindfolded and really not expecting this type of stimulation.

The nervous system is constantly trying to minimize the amount of free energy or surprise. Surprise is "good" because it leads to continuous refinement of the brain's models of the world, but the goal in the long run, is focused on keeping the unanticipated as close to a minimum as possible.[15] Being swamped with an excess of surprise would be chaotic. One of the most obvious ways to quench surprise is to anticipate and ignore self-generated activities. In the brain's model building efforts then, there is a great benefit in including a self that is situated in a world outside of or beyond that self. A "self" appears to come along for a free ride, as the brain attempts to build useful "hallucinations" of an objective world.[16] Self and world, self and not-self—these distinctions are intertwined at root, so that positing one implies the existence of the other.

How Sure Are You That You Are Your Body?

In holding our hand before our eyes, in looking at our familiar face in the mirror, there is an obvious sense of ownership. We do not question that we are this body in a very direct way. This body belongs to us, is us. Beyond "our body," the world "out there" begins. A fun way to explore your boundaries, from the inside-out perspective, is offered in In-Sight 11.2.

In-Sight 11.2

When we look at ourselves in the mirror, we confront our presence as a physical fact, similar to other objects, both animate and even inanimate ones.

Here, we want to intentionally remove reliance on the visual impression of the body, or any memories or conceptual knowledge about the body, and rely only on the direct, nonconceptual sense of the body's boundaries exclusively from what is immediately present in consciousness. This is not about visualizing your body's contours, but about staying with the bare givens of the experience itself.

Assume a comfortable seated position and close your eyes. Sticking with direct experience only, can you feel where your body ends and the outside world begins? What makes up this boundary: is it fluctuations in temperature? A palpable texture? A pulsation or tension? What happens in your experience when you try to pinpoint the exact contours of the edges—are they sharp and jagged or blurred and fluid? Does your sense of being a definite "I" localize within the body, or does it spill over to occupy an expansive, nondistinct space, encompassing areas both inside and outside of the body?

Next, draw your attention to your hand. Do not *think* about your hand, but feel your right hand, allowing your attention to expand throughout the interior space of the hand, like slipping into a glove. Can you feel a boundary between where the skin of your hand is contacted by the outside world? Does the edge feel smooth or sharp?

Next, tense all of the muscles in your body, and hold this tension sustained for a few moments. What happens to the sense of the boundary between your body and the outside world during the tension. Can you feel the boundary getting thicker? After a few moments, allow all of your muscles to become completely relaxed, gently letting go of all tension. Try to feel into your boundaries now, and see if they have reorganized.

When we turn to what neuroscience teaches us, we learn that this feeling of having a body is based on another model that integrates signals arriving from multiple senses. In Chapter 7, we read about how malleable "dream bodies" are. But even during waking hours, the body that we so intimately identify with is far from the reliable anchor for personhood that we assume it to be. The sense of owning our bodies can be drastically altered within about a minute or two of delivering carefully choreographed experiences to the brain.

The classic example of this is the rubber hand illusion. Experimenters create this illusion by having a subject look down at a realistic looking rubber hand placed on a desk.[17] Meanwhile, the subject's real hand is hidden behind a screen. Next, the experimenter provides two sets of synchronous cues: the subject's real hand is stroked with gentle tactile sensations, like a brush applied to the back of the hand, in concert with the subject's watching a brush being moved across the fake rubber hand. Within a surprisingly short amount of time, the subject becomes convinced that the rubber hand is in fact an extension of their own body. When the experimenter approaches the fake rubber hand with a needle, the subject squirms. A subject can even embody this fake hand to the point of feeling real pain when the rubber hand is pinched or cut. These findings suggest that the combination of different sense modalities—in this case vision (seeing the rubber hand being stroked) and touch (feeling the actual touch on one's real, concealed hand)—are combined into a single model or representation of the body.

Bigna Lenggenhager, Tej Tadi, Thomas Metzinger, and Olaf Blanke have relied on virtual-reality technology to push the rubber-hand illusion even further—to the point that, for one study, they induced people to embody entirely fake virtual bodies.[18] This required outfitting a subject with a head-mounted display that presented two incrementally adjusted images to the left and right eye. The slight disparity created a compelling sense of being in a 3D room. A video camera captured the subject's every movement from behind and projected this image to appear about two meters in front of them in the virtual room. The subject's real back was then stroked, either synchronously or asynchronously, while experimenters provided visual feedback of that subject's virtual self being stroked. When the sense of touch and vision coincided, a subject felt that their conscious self was located in the virtual body in front of them. The full-body illusion worked just as well when synchronous strokes were paired with the body of a mannequin. In terms of what this conscious experience might have felt like from the first-person perspective, it was as if one were being pulled out and slipped into the body of a mannequin. The illusion failed when the visual and touch sensations were delivered asynchronously or when the experimenters projected a tall rectangular block bearing no relation to an actual body. The appearance of the rectangular block apparently stretched the brain's predictive models to such an extent that it was no longer possible to construct a valid model of bodily selfhood.

The significance of such experiments is their ability to demonstrate that the sense of bodily ownership is a dynamically constructed model. As Thomas Metzinger puts it, "You don't need to do anything to achieve this effect. It seems to be the result of complex, dynamic self-organization in the brain. The emergence of the bodily self-model . . . is what you experience as your own body and your own limbs."[19] The body that we feel is ours during waking hours is as much the outcome of a model-building process as the dream body.

Bodies, from the Outside and Inside

The body is not just what you and others see. In addition to the external senses, like sight and touch, there is also the realm of interoception, a sensing from within the interior of the body. Usually, the only time that we become aware of interoceptive signals is when they begin to deviate from expected values, such as when your heart beats irregularly.

Interoception is an important, though often overlooked, sense, encompassing all of the rich, bubbling medley of sensations and movements that happen below the surface of the skin. Collectively, these sensations constitute what the developmental psychologist Margaret Mahler called the "inner core of the body image."[20] Mahler's work suggested that this inner core dominates one's sense of being a self long before there is an awareness that we are also bodies that can be seen in mirrors or by other people. This inner core is the kernel of minimal selfhood. Contemporary neuroscientists like Anil Seth[21] and A.D. Craig[22] have recognized the role of interoception in establishing these most fundamental bases of selfhood.

The contemplative psychologist Karlfried Graf Dürckheim spoke of the distinction between the body that you *are* and the body that you *have,* which contains parallels to this outside-inside nature of the living body.[23] The phenomenological tradition of the early twentieth century, in its turn, distinguished between the actual experience of the lived body (*Leib*) in contrast to the body as a mere physical object (*Körper*). The outward facing aspect of the body is the one that you see when you encounter your face in the mirror and recognize it as being you. The representation of the body's interior appears to be closer to the minimal/embodied self—the sense *that* you are. The second, external body representation, by contrast, seems to serve as the primary way to define your boundaries, your sense of where "you" end, and the world begins. Eventually this sense of the body as an object, like other objects, becomes a vital aspect in the creation of your own unique narrative/autobiographical self.

In looking for the mechanisms and processes that give rise to selfhood, experiments on phenomena like the rubber-hand illusion have mostly focused on the role of integrating and binding sensory signals into a coherent model. Yet we would be remiss if we were to leave out what is arguably an even deeper core of embodied selfhood–the spontaneous bubbling up of an intention to move, to propel through space. Movement enacts a process of world discovery, and this entire process of discovery is shot through with a sense of purpose.

Impulses to explore and act, to move into a world full of stimuli that have special meaning for us, spontaneously emanates from deep down within the body's interior. In line with a venerable tradition of neurobiologists, including Walter Freeman[24], Jaak Panksepp[25], Rodolfo Llinás[26] and many others, it is indeed possible to locate the earliest kernels of selfhood in these ancient neurodynamic energies and movement potentials animating the behavior and psyche.

Panksepp, for example, has argued that the true anchor of embodied selfhood and agency lies in what we might envision as a primordial coordinate space of possible bodily actions.[27] For Panksepp, these processes are neurally rooted in some of the oldest regions within the brain stem, where cells maintain a map of the entire body, uniting both its internal (visceral) and external (skeletomuscular) facets. These brain regions funnel signals that direct the body's movements in space as well as all of the various inner movements (like changes in blood flow or movements of the gastrointestinal tract). These primitive brain structures endow biological organisms with what Panksepp called a simple ego-type life form (or SELF).

Simple ego-type life forms, as intelligent and hierarchically organized sets of Markov blankets, maintain and protect their own boundaries precisely by moving and sensing the results of their movements. We can use these neurological facts to modify Descartes's famous "I think, therefore I am" dictum to read "I move and feel, therefore I am."

As brain evolution continued to add more complex structures, these distinct components of being a definite self began to merge and coalesce into higher-order models that pooled each of the corresponding elements.[28] An integrated sense of the body—both as sensed from within and experienced as a distinct object—normally come together and cooperate to impart a fairly integral and continuous sense of being the person that you are. The minimal/embodied rudiments of selfhood link up and become fairly coextensive with more conceptually elaborated narrative versions of selfhood. As we are about to see, however, specific components of selfhood can also come undone in various situations.

Varieties of Selflessness

Can there be such a thing as a selfless state of consciousness? The case of Bernadette Roberts, as well as Jill Bolte Taylor's stroke, provide compelling phenomenological evidence that the ordinary sense of self can undergo dramatic alterations. Given the variety of selves, a wide spectrum of conditions involve diminished or absent selfhood. Raphaël Millière, an Oxford philosopher who has extensively studied such phenomena, pooling findings gathered from studies of sensory deprivation, dreams, meditation, drug-elicited states, paralysis and different types of clinical conditions, believes that there is substantial evidence to support at least six different types of selflessness.[29] These can range from the dissolution of the narrative self to more unusual cases where there is an absent sense of being an embodied person and even ones where one's thoughts are no longer perceived as belonging to any one in particular. Here we will focus on just a few salient examples of selflessness that might provide us with deeper insights into the mysteries of consciousness.

Drug-Induced Ego Dissolution

"'I' was no more, blasted to a confetti cloud by an explosive force I could no longer locate in my head, because it had exploded that too, expanding to become all that there was . . . Whatever this was, it was not a hallucination. A hallucination implies a reality and a point of reference and an entity to have it. None of those things remained . . . every touchstone that tells us 'I exist' was annihilated, and yet I remained conscious."[30]

Author and journalist Michael Pollan's dramatic account of an experience evoked by a potent psychedelic known as 5-MeO-DMT exemplifies one of the hallmarks of the psychedelic drug experience: a loss of perceiving oneself to be a distinct self. The phenomenon of drug-induced ego dissolution, with the loss of self-boundaries, can be accompanied by blissful feelings of unity or merging of the self with the universe or a higher power. It can also be an experience filled with apprehension and anxiety that occurs as the sense of self recedes.[31] Often, psychedelic experiences are a combination of both types of

experiences, which shape-shift and flow seamlessly during a psychedelic session.

During drug-induced ego dissolution, a psychedelic voyager often
retains a glimmer of "being here now" and may even retain awareness of residing in a body immersed in a sea of vivid colors and
sensations. What is lost or drastically reduced under the influence of
psychedelics is the narrative self, or more accurately the constraints
imposed by the narrative/autobiographical self that is rigidly anchored
in stories and events located at discrete points on a past-present-future
timeline. In neurodynamic terms, the large-scale networks that instantiate this more abstract sense of self lose their capacity to entrain more
local assemblies of neuronal populations. The top-down predictions
that constrain lower rungs of predictive hierarchies become relaxed.

Psychedelic substances replace the narrative self, for however short
a time, with a sense of full engagement with stimuli in the present moment. The stories that we self-narrate can mire us in the past and are
rarely if ever questioned, leaving little room for alternate ways to think
about ourselves and the world. Psychedelics can permit access to unfiltered experiences that challenge tired, stale, and dysfunctional views
of our past, current, and future realities. Ideally, new ways of seeing
the world can be incorporated into more flexible and adaptive models
of the self and the world. The potential also exists, however, for frightening experiences of self-loss to arise during psychedelic experiences.
The ability of psychedelics to alter both the narrative and embodied
nature of selfhood seems to be a central component in their capacity
to sustain transformative insights.[32]

Neuroscientists have used brain imaging to reveal what occurs
within the brain as psychedelic experiences of (narrative) ego dissolution unwind. On a neurobiological level, psychedelic substances
disrupt the connectivity of the default mode network with regions of
the brain associated with how self-related information is processed.
Yet these changes occur simultaneously with increases in both global
connectivity of brain regions and with greater integration of activity
among brain areas, which together produce feelings of ego dissolution. Thus, the powerful subjective experiences of connectedness
or unity that can accompany ego dissolution mirror alterations in
connectivity and integration of neural activity across different brain
regions.[33]

Depersonalization

Depersonalization is a condition marked by profound alterations in the sense of self and reality. People who experience depersonalization feel split off from essential aspects of their selfhood—from their emotions, thoughts, sensations, and actions—and this often gives way to a disquieting sense of unreality and emotional numbing. Depersonalization can extend to feelings of being disconnected from the body, as in out-of-body experiences, or to the sense of a loss of ownership of a part of the body—perhaps a hand or arm—or in extreme cases the entire body. As philosopher Jennifer Windt comments, consciousness can be maintained even in the absence of bodily ownership, such as in dreams in which individuals "experienced themselves as a disembodied point or freely moving center of awareness."[34] Thus, a lack of bodily awareness, which marks depersonalization, can cut across dreams and waking life.

Dreamlike states are present not only in depersonalization, but also in derealization, which is typified by feelings of unreality or a foggy dreamlike state that arises in relation to surroundings, objects, and people. In fact these two conditions are so enmeshed that in the latest diagnostic manual of psychiatric disorders (DSM-5), they are united in the diagnosis of depersonalization / derealization disorder when feelings of inhabiting a secure "reality" and stable sense of self are so negative and persistent that they interfere with daily functioning. As noted by Daphne Simeon, a psychiatrist, and Jeffrey Abugel, a medical journalist who himself experienced depersonalization, "people with chronic depersonalization are . . . living everyday with the fear and unreality of a dream state come true" and they are "never quite sure who they are in a sense."[35]

Depersonalization / derealization often occurs when sleep and waking states of consciousness overlap, although in a decidedly different way from lucid dreaming. In lucid dreaming, a conscious perception of reality intrudes into the dream. In depersonalization / derealization, quite the opposite occurs: dreamlike states infiltrate daytime reality. The line between dreams and waking life can become so blurred that dreamlike experiences find their way into daily life, shading into feelings of unreality and dream-reality confusion even while we are awake. This can create a jarring sense of disconnect between dreamlike experiences during waking hours and the demands of mundane life. In sum, experiences related to depersonalization underscore the linkage and overlap of states of consciousness across sleep and wakefulness and how we could come to have the subjective sense that, at least in certain circumstances, we are literally "dreaming reality."

Dissolving of Self Boundaries in Long-Term Meditators

The experience of selflessness represents the apex of certain contemplative paths. In some forms of Buddhist meditation, for example, a profound and direct realization of selflessness can lead to a lasting seeing-through of the "self illusion." Meditative practices, by stilling the mind's attachments to the past and fantastical simulations of affectively laden future scenarios, might denature the neuronally "sticky" bonds that hold more abstract, narrative models of self intact under ordinary circumstances.[36]

Long-term, expert meditation practitioners offer unique opportunities for neurophenomenological studies of selflessness phenomena. In a series of scientific articles, Yochai Ataria and colleagues reported on an individual who, after forty years of dedicated Buddhist practice and over twenty thousand hours of meditation practice, could flexibly dissolve his sense of perceived boundaries between self and world.[37] There was some evidence that his prolonged training had actually changed specific aspects of his brain physiology. For instance, his DMN had become chronically underactive relative to that of individuals with less extensive training.

This study revealed three distinct forms in which the conscious experience of self and world could become organized. The most familiar such state was the one in which the sense of being a conceptually defined, narrative/autobiographical self was firmly intact. When this experience of self-dominated consciousness, it was associated with a sharp separation between a self located "in here" and the world "out there."

An intermediate stage of dissolving boundaries was experienced as a progressive increase in the porousness of the boundary between self and world, so that "it's like as if I'm in a way made of air and airiness extends out" and "the voices outside are kind of inside, I include them within the bubble . . . sound is known, but I can't tell you really what it is, it's included in the immediate bubble, the sound is inside it some-where, but I can't tell you exactly what it is." While there was still some sense of having a body, the body was experienced as an expansive web.

Finally, with more completely dissolved boundaries, there was a "sense of dropping, it's like falling into empty space . . . I dissolve into the world and where you have the boundaries and the self as foreground and the world as background or as ground, here there isn't a ground, there isn't a foreground, in a way the background is everything, although not identical to the world, I'm not separate from background . . . there's no personal point of view, it's the world point of view, it's like the world looking, not ME looking, the world is looking."

This dissolved state was associated with a unique signature of functional brain activity, characterized by two distinct processes. The first set of neural changes involved a suppression of neural mechanisms in the posterior part of the DMN, which normally helps to integrate a conceptually elaborated sense of a distinct self. The second pattern involved a change in neural activity in the parts of the brain that typically function to yoke the sense of self to one's perception of the body as an object. As chronic neural activity in both of these locations is reduced, one's conscious experience is of increasingly dissolved boundaries between self and world.

The Vicissitudes of Ego Dissolution

Self-loss experiences clearly come in many varieties. Consider Bernadette Roberts. As she approached the brink of what would eventually become a transformative and positive experience of self-loss, she felt an impending sense of annihilation of her self boundaries, which threatened to overwhelm her. Such negative experiences can accompany ego loss in psychedelic and other states too, such as those produced by meditation, and they represent the flip side of blissful experiences of unity and oneness.[38]

Psychologist David Yaden and his associates identified two aspects of self-transcendent experiences.[39] The first is what they called an *annihilational component,* in which the bodily sense of self, self-boundaries, social boundaries, and self-salience diminish appreciably or dissolve entirely. People who experience self-loss, for example, in some cases of schizophrenia, during psychotic episodes, and when loss of contact with the self and reality occurs in cases of depersonalization/derealization disorder, exemplify the dark side of the annihilational component of self-loss. Yet, on the bright side, Yaden and his colleagues contend that people with excessive self-focus, such as depressed individuals who ruminate, feel guilt, or experience anxiety about future events, could actually benefit from decreased self-focus. In some cases, relaxing a tight or a habitually rigid grip on the self's boundaries can be beneficial. By accessing the property that the psychiatrist Daniel Siegel refers to as *ipseity,* so characteristic of the minimal/embodied self, one can experience a sense of emotional liberation from staid, old strictures.[40]

Yaden and colleagues identified a second aspect of self-transcendent experiences that they called "relational" that is associated with a sense of connectedness or unity, which extends beyond the self and is linked with mystical experiences. Research led by Anne Taves suggests that mystical experience may be a variant of ego dissolution, as relational experiences are often reported in conjunction with mystical experiences in which the sense of self can fall away entirely, creating a distinctionless sense of unity with one's surroundings.[41]

Whether a person will predominantly be conscious of one or the other component of self-transcendent experience is multidetermined. Psychedelics provide a good example. Given the multitude of potential responses to psychedelics, it is best to think of experiences of ego loss as falling on a continuum. On one end are those that are peaceful and blissful, at the other end are ones that are distressing and even terrifying, and in the middle are neutral experiences. Even the same person across sessions—or within the same session—can respond to psychedelics very differently and be located at different points on this continuum at different times and under different circumstances.

When feelings of dissolved self-boundaries are interpreted as an increase in connectedness with the social and natural world, there is a more positive experience.[42] At its height, the sense of connectedness and unity fosters a visceral sense of our participation in a shared humanity that transcends social and cultural boundaries and helps us to feel aligned with a universal, all-pervasive dimension of togetherness.

This intuition appears to be the very stuff of mystical experiences, reported in all human cultures throughout the world. Even glimpses of such experiences appear to be profoundly transformative for many people. The latest research suggests that the clinical effectiveness of different psychedelics in treating disorders like depression is positively predicted by their varying abilities to evoke these mystical-type experiences.[43]

The Center That Does Not Hold:
Buddhist Perspectives on No-Self

There is a tendency to link contemporary scientific discoveries concerning the constructed selfhood to frameworks borrowed from contemplative practice, particularly to the teachings of various Buddhist schools on the concept of no-self. In engaging upon any such comparisons, we have to remain on guard against committing two possible errors.

The first error is to ignore the sophisticated and nuanced arguments about the nature of selfhood and just what no-self actually means. The Buddha's doctrine of *anatta* (no-self) is among the most subtle and complex of topics, which practitioners and scholars have debated for hundreds of years.

The second error is the treacherous attempt to compare technical terms that derive from radically divergent worldviews. As the scholar Brian Lancaster has capably documented, any attempt to divorce the Buddhist teachings on *anatta* from the code of ethics and morality that underpins the practice of Buddhism does unintended violence to the depth of meaning behind concepts like no-self.[44] One result, for example, is that the Buddhist teachings of no-self have sometimes been coopted to support distinctly modern forms of a prevailing cultural nihilism that would seek to reduce these very complex teachings to overly simplistic conclusions, such as that the self is an "illusion," "a myth," and that "ego dissolution" is, *a priori,* a goal to strive for. Unmoored from proper contextualization within a larger worldview, the conclusion that the "self is an illusion," which needs to be annihilated, might lead to an ethical *laissez faire* attitude. From the traditional perspective, this would be the exact antithesis of the original Buddhist teachings.

Since the newly emerging field of contemplative science has drawn most extensively from Buddhist wells, and since this ground is fairly well trod by now, we will limit our discussion of it. For our purposes, perhaps the most important things to note concern the scientist-like nuance that the Buddha introduced on this topic. For example, in the scriptures contained in the Pali canon, the oldest bundle of texts which contain written records of the Buddha's discourses, we find him expounding on different aspects of the self, including the "gross acquired self" and the "mind-made acquired self."[45] The former term refers to that very compelling dimension of our experience where we implicitly identify ourselves with the physical body. The "mind-made acquired self" encompasses the inner world of thoughts, running commentaries, memories, and daydreams that, unlike the physical body, might be said to belong to a more subtle, mental sense of selfhood. Just as we learn to take the body as being "me" or "mine," so too do mental phenomena give rise to a compelling sense of "me."

The historical Buddha himself derived his teachings by undertaking a personal search following the trail of suffering (*dukkha*) and conducting a very deep and interior inquiry into its origins. Quieting of the mindstream allowed him to peer into the depths of mental life. Rather than discovering an independently existing and substantive thing called self, he discovered the arising and passing away of different temporary aggregates forming constituent elements of phenomenological experience. The absence of an enduring, permanent, substantive, fixed entity or thing called self leads us, then, to the doctrine of *anatta*—no-self.

Although the full meaning of Buddha's doctrine of no-self is unlikely to be resolved, and its interpretation will continue to be debated, some scholars have strongly challenged the simplified (but popular) conception that it simply points to the self as an illusion. Properly speaking, the Buddha seems rather to have been disputing the acquired narrative of selfhood, which for most of us becomes incredibly rigid over the course of a lifetime. One scholar, having studied the Pali canon in detail, concludes that the Buddha "does not say simply that the self has no reality at all, but that certain things, with which the unlearned man identifies himself, are not the self and that is why one should grow disgusted with them, become detached from them and be liberated."[46] This would suggest that it is that solidified ego, the conceptual mega-construct that attempts to constantly aggrandize itself, often at the cost of others, which has no true existence and that constitutes the no-self of Buddhism. We will now see what, if any, parallel to these teachings exist in Christianity.

The Perspective of Ancient Christianity

Christian teachings hold personhood as foundational.[47] Ultimate reality is understood as intensely personal and loving, rather than as an impersonal boundlessness into which the contemplative's ego dissolves. At the same time, however, the rich tradition of ancient, contemplative Christianity makes critical distinctions between different senses of self. An important distinction is made between a surface definition (a "false self") and a more authentic, deep, "true self."[48] The surface or false-self includes the all-too-familiar perception of being a *separate* self—an individual—who is set off, positioned over and against a world out there. The false self is grounded in the feeling of being separate and self-enclosed and confronts the outside world from an egocentric frame of reference. Everything in experience is carved up into what is "mine" versus "not mine." In a nutshell, this inveterate (but false) self is that experience of an "I" that ceaselessly desires and fears things and other people in the outside world.

The false self subtly insinuates itself as being a core component of reality, giving rise to what the Christian spiritual masters called *philautia*—basic selfishness.[49] *Philautia* comes to afflict perceptions, thoughts, emotions, and behaviors across many levels, from the conscious down to the subconscious mind. According to the experience of the Christian contemplatives, when the mind's functions are so thoroughly penetrated by *philautia,* our experience of self and world becomes grossly distorted. Characteristic of this mode of perception is the tendency to experience everything and everyone through the prism of personal wants and needs, such that the richness of reality is reduced to a measly trickle that pours through our sensory portals, resulting in a massive restriction of potential.

Philautia splits the categories of self and world into sharply experienced dualities, through gluttony, lust, greed, envy, anger, and other powerful feelings. The world is then entirely encoded with different shades of pleasure and displeasure corresponding to one's momentary physical and psychological needs. What is considered neutral, relative to one's state, is outright ignored. Unrestrained desire perverts perception from its original, pristine, and compassionate experience of reality into a kind of personal fetish.

Echoing the metaphors of many sages from the Far East, the early Christian writers repeatedly referred to humanity's being in a state akin to sleepwalking, seduced by fantasies, mirages, and illusions. Even in the midst of ordinary, waking consciousness, one then inhabits a world that is like a subjective hall of mirrors, "just as a child, young and guileless, delights in seeing a conjuror and in his innocence follows him about."[50] Saint Gregory of Nyssa writes that "nothing in life appears as it is; rather, life shows us some things in place of others according to our deceived imaginations . . . hiding itself under the illusion of appearances," and Saint John Chrysostom compares this mode of existence to "dust that the wind scatters . . . a shadow, a wisp of smoke . . . the leaf that is the plaything of a puff of air . . . a dream, a sound that passes . . . a light breeze that vanishes . . . the flimsy feather that is caught up in flight."[51] According to these teachings, it is pleasure-seeking and self-love that are the twin engines perpetuating this misperception of reality. Viewed from this perspective, Christianity and Buddhism would seemingly agree that the individual, when defined as this separate, independently existing self, is fundamentally unreal.[52]

In the teachings of Christianity, the false self, under the sway of *philautia,* is constructed over the course of a lifetime. Human beings have a strong penchant to fall under the spell of the false self, which is taught to engender a "fallen" version of reality. The far-reaching ability of philautia to distort the world and people around us, from this perspective, is not only characteristic of striking examples of flagrant narcissists, but remains operative to a lesser or greater degree in all of us much of the time.

The path of Christian spirituality is one of a radical and deep transformation reaching down into the very roots of our being, restructuring human consciousness from the ground up. The Scriptures are full of references to the need, through much tribulation and ceaseless struggle (an "unseen warfare"), to put to death the false self. This is a form of "psychological death"—what many mystics of the past have referred to as "dying before we die"—after which we can be liberated to come truly alive.

In its initial stages, this putting to death of the false self requires an unswerving commitment to cutting off various addictions, compulsions, and obsessive and intrusive thoughts that reinforce the false self.[53] Undoing of the false self entails a conscious decoupling from deeply ingrained habits of thinking, feeling, and acting. The contemplative cultivates constant watchfulness over one's interior states and is ever on the lookout for the extremely subtle and rapid ways in which perception shifts into its selfish, egotistical mode. Conditioned patterns refuse to die without putting up a life-or-death struggle, necessitating various spiritual exercises (ascesis) including interior vigilance, ceaseless prayer, increasingly deeper engagement with the reading and assimilation of the Scriptural teachings into one's psychological life. When ascesis is combined with regular participation in the purifying sacraments offered by the Church, it functions to effect a very gradual cure of the underlying malady of *philautia*.

If the goal of the Christian contemplative is a mortification of the false self, this is not done so as to precipitate an undifferentiated form of ego dissolution. Rather, the ultimate psychological death of the isolated, false self is what allows for the true, deep self to be revealed and become the new center of gravity in one's consciousness. According to the teachings of the Christian East, every person bears an unmistakable stamp of divinity, a sacred and irreplaceable facet reflecting the brilliant radiance of a reality that is inherently good, loving, and nurturing. The true self is believed to be a unique and unrepeatable window that opens into an experience of that divine reality. From the Christian perspective, the false self stands between us and the experience of a divine reality, an abiding sense of unity and goodness. The recovery of the true self is believed to be assisted by a lifelong dedication to contemplative ascesis.

People might experience brief flashes and glimpses of their deep self at different moments. During an extended vacation in an exotic environment the veils might temporarily lift to disclose a reality far more mysterious, benevolent, and transcendent than what is typically disclosed to consciousness. Without serious and prolonged devotion to the path of spiritual transformation, however, these infrequent episodes usually dissipate back into the more familiar orbit of the false self. For a contemplative, the psychological center of gravity eventually needs to shift from the orbit of the false self to that of the true person, and following the gradual process that will accomplish that decisive shift requires enormous amounts of conscious effort.

How does perception operate from the perspective of the true self? Rowan Williams, the former archbishop of Canterbury and theologian, points out that abolishing the false self does not result in experiencing oneself melt into an undifferentiated background. A unique and distinct point of consciousness continues to exist, but now with a consciousness of interconnectedness with all of life. Contemplation "might be understood as the discipline of opening the mind to a 'world' in the full sense."[54] Reality discloses itself as a luminous, living network of here-and-now interactions, encompassing the entire cosmos. The "foreground"—a personal nexus of consciousness— enters into a dynamic relationship or dance with a beneficent and caring "background."

The bedrock of what is real is a pulsing ecological web of energies that call for appropriate responsiveness in dynamically changing situations. Christianity teaches that the sense of distinct personhood is mutually compatible with a consciously experienced feeling of unity. "Our distinctness," Rowan Williams writes, "becomes not a solid identity but a unique 'point' . . . from which the communication of life radiates."[55] A person is an integral and organic component embedded within a more expansive network of living interactions. There is no other, insular self, except as a concept or narrative, something really like an "illusion."

Rigid Boundaries, Open Boundaries, and Well-Being

We will now offer a few final reflections on selfhood and how it relates to our well-being. From the evidence we have reviewed, we can discern a continuum of experiencing self and world. On one end of this continuum is a fairly rigid, dualistic or bipolar organization of consciousness in which the boundaries of the self are rigid and sharply contrasted with the world beyond a "skin encapsulated" self. This mode of consciousness is the default setting for most of us. When, however, the self boundary becomes too habitual and inflexible, it can cause substantial mental suffering.[56] On the other end of the continuum, we have an increasingly permeable sense of a self boundary, even to the point of virtual disappearance, as in experiences of drug-induced ego dissolution. The ego dissolution facilitated by ingestion of a psychedelic substance presumably loosens the narrative self, while leaving intact at least some version of embodied selfhood. In fact, psychedelics, like certain forms of sensory isolation, might enhance interoceptive abilities, allowing the ability to remediate habitual and restrictive forms of narrative selfhood.[57] The collective Christian contemplative experience attests that the deeper and more authentic dimensions of selfhood are entirely compatible with profound experiences of unity with all of existence, indicating that self boundaries can become more open to the world, without leading to self-annihilation and absorption into nothingness; personal being and communion are mutually compatible.

The psychologist, Stanley Keleman suggested that to be fully alive means to be capable of letting go of overly rigid self boundaries, which are expressed somatically and psychologically.[58] This frees us up to repeatedly dissolve and reform our edges in a dynamic fashion as the situation requires, such as when we are called to become assertive or stand up for a cause. When give and take is called for in a collaborative effort, then we can relax self boundaries, leaving them more flexible for subsequent reformation. According to Keleman, the repeated mini dissolutions of self boundaries, which we undergo throughout life, prepare us to accept the ultimate challenge of facing death, the final form of ego dissolution, at all levels from narrative to the minimal, with greater equanimity.

Trauma can precipitate a chronic freezing and painful contraction of self boundaries. In terms of the embodied self, a person can become locked into a state of physiological hypervigilance and uptightness. Traumatic experiences can lead one to experience the body simply as an external object, a "thing," similar to other inanimate objects. At the level of narrative selfhood, traumatic memories seem to function as psychological "black holes"—centers of gravity, so to speak, that completely pull in one's sense of identity and lead it to become inordinately defined with respect to elements of the trauma itself.[59] As a neurological parallel, early life traumas are capable of imprinting residues in spontaneous neuronal oscillations, within brain regions that are intimately involved in maintaining the autobiographical self.[60]

These tighter self boundaries following trauma are motivated by the mind's and body's attempts to heal, using a stiff boundary as a protective wall against an external world perceived to be hostile. Yet, these defensive attempts can backfire over time and produce prolonged suffering. The chronically maintained thickening of the self boundary, the gross- and mind-acquired self, is what the Buddha identified as the root origin of our suffering.

On the other hand, the ability to fully let go of self boundaries requires what the philosopher Jean Gebser has called *Urvertrauen*—primordial trust.[61] This refers to a radical openness and receptivity to being affected by the world and entering into deep communion with it, with no conditions required. Underlying this trust is an unshakeable belief in the fundamental beneficence of life as an irrepressible impulse, a viscerally felt confidence that, in the words of Julian of Norwich, "all shall be well, and all shall be well, and all manner of things shall be well," a paradoxical truth that can persist even in the face of shattering tragedies.

12

The Emptiness of Being

Some states of consciousness bear little resemblance to ordinary waking experiences. A few quotes may serve to hint at the states probably the furthest removed:

> "no imagery, just blackness: yet the blackness was pregnant with the unknown. . . . I don't see anything . . . don't feel anything. While nothing is happening, everything happens . . . absolute serenity."[1]

> "no symbols are encountered, visual or otherwise . . . all awareness of the self as body or special entity leaves . . . characterized by peace, silence."[2]

> "I remember regaining awareness before being able to see, hear, or feel anything. Incidentally, that period of darkness, silence, and lack of sensation was probably the most peaceful experience I've ever had."[3]

The first two are vivid accounts of imageless, lucid dreams. The last one describes a state induced by a general anesthetic. The psychologist Wilson Van Dusen, recounting his self-experiments with a mixture of nitrous oxide and oxygen, wrote that "at the lowest stage of unconsciousness, in which breathing itself stopped, I had tremendous cosmic experiences that I was very disappointed to leave by someone bringing me back to life."[4]

In experiences of ego dissolution, boundaries between the "self" and the "outside world" become translucent or disappear. Now we are ready to peel back yet another layer and look at a perplexing netherworld of conscious experiences, where it seems that both the contents of consciousness and the subject for whom they appear are reduced to a bare minimum or even disappear entirely.

What is left when the contents of experience are so reduced that they seem to vanish? We can imagine it as a perfect vacuum, but a vacuum that is aware, providing the medium for each and every scrap of experience that we could ever conceivably become aware *of*. Even the perspectival structure of experience—the phenomenological fact of having a specific point of view—seems to be missing this far down the consciousness spectrum. In professional jargon, this is a minimal phenomenal experience (MPE).[5]

While modern science knows little about MPEs, the scholar Robert K. C. Forman has argued that studying them might do for consciousness studies what the study of single-celled organisms did for biology.[6] By isolating the most minimal experiences, we could gain an unprecedented window into the origins of consciousness, just as in the most simple life forms can be found the mysterious keys to multicellularity.

By and large, modern Western philosophy and science have tended to neglect the study of MPEs, mostly because of a presumption that consciousness can never be without content. How could one be aware of nothing—and if one was, would this not already be *something?* Consciousness researcher Talis Bachmann contends that, even in moments like those described in the quotes above, the very feeling of what is lacking is still a form of content.[7]

Yet spiritual traditions from around the world have long assured us that MPEs are real, empirically verifiable states of the human mind.[8] Buddhism, for example, speaks of the "clear light of awareness" as the most subtle form of consciousness, which supposedly dawns for each person at the moment of physical death, when the senses collapse into a state of utter quiescence.[9] The famous *Yoga Sutras* of Patanjali contain systematic instructions for stilling the mind's fluctuations so completely that only a pure awareness remains, unadulterated by any specific content.[10] The Christian tradition, using its own distinct terminology, contains descriptions of similar states of consciousness.[11] Mystical traditions in Judaism and Islam offer deeply reverent descriptions of MPE-type phenomena.[12]

In this chapter, we will see that the Indian and Tibetan philosophical systems in particular contain many descriptions of very subtle MPEs. In both systems, these dimensions of the mind can be observed by entering into profound states of meditation and interior silence. Today, as neuroscience begins to make its fledgling forays into this veritable Antarctica of the brain/mind, we stand at the threshold of a new frontier of consciousness studies, where neurobiological hypotheses concerning MPEs can be compared with the experiential findings of the great mystics.[13]

The Mystery of White Dreaming

In 1946, the psychoanalyst Bertram Lewin wrote a paper in which he drew attention to something he called a *blank dream*.[14] Such dreams are devoid of all visual content, presenting an empty screen that ordinarily functions as the backdrop of dream narratives. This dream screen, according to Lewin, is always present as the context for all dream events, but it normally goes unnoticed as we become utterly mesmerized by dream characters and storylines.

"The dream screen is not often noted or mentioned by the analytic patient," Lewin wrote, "and in the practical business of dream interpretation, the analyst is not concerned with it." Yet, under some situations, the screen alone showed up in the dreams of his patients.

For a long time, the dream screen went largely unnoticed by sleep psychologists. It was not until about three decades after Lewin's paper that such a phenomenon was independently discovered and found to account for approximately 30 percent of the dream reports from dream diaries.[15] These were dreams that were completely lacking in any specifiable content.

A more meticulous experiment conducted recently reported similar figures.[16] The researchers used what is known as a serial awakening paradigm, by which an automated computer sound went off at fifteen- to thirty-minute intervals as participants were monitored in a sleep laboratory. Immediately upon awakening, and across a total of forty-four separate nights, each participant was instructed to lie in bed, with closed eyes, and to respond to specific prompts, such as "What was the last thing going through your mind prior to the alarm sound?" Based on more than seven hundred responses from all different sleep stages, this study reported that dreams with no actual content (as it termed them, *white dreams*) made up around 30 percent of all dreams—a similar proportion to that found in earlier work. Another 20 percent of the awakenings prompted descriptions of dreams in which there had only been vacuity (absence of experience, rather than experience of absence), and the remainder described narratively rich dreams.

Contentless dreams can occur during any part of the sleep cycle, but they seem more likely during specific portions of the night. Awakenings during the early portion of the night, when so-called slow-wave or NREM sleep dominates, elicit white dreams close to 40 percent of the time.[17] In the traditional understanding, these sleep stages would have been classified as periods when consciousness was presumed to be absent.

White dreams remain shrouded in mystery. Some researchers have assumed that such dreams reflect nothing more than forgetting. Plenty of us have had the experience of being convinced that we were in the midst of a compelling dream story, only to have it disappear into a fog of forgetfulness the moment we wake up. There is little doubt that at least some, and maybe even most, white dreams *are* simply forgotten dreams that were rich in all kinds of content that we later fail to recall.

Some scientists have proposed that white dreams do, in fact, contain some content of a very weak perceptual quality. Support for this hypothesis comes from a study using an electroencephalograph to compare the amplitude of high-frequency neural oscillations in sensory brain regions during periods of no dreaming or during content-rich dreaming to the amplitude in periods associated with reports of white dreaming.[18] In white-dream periods, the oscillatory amplitudes were stronger than in the definite periods of absent dreaming, but these oscillations were weaker in strength than those of florid dreaming. The reduced neural activity in posterior sensory brain regions likely indicates perceptual representations of diminished vividness, greater ambiguity, and less stability.[19]

Fragmentary instances of white dreaming are also frequently reported during sedation with general anesthetics, when participants are unresponsive to verbal prompts.[20] To complicate matters further, it is now widely recognized that even after-the-fact reports of unconscious oblivion from patients who were sedated during surgical procedures are not always true. There have been surgical patients, for example, who denied having been awake at all during general anesthesia but who had actually carried on conversations using hand signals during their procedures, indicating there had likely been at least a minimal form of consciousness.[21] So, while many white dreams with little or no content may be simply hazy in memory, scientists cannot rule out the possibility that some portion of them really are high-fidelity reports of conscious experiences of a "void."

Dream researcher Jennifer Windt has suggests that white dreams may be a transitional state of consciousness between periods of deep, dreamless sleep, and dreams richly populated with content.[22] She holds up white dreaming as a leading MPE candidate, a prototypical state in which all sense of being a self extended in space disappears, leaving only a sense of the passage of time elapsing in an ever-present "now." Also interesting are cases of so-called subjective insomnia, a condition in which people report that they are awake even though their brain activity indicates they are actually asleep.[23] The two phenomena may be closely related.[24]

Windt has outlined a few strategies for future researchers to increase their sensitivity to white dreaming and discover genuine examples of this state of consciousness.[25] One is to awaken test subjects in a sleep laboratory at regular intervals but pose inquiries that do not exclusively emphasize experiential content. For example, instead of asking *What was the last* thing *going through your mind?*—a question bound to direct attention to the contents of consciousness—a researcher might ask *What, if anything, were you feeling prior to awakening?* This might elicit responses about more direct and immediate forms of experience that pertain to the basic qualities of *being* rather than *doing*. For instance, the preceding sleep period might have been experienced as deeply restful and relaxed, or tense and troubled, or perhaps groggy—all of which are dimensions of experience with no "content" as such, and unlikely to be reported to a questioner probing only for dream content. Another potentially productive strategy is to have subjects generate estimates of elapsed time—providing indications of whether an experience of pure temporality, independent of characters or a story, was present during certain parts of the sleep cycle. Readers who are strongly committed psychonauts might find the self-experiments described in In-Sight 12.1 helpful to developing an intuitive sense of white dreams and MPEs.

In-Sight 12.1

As a word of caution, only the most committed psychonauts will want to explore this particular exercise. You will need to set up your phone or other digital device to have repeated alarms sound at roughly twenty-minute intervals as you are drifting off to sleep. At each of these awakenings, quickly record a spoken account of what you were experiencing immediately prior to being roused by the alarm. Inquiries suggested in the text could be used to guide these quick summaries. For example: *Do you recall any feelings right before awakening? Do you have any sense of how much time has elapsed since the prior awakening? Can you recall any dream content or not?* Review your audio recordings or jotted notes the next day and try to detect any experiences resembling the descriptions of white dreaming given above.

Entering the Antarctica of the Mind

Lucid dreaming offers a potentially rich inner laboratory for exploring MPEs. Stephen LaBerge, a pioneer in this area, has cataloged a unique class of lucid dreams that unfold in minimal perceptual environments.[26] A lucid dream world may begin as rich in content, and then become progressively stripped until the dreamer is a boundless point of consciousness floating in empty space. As the lucid dream shifts from being perceptually minimal to sinking into an imageless void, we gradually enter the realm of pure MPEs.

Some talented lucid dreamers have made a concerted effort as a community to map the furthest reaches of these imageless lucid dreams, past the point where any sense of being a self extended in space evaporates.[27] One prolific lucid dreamer, Robert Waggoner, refers to sojourns through empty dream spaces as passages into the "gray state." He defines the gray state as the imageless backdrop that lucid dreamers can become aware of when visual imagery disappears. Although many lucid dreamers assume that, at that point, the dreaming process has ended and they choose to wake up, Waggoner found that it was possible to remain in that process and so enter into what he called "clear light" dreams.[28] He describes these dreams as being memorable for having no action, no story line, no characters, and no content except for a luminous expanse.

Fariba Bogzaran, a visual artist, has explored and written about these unique, contentless lucid dreams, which she calls *hyperspace lucidity*.[29] Her term encompasses an entire family of content-free lucid dreams, and emphasizes the profound stillness frequently described, where all sensory objects disappear—including the sense of inhabiting a body—but where sensations of subtle energetic currents and feelings of tranquility can remain. The subtle currents indicate that at least some very residual form of minimal/embodied selfhood lingers. Bogzaran's work has been at the intersection of imageless dreams and a style of modern painting known as Lucid Art, which she established with fellow artist Gordon Onslow Ford. In their artistic depictions of these states of consciousness, often the only figural elements are simple geometric shapes.

The difference between "ordinary" lucid dreams with content and imageless lucid dreams may be the collapse of any duality or perceived otherness between the dream self and the dream world.[30] Similar experiences are reported by advanced meditators during waking, involving a consciousness that contains little or no cognitive and perceptual content, but is marked by pervasive feelings of peace and well-being.[31]

It is clear that the presumption that consciousness is simply absent during dreamless sleep and contentless dreaming merits serious reevaluation.[32] In Chapter 7, we described experiments that have opened up lines of communication between lucid dreamers and experimenters, in real time.[33] It is unclear whether these communications across observers in distinct states of consciousness can be extended to the experiencing of lucid *dreamless* sleep, but the possibility of it would offer modern science the clearest insight yet into these very subtle layers of the mind.

Sleep researchers will not be the only ones to help guide these advances in our knowledge. Medical science has been making major contributions to such efforts as well, as evidenced by the research of people like Adrian Owen.[34] Neurologists and other professionals working with patients who suffer from brain injuries that impair consciousness have recognized a need for updating the classical and simplistic notions of their predecessors. It used to be commonly believed that consciousness varies along a single dimension of arousal, from unconsciousness on the one end (coma) to bright wakefulness on the other. According to this scheme, the various stages of sleep, anesthetic sedation, to alert wakefulness, can be localized along this one arousal dimension.

As we learn more from patients with conditions such as locked-in syndrome (where there is full awareness with no capability to move or communicate except through eye movements), vegetative (intact sleep-waking cycles but no signs of awareness), and minimally conscious (inconsistent but clear behavioral evidence of basic forms of awareness) states, there is a need to expand upon the single dimensional model.[35] The range of different states of consciousness is incredibly vast and probably varies simultaneously along many dimensions. One such dimension captures content, or the sheer number of things or objects that we are aware of. When all objects for experience disappear, it seems to become much more likely that a sense of self may go along with it too. Self- and world-models seem to be inextricably intertwined so that when there is no content to experience, the experiencer slowly becomes unsustainable as well. But, as odd as it might sound, it seems that consciousness can survive the disappearance of both the stuff of experience and the experiencer.[36]

Another dimension of consciousness captures a feature that scientists have called *global availability,* the extent to which the things that we become conscious of can be used to actively guide our behaviors.[37] If you are fully awake, and not suffering from serious brain damage, you perceive a world full of stuff that is used to guide and course-correct your movements in space—you see a coffee cup on your desk, and this information is then used to program the reaching of your hand to grasp it. In this scenario, the virtual-reality simulator we have discussed previously is doing the job that it evolved to perform, by serving as a useful interface between the body and the outside world. But in some instances you may simply have a bare sensation of existing or being alive, without using that feeling to *do* anything in particular beyond itself—experiencing the most minimal form of embodied selfhood. This is that fine distinction between doing and being that seems to vary depending on where you are, at any given point, along the consciousness spectrum. The feeling of *being* may not even be capable of verbal expression, outside of a vivid form of knowing. Psychologist William James called it "plain, unqualified actuality, or existence, a simple *that*."[38]

Right now, the science of consciousness really does not know just how many of these experiential organizing dimensions exist, but new efforts are being expended in this direction. It is conceivable that another dimension has to do with how accessible our experience is to verbal access, and how lucid or dim the experience is.[39] A lucid, imageless dream, then, would be quite distinct from the dull state occupied by a patient who is minimally conscious as a result of brain injury. You could imagine that the many varieties of experience occupy an extremely large state-space, with any particular experience representing a kind of local pocket within a vast expanse of possibilities.

From this new vantage point, the status of MPEs, which previously might have been dismissed as a *bona fide* phenomenon, becomes increasingly plausible. That modern science currently lacks an adequate map to guide these explorations can be seen as either a glaring loss or an exquisite opportunity to learn from other traditions in which the careful study of experience has flourished for centuries.

The Absolute Zero Point

Descriptions of MPEs can be readily found in reports from awake people during very deep stages of meditation or during extreme sensory privation. Floating in the darkness of John Lilly's isolation tank, the Zen priest and anthropologist Joan Halifax described losing her sense of body boundaries and time, and entering "a continuous void that was not boring, yet empty, not engaging, yet full" with "total peace." The transpersonal psychologist, Stan Grof, referred to an "absolute void" that he likened to "consciousness of interstellar space" with a feeling of "timelessness."[40]

These phenomenological reports have been replicated in contemporary research studies on sensory deprivation. Some research subjects report losing all sense of time and space. One of them describes a sensation that "I . . . just flew away into nothing," so that they were unable to "feel whether I was in the tank or where I was." Another one describes an experience of "losing time and space."[41]

In his autobiography, John Lilly describes an absolute zero point (which elsewhere he calls "the quiet center"), a unique state of consciousness discovered in his long periods of floating in utter darkness and silence.[42] He describes this point as a "completely black, completely silent, empty space without a body," and a position from which he could then transition into other subjective experiences.

Additional examples along similar lines can be gathered from the study of psychedelia. As mentioned in Chapter 11, apparently unrivaled in its capacity to evoke states of ego dissolution is 5-MeO-DMT, the psychoactive secretion from the Sonoran desert toad. The same substance can elicit both an overwhelming barrage of sensations (for instance, feelings of simultaneously experiencing every possible thought that could ever occur), and a vacuum of all objects and selfhood. It frequently happens that the second stage is reached after a preliminary passage through the overload phase. The sudden flooding of consciousness with a massive influx of content appears to be a potential trigger for an MPE event.

Perhaps the most authoritative source on the phenomenological effects of hallucinogenic tryptamines like 5-MeO-DMT is the collection of systematic observations made by chemist Alexander Shulgin, a doyen of psychedelic pharmacology, and his wife, Ann Shulgin. They extensively recorded the precise dosages, routes of administration, and psychopharmacological properties of hundreds of phenethylamine and tryptamine derivatives. Shulgin recounts how a person who took a large dose of 5-MeO-DMT quickly passed into a "coma-like state" that the man could subsequently describe only as blissful.[43] Research surveys suggest that the vast majority of people who take 5-MeO-DMT have intense experiences that involve transcending all sense of time, space, and self.[44]

Such experiences bear many similarities to what John Lilly called the absolute zero point. This apparent convergence hints at the possibility that there might be several different routes to inducing minimal states of consciousness, ranging from specific meditative techniques, to lucid dreaming, ingesting mind-altering substances and prolonged sensory deprivation. In apparently quite rare cases, there have been reports of spontaneously occurring episodes of minimal consciousness.

The nineteenth-century English poet J. A. Symonds appears to have been subject to such self-described "trances," often anticipated by sharp changes in mood. Symonds's episodes "consisted in a gradual but swiftly progressive obliteration of space, time, sensation, and the multifarious factors of experience . . . [I]n proportion as these conditions of ordinary consciousness were subtracted, the sense of an underlying or essential consciousness acquired intensity." Symonds then describes how the world became "without form and void of content," leaving nothing behind but a sense of "keenly sentient being." Symonds would just as spontaneously regain ordinary consciousness in a process that he compared to "awakening from an anesthetic influence."[45] Robert Forman has gathered numerous examples of spontaneously occurring MPEs.[46]

It is exceedingly difficult to say whether these sundry examples point to an equivalent state of consciousness, but it does appear plausible to suggest that there is at least a family of minimal conscious states with numerous shared features. One way to address these tantalizing questions is through the use of computational analyses of written descriptions of these different states of consciousness that quantify the similarity in subjective experiences induced by psychedelics and lucid dreaming and between psychedelics like DMT and near-death experiences.[47]

Paradigm Shifts

It is only recently that some cognitive neuroscientists have begun to address the existence and nature of MPEs; most research programs have instead concerned themselves with waking states of consciousness that are rich in content.[48] Consciousness has been assumed, *prima facie*, to be entirely absent in states without sensory representations, or when stripped bare of all images and thoughts. This has not been the case, however, in the world's great religious traditions based on contemplation. For example, Indo-Tibetan meditation schools have worked with these more unusual states of consciousness for millennia, amassing generations of knowledge about them.⁴⁹ Consider Advaita Vedanta, a prominent strand of Indian philosophical thought the main exponent of which was the sixth-century sage Gaudapada. According to his writings, there is an empty, luminous awareness that can be realized in deep, dreamless sleep.⁵⁰ Although it goes largely unnoticed, adherents of Advaita Vedanta believed that this contentless consciousness was actually the ground upon which waking and dreaming experiences rested. It is intriguing that scientists may now be able to test some of these daring suggestions.⁵¹

Patanjali's Yoga Sutras

The aforementioned *Yoga Sutras* of Patanjali, written around 200 CE, are centrally concerned with MPEs. What follows will necessarily be an abbreviated summary of Patanjali's thought.⁵²

Patanjali probably would have agreed with John Lilly's contention that the brain is a sophisticated biocomputer loaded with numerous programs, some of which are part of our biological heritage and others of which are created during the course of life. As the various programs play out, they subject the mind to many different agitations or fluctuations of consciousness. In Patanjali's yoga, all of these fluctuations are vortices or wave patterns that play out on the mind's surface. Patanjali referred to them as *vrtti,* and categorized them into five basic kinds.

The first kind of *vrtti* are the *objects of perception,* the stuff out of which most of our experiences are composed, such as the almost endless variety of things that we can perceive through the senses. Although Patanjali held that these were no more than a type of ripples that form within an all-pervasive field of awareness, we normally take them to be the sole ingredients of an objective reality. The second kind of fluctuation consists of many forms of *perceptual error,* including hallucinations. The third closely related form of mind fluctuations are *imaginative* forms, as when we close our eyes and visualize a plump lemon. Sleep, in Patanjali's system, is considered to be a distinct form of *modification of consciousness,* and the last of the *vrtti* is *memory.* Taken together, these five forms of agitation make up virtually the entirety of our conventional mental life, determining the content of our conscious lives. Adding further momentum to the whirling *vrtti* are the instincts attributed to our evolutionary heritage (*vasanas*).

On top of these five fluctuations, one of the most basic programs we come equipped with is the one that gives us a unique perspective on the whole virtual-reality show and endows us with the very convincing sense that we are separate individuals. This is the ingenious trick by which we become ensnared through grasping and wanting to hold on to things that we like, while avoiding loss and pain. According to Patanjali, the more that these successive wave patterns play themselves out in awareness, the more enmeshed we become, and the cost is inevitable suffering.

On the basis of this diagnosis, Patanjali goes on to advance one of the most thoroughgoing and radical manuals for dismantling the virtual-reality simulator that runs ordinary consciousness. His systematic yogic system prescribes a methodical set of psycho-physiological exercises, involving instructions on the control of posture (*asanas*), breath (*pranayama*), and attention (*ekagrata*). The entire goal of Patanjali's Yoga is to completely still the mind's ceaseless *vrtti*. To accomplish this end, one adopts body positions in which movement is frozen, breath is suppressed to the point of hibernation, and attention is withdrawn from the objects of sensory experience until it settles to a one-pointed focus. At the *pratyahara* stage of Patanjali's yoga, the yogi's senses are systematically withdrawn from the external world, akin to a turtle's retracting its head and limbs into its protective shell. In contrast to being immersed within a John Lilly–style sensory deprivation tank, the yogi achieves this state primarily by using very fine control of concentrative powers.

According to Patanjali, at the moment when all the whirling of the *vrtti* has been extinguished, pure consciousness reveals itself through what he called the "seedless" *samadhi*. This is consciousness without any object, shorn of all images or symbols. All sense of being a separate self having experiences is arrested as awareness shines forth in its naturally pristine form, totally emptied of its usual partitioning into the content of experience and an experiencer. Patanjali characterizes this empty consciousness as being intrinsically luminous but having no object or subject structure. For the yogi, this experience of the seedless *samadhi* offers a penetrating experiential realization into how this empty luminosity corresponds to an ultimate reality, which is ordinarily misconstrued by the whirling *vrtti* to conjure up an illusory reality that creates a multitude of desires and fears. For the Patanjali and those who followed his teachings, the entire point of yoga was to achieve a lasting transformation of one's consciousness by fostering a union with pure consciousness. Indeed, the word *yoga* refers to becoming yoked to that empty, disinterested witness consciousness.[53] Patanjali believed that only by achieving this permanent union could we become liberated from enslavement to addictions and cravings.

Entering Samadhi

In Hindu and many Buddhist traditions, *samadhi* is the contemplative instrument that is used to transport us into the deep space regions of the mind. In one definition, *samadhi* is a state of exquisite, single-pointed concentration. While there are many different classes of *samadhi,* with differing interpretations in Hindu and Buddhist meditative approaches, in this section we focus on the Theravada Buddhist tradition, which continues to be practiced today, especially in Thailand, Sri Lanka, and Cambodia.

Samadhi in that lineage is usually analyzed in terms of four stages (*jhanas*) of meditative absorption, each one deeper.[54] The achievement of each *jhana* serves as a stepping stone to the next one. Texts (sutras) belonging to the Pali canon, the collected scriptures of Theravada Buddhism, as well as the commentaries on these, contain many beautiful, poetic descriptions of the stages of absorption.

Each *jhana* possesses unique properties called *jhana* factors, which are peeled away as one progresses through the stages.[55] The traditional instructions for achieving the *jhanas* are, first, to seek a solitary place where it is possible to remain in sensory isolation for long periods. The meditator then withdraws attention from external sensory objects and focuses single-pointedly on the breath. Having single-pointedly focused attention on breathing, and thus having accomplished seclusion from both the outside world and inner distractions, the meditator will have arrived at the first *jhana*, characterized by rapture and pleasure suffusing the entire body. Achievement of the second *jhana* involves the cessation of all manner of discursive thinking. By the third *jhana*, bodily rapture has disappeared, leaving only the pleasure of a completely alert and clear mind. The fourth *jhana* involves the disappearance of pleasure too, leaving a state of utter equanimity, a neither pleasant nor painful feeling. At this point, what remains alongside the perfect equanimity is a poised, one-pointed awareness.

The "formless" *jhanas* lie beyond the fourth meditative absorption. The classical texts describing these *jhanas* contain allusions to some of the subtlest forms of phenomenology. The names of the stages can indirectly point to the largely ineffable nature of these experiences: they are known as the bases of boundless space, boundless consciousness, nothingness, and neither-perception-nor-nonperception. A common consensus is that the sense of being embodied disappears in the formless *jhanas,* suggesting a total disappearance or at least a radical suppression of the minimal/embodied self-representation. Having attained the final stage of absorption, the meditator is said to abide in a state of aware emptiness. Buddhist textbooks of meditation include mention of having attained the "station of nothing whatever" and even passing beyond "the field of nothing."

Tibetan Sleep Yoga Revisited

Tibetan Buddhism contains highly detailed maps of experiential inner spaces that are largely unknown to the contemporary science of consciousness.[56] Not only are such states detailed with great precision, but there are also systematic instructions on inducing such radical transformations of consciousness. The practice of Tibetan sleep yoga is particularly germane. While Tibetan dream yoga deals extensively with lucid dreams (as we learned in Chapter 7), sleep yoga, or luminosity yoga as it is also called, is a more advanced practice related to the deeper, slow-wave (NREM) stages of sleeping that are empty of content and that modern science would treat as lacking consciousness.

Tibetan Buddhism draws an explicit distinction between the "sleep of ignorance" and what it calls "clear light sleep."[57] This distinction honors the broader one that Vajrayana makes between coarse, subtle, and very subtle types of mental activity. The "sleep of ignorance" is what happens for the majority of us, especially without prior meditative training. We tend to notice only the coarse manifestations of the mind, where attention can engage with vivid imagery or thoughts. Unless we are in the midst of a dream, we have no apparent conscious recollection of dreamless sleep. Even in cases where there is a residual awareness, as is the case with white dreams or imageless lucid dreams, these experiences are described as those of a dark void.

By comparison, during "clear light" sleep, accomplished by the Tibetan masters only after prolonged engagement with the practice of sleep (or luminosity) yoga, consciousness remains bright, lucid, and vivid, even without any dream content. How can we describe these apparently paradoxical states of consciousness? As Tenzin Wangyal Rinpoche describes it, "there is no film and no projection. Sleep yoga is imageless. The practice is the direct recognition of awareness by awareness, light illuminating light itself. It is luminosity without images of any kind."[58] The clarity aspect is believed to reflect the spacious emptiness of awareness, while the light emphasizes that this spaciousness is keenly sentient, rather than dark and void. These twin dimensions of clarity and spacious emptiness are believed to be as inseparable as wetness and water. As Tenzin goes on to say, it is next to impossible to put this experience into words, since language presumes the structure of a separate subject and object, and therefore "the only way to know the clear light is to know it directly."

The Stillness of Christian Prayer

The philosopher W. T. Stace devoted a good portion of his scholarly efforts to studying *introvertive* mysticism, which he defined as a consciousness of undifferentiated unity, marked by a total absence of perceptual or symbolic content.[59] He found abundant descriptions of this state not only in the Far East but scattered throughout the Christian mystical tradition. Medieval mystics Jan van Ruysbroeck, Meister Eckhart, Saint Teresa of Avila, Saint John of the Cross, and the unknown author of *The Cloud of Unknowing* all agreed that the highest stages of contemplation entailed the extinction of mental chatter, including perceptions, memories, and imagination. Saint John of the Cross writes about becoming "emptied of all . . . imagined forms, figures, and images," so as to "rest without engaging in any particular meditation and without positing acts and exercising the faculties of memory, understanding, and will."[60] This extreme quiescence, in which all cognitive and motivational tendencies came to a stop, could culminate in an ineffable state of communion with what was understood to be ultimate reality. Poetic descriptions of that highest form of Christian contemplation often mentioned a darkness, emptiness, nothingness, silence, dazzling obscurity—yet a darkness in which there is a personal encounter with an eternal light.

Introvertive mysticism is prevalent in the hesychastic tradition of the Eastern Orthodox Church.[61] A central tenet of these teachings is that attention is ordinarily completely magnetized by the senses. Hesychast texts characterized this outward dispersal of attention as movement along a linear path, as when an arrow is shot at a target. The Christian hesychasts taught that linear attention engenders a state of mind similar to a waking hypnosis, wherein external appearances trigger ways of reacting to the world that are based on associative learning and habits.

The hesychast strives to reverse this linear "leakage" of attention by exerting a great effort keeping attentional resources unencumbered by involvement with images drawn from perception, imagination or thoughts. Attention is withdrawn from external objects to rest within the interior, forcing attention to take a "circular" rather than its usual, linear route.[62] Hesychasts used the Jesus prayer, consisting of a short phrase, repeated over and over, to anchor attention, and to gradually begin to impose greater degrees of integration over one's habitual ways of attending and reacting to external events. As the contemplative begins to engage in this form of inner prayer, attention starts to shift focus from its starting position (generally, around the head) down into the heart and gut areas. Visualizing one's mind slowly draining from the head to the heart was sometimes recommended as an aid to prayer.[63]

In the Christian understanding, the heart is more than just an organ—it is the epicenter of a person, the locus from which one's will, desires, feelings, and thoughts originate and radiate outward.[64] Attention is drawn from its usual wandering and fragmentation and gently directed, via the breath, to focus on more and more inner dimensions of this personal epicenter.[65] From the perspective of the material reviewed in Chapter 11, we could characterize the deeper stages of prayer as increasingly embodied once the more conceptually dominated narrative models of selfhood are rendered silent in favor of a more minimal flavor of selfhood.

The Jesus prayer was traditionally pursued by adopting specific bodily postures, usually seated on a low stool, and slowing the breath, in environments with minimal sensory distractions. One-pointed concentration, silence, isolation, darkness, and physical immobility have been traditionally important in this style of prayer. Deeping of prayer was regularly accompanied by stirrings of mental images and bodily impulses. The elders of this tradition offered extensive instructions on ways of minimizing involvement with thoughts, feelings, and impulses so as to achieve a one-pointedness of attention during prayer. "From whatever side a thought may appear," reads an instruction in prayer, "the mind immediately chases it away, before it has had time to enter, and become a thought or an image."[66] Spontaneous flights of mind wandering are eventually completely overcome to reveal a deeper tranquility. Over time, attention became steadily withdrawn from its usual dispersions to rest in a more pure, detached state. In contrast to its usual absorption into the contents of consciousness-sensations, thoughts, memories, fantasies, and feelings-the Christian hesychast cultivated a reservoir of "unbound" attention.

The deepest stages of prayer entail *hesychia*, a Greek term denoting complete inner silence or stillness. In the absence of images or concepts, as all stirrings become quiescent, what remains is a detached witnessing and reality is finally encountered without being distorted by personal fears or desires. The inner quiet of hesychia is not a passive trancelike state that the contemplative slips into: it requires a dynamic tension between receptivity and an active alertness to avoid becoming mesmerized by mental phenomena.

Although hesychia is an extremely advanced stage of Christian prayer, it is not in itself the end goal for a contemplative, but rather serves as a prerequisite toward a unitive experience of God. These far reaches of the contemplative journey are shrouded in mystery. As commonly taught, they occur within a paradoxical "luminous darkness," beyond images, words, and thoughts, and apparently beyond the sense of being confined within a body.[67] The classic texts frequently allude to an experience of a light that is both material and nonmaterial, and that cannot be circumscribed as either inner or external. This expanseless luminosity is often described as "noetic," or conveying a definite sense of knowing-through-identity that cannot be summarized in words or formulas.[68] Alluding to this state, the monk Evagrius declared that "Blessed is he who has reached, during prayer, unconsciousness which is not to be surpassed," and

Saint Isaac the Syrian wrote of the "secret place of the thick darkness of Your glory."[69] This darkness, however, is not a state of nescience, not an experience in which one's personhood is abolished, but rather a heightened state of personal communion with ultimate reality that defies all conceptual and image-based representations or linguistic descriptions.

Toward a Neurobiology of MPEs

The evidence collected in this chapter suggests that, at the very least, there is a broad family of phenomenological experiences that involve the presence of consciousness with very low arousal and minimal or no content. Reference to MPEs within the scientific literature, such as there is, concerns itself solely with the phenomenological dimensions of that experience and stays silent on the metaphysical significance and interpretations that are important to the world's great religious traditions. We remain very far from having a detailed neuroscientific account of MPEs. The psychiatrist and philosopher Georg Northoff has advanced an empirically based spatiotemporal model of consciousness that provides a useful way to approach the topic.[70]

According to him, the brain mechanisms associated with the presence or absence of consciousness (consciousness-as-such) are at least partially independent from those correlated with the specific *contents* of consciousness. He reviews an impressive amount of evidence that there is a characteristic pattern of spatiotemporal integration evident in brain activity—a pattern which. when it is present, appears to be associated with instantiating consciousness-as-such. In other words, a characteristic mode of spatiotemporal organization within the brain provides a neural prerequisite for the rise of any content-based manifestation of consciousness whatsoever.

Spatiotemporal webs of spontaneous brain activity that serve as neural prerequisites of consciousness can be, and have been, readily quantified using a wide variety of mathematical measures. Awareness of specific contents of consciousness occurs when neural traces of sensory signals successfully mesh and interact with the spatiotemporal webs of spontaneous brain activity. Consciousness-as-such can be understood as the pure (empty) potential for the arising of concrete experiences.

These two processes—the prerequisites and contents of consciousness—are seamlessly intertwined in the normal course of things. Nevertheless, Northoff hypothesizes that in some situations, such as very deep meditative states or sensory deprivation, content-based integration dips to a minimum, excluding even body-based inputs. As a result, it becomes possible to have experience of the neural prerequisite for consciousness-as-such. In addition to contexts like meditative absorption and profound sensory isolation, MPEs can apparently be provoked by a heavy barrage of stimuli that overwhelm the brain's content-based integration mechanisms, pushing them beyond their limits.

Contemporary investigators are now beginning to formulate hypotheses about specific neuronal mechanisms that could be associated with minimal consciousness events. Thomas Metzinger has made one such adventurous prediction.[71] At the core of his proposal lies the highly interconnected network of brain cells known as the Ascending Reticulothalamic Activating System (ARAS). The ARAS is critically involved in monitoring critical life functions and sends antenna-like projections that ignite the rest of the brain.[72]

The ARAS functions like a dimmer switch and provides the brain with a way to regulate its own arousal. Being a sophisticated prediction engine, the brain needs to explain away activity originating from the ARAS, one of its strongest sources of incoming activation. Metzinger thinks that the experience of an empty PCE may be a best guess the brain is able to come up with for explaining away the ongoing neural signaling coming from the ARAS. The empty but sentient space then could be the phenomenal experience of this model—a virtual simulation of the contentless ARAS signal.

Metzinger's ideas bear a resemblance to the hypothesis advanced by Ravinder Jerath. Jerath, a medical doctor, sees a role for the entire body and not just the ARAS in creating a global backdrop for awareness. According to him, ongoing electrochemical activity within the body as a whole, including the brain, is responsible for creating an internal 3D "dark space."[73] This space is a naked coordinate system that provides a canvas to be filled in by specific contents. It is the metaphorical Antarctica of the mind.

Once it has been clothed with specific sensory qualities, the background "dark space" becomes a private 3D virtual interface for the experience of being a self-in-world. Those experiences of being a self-in-world are the stuff of our life, whether we are awake, lost in reexperiencing old memories, dreaming, thinking, or whatever other experience we can imagine. Not being filled with content, the internal 3D "default space" is extremely unlikely to be the focus of awareness. Most of the time it is far too subtle for us to notice it, echoing Lewin's observation that we fail to notice the dream screen.

Neuroscientist and philosopher Antti Revonsuo likewise speaks of an *a priori* existing "empty phenomenal space as such," which he claims "resides at a level that is . . . nonphenomenal (or, more accurately, *sub*phenomenal). The coordinate system merely allows phenomenal features to be realized in a spatially unified and organized form. We never experience the contentless coordinate system as such." Revonsuo speculates that if it entered consciousness, that empty phenomenal space would result in "the subjective sense of pure spatial extension: an experience of pure, empty space." But, he goes on, "it is doubtful that such experiences of pure empty space ever happen."[74] While this is probably true most of the time, it may not be true in principle. Positive experiences of boundless empty space are not infrequently reported in various types of MPEs.

Jerath speculates that deep mental stillness, achieved through very advanced stages of meditation and prayer might be a sufficient trigger to induce the MPE-type reports that are prevalent in the world's spiritual traditions.[75] The "dark space" can suddenly come into the foreground when attention is withdrawn entirely from contents of consciousness, somewhat like how turning off a loud fan that has been active for a long time can suddenly bring into one's focus the pervasive silence that was ever-present in the background. General anesthesia has the potential to precipitate similar experiences through a chemically induced quieting of the brain's sensory filling-in.

Settling into deep meditative stillness requires not only fine attentional control to resist sensory distractions but also the quieting of internal mental chatter. As a neuronal correlate of this quieting, a recent study performed brain imaging of an expert Tibetan Buddhist meditator with over fifty thousand hours of meditation experience who was able to enter into an MPE-type experience during brain scanning. The researchers discovered that a brain signature of this state was associated with diminished cross-talk among nodes of the brain's internally oriented default-mode network (DMN).[76] This reconfiguration of a major brain network might be tracking the dissolution of more narrative representations of selfhood. Another study has provided the first evidence of widespread electrophysiological changes occurring during meditation-induced cessations of waking consciousness.[77] Perhaps such studies will do much to convince the larger scientific field that MPEs are replicable states of consciousness, much as the early psychophysiological studies accomplished for lucid dreaming.

Because Jerath's model takes into account the role of the body as a whole, and not just the brain, in shaping the "dark space" of consciousness, it can help to explain why voluntary breathing practices have been so important to inaugurating experiences of consciousness without content. As we have seen, the breath is commonly restrained, and even slowed to an absolute minimum, in several different religious traditions, ranging from Patanjali style slow yogic breathing to the Christian hesychasts. Jerath suggests that voluntary slowing of the breath alters the bioelectrical excitability of cells in the body and can enhance the synchronization between breath and heart rhythms.[78] The increased coherence between breathing and heartbeats modulates autonomic and central nervous system activity to promote profound states of calmness. The physical immobility which one often finds in these contemplative practices further reduces the amount of additional sensory information that needs to be explained away, and further amplifies a model of the "dark space." Cessation of movement alleviates the burden of error prediction that the brain is required to perform.

The discussion of MPEs has taken us to the most subtle dimensions of the consciousness spectrum from which fully elaborated self-in-world models sprout. Here, we approach an incredibly thorny set of questions that blur the line between phenomenology and ontology. It is now time to venture into areas where we probe into the furthest limits of what ultimately may lie underneath all of the various self-in-world models and whether we can approximate a more fundamental encounter with reality than what is suggested by the metaphors of virtual-reality simulation.

13

Nonduality and Reality
Beyond the Dream

In the last two chapters, our sojourn took us to the veritable hinterlands of the mind, those indistinct and ethereal far reaches of the consciousness spectrum where self and world boundaries shade off into nebulous horizons, or disappear altogether. Along the way, the story has drifted farther and farther away from the seemingly solid ground that makes up the ordinariness of routine experience.

According to the view expounded throughout this book, the nonordinary dimensions of consciousness, although they appear to be quite distant from "normal" waking experiences, are not just curious sideshows. Rather, they offer an opportunity to study the foundational dimensions of consciousness.

Now we arrive at the most subtle aspect of experience: consciousness itself, the ubiquitous manifold on which all other states of consciousness rest. It is a paradox that although consciousness permeates all other states of mind we are frequently unaware of its presence, as a fish is oblivious to the water in which it swims. Traditionally, neuroscience has tried to understand consciousness by focusing on the specific contents of consciousness, such as the experience of seeing. As to "conscious experiences in their entirety, irrespective of their specific contents," this is a domain mostly unexplored by researchers.[1] For this reason, rather than beginning from any particular scientific theory, we will begin this final chapter by drawing from various ancient contemplative schools. We will intentionally stretch the discussion of this topic to the utmost limits of epistemology (*how do we know what we know*), ontology (*what is actually real*), and metaphysics (*what are the first principles upon which everything else rests*). The depths of phenomenology will take us also to the edges of ontology.

This chapter is the most speculative one, scientifically speaking. But we will unabashedly allow caution to fly to the wind for the sake of laying out some radical and intentionally provocative suggestions. The edifice of contemporary science is built on the accumulation of "objective" knowledge, acquired through empirical induction and logical deduction. Hence, the prospect of relying on phenomenology to make any inferences about ontology is considered illegitimate, an intellectual *faux pas*.

The ability to be prodded, cut, and measured has frequently been a prerequisite for something to qualify as a real phenomenon (to have ontological status). Yet an explicit aim of this book has been to nudge scientific theories of consciousness into reevaluating the extent to which this epistemic bias, along with a tendency to treat waking consciousness as the normative reference, is more of a cultural and historical artifact than a basic truth. For many of the great spiritual traditions we talk about here, there is a predilection to treat a certain kind of contemplative phenomenology as inseparable from questions of ontology.

Ultimately, our intention is to encourage a willingness to challenge and question the long-held but often unexamined presuppositions that are endemic to modern scientific theories of consciousness. In our opinion, such an open-ended and curious attitude is critical for advancing this entire field of studies and coming to a more balanced assessment of the latent potentials of the human brain/mind.[2]

First- and Second-Order Realities: The Virtually Real and the Really Real

The various strands of religious thought woven throughout this book share a common premise: our ordinary way of experiencing obscures deeper insights into the true nature of our mind and the world around us. For example, in the Vedanta tradition, ordinary experience is understood as clouded over by *avidya*—an ignorance that operates outside of the confines of awareness—to both conceal and radically distort awareness of the self and world.[3] This misinterpretation of reality produces what the Buddhist scholar Reginald Ray calls a second-order reality.[4] Second-order reality is a step removed from the objective world. It is a heavily edited, filtered, and censored version of the underlying base or first-order reality—self and world as they *really* are, in their relatively undistorted state. Second-order reality is warped by deep-seated psychological biases that prevent direct contact with what is truly present. As an analogy, consider a sleeper who hears a trickle of water while dozing: this may get incorporated into a hallucinatory dream narrative as the sound of paddling in a canoe. The objective sound of rainwater is thus distorted into a hallucinatory, subjective impression. We can compare this example to how one's desires and fears, thoughts, and beliefs shape the subjective impressions of objective happenings during waking hours, too, leading to a highly idiosyncratic experience of reality.

Second-order reality is the dreamlike production and construction we have considered in earlier chapters. According to Buddhist psychology, one of the defining characteristics of second-order reality is that it imposes a deep fissure into the experience of self and world. This happens when experience is filtered through a lens that splits a holistic reality into the apparent but artificial polarities of self and not-self. In previous chapters we have used the word *nondual*. The term will also be used throughout this chapter to refer to a form of experiencing that entails no splitting of underlying reality into the separate categories of me versus world. On top of our habit of organizing experience using a dualistic frame, the Buddhist tradition teaches that feelings impart yet another pair of distortions by virtue of which we filter reality through the twin prisms of desire and fear. Subjectively neutral aspects of reality tend to slip away and get edited out from the experience of second-order reality.

Like Buddhism, Christianity refers to a basic ignorance that causes one to misinterpret the true nature of reality.[5] The seemingly ordinary state of waking consciousness was understood by the early Christian tradition to be similar to a hypnotic sleep, maintained by a deep seated tendency to experience oneself as separate and then to succumb to the pull of sensory appearances and images drawn from memory and imagination, all molded according to a need for the gratification of self-centered desires.[6] One of the great fathers of the Christian Church compared this condition of humanity to repetitively treading a circular arc defined by an absence of any higher meaning beyond having a consciousness that is passively captive to seeking after pleasure and avoiding pain. "With our eyes blindfolded, we walk around the mill of life," wrote St. Gregory of Nyssa, "always treading the same circular path and returning to the same things . . . appetite, satiety, sleep, waking up, emptiness, fullness. From the former of each pair we constantly pass to the latter, and we never cease to go round in a circle."[7] For Buddhists, too, to be stuck in the suffering of samsara meant constantly going around in circles. The mind, in this condition, operates passively, beholden to inexorable laws of association that passively condition one's mind and body to slip into the same old grooves and ruts inscribed into the mindstream by one's biological inheritance and life experiences.

There are numerous points of contact between the world's great spiritual traditions and modern science on this fundamental premise: much of what we naively take to be consensus reality is a tightly constrained hallucination. Contemporary fields of research, like microphenomenology, agree that a combination of largely automatic cognitive and emotional biases cloud experience and need to be overcome to establish a more direct form of contact with the true nature of the mind and the external world.[8]

But is it possible to have experiential access to what lies underneath (or beyond) the dreamlike second-order reality? Can there be a more direct experience of self and world as they *really* are? At least since Immanuel Kant, science has believed that human epistemic structures (organizing how the mind comes to know itself and the world) are inescapably subjective. Kant believed that the world of *noumena,* or first-order reality, is fundamentally inaccessible.[9] Contemporary cognitive neuroscience of consciousness provides many details about how second-order reality is constructed by the brain, but it generally stops short of pondering the possibility of alternative forms of brain/mind organization that are less distorted.

The world's spiritual paths teach that to "awaken" from or transcend experiences of second-order reality requires a lifelong path of engagement with practices such as meditation and prayer. Such practices transform the deepest rooted epistemic structures and fundamentally alter how one encounters reality, effecting a lasting change in how one perceives, feels, thinks and acts. In contrast to Kantian notions of the inaccessible noumena, most spiritual paths seem to be arguing that it is possible to be in communion with that underlying ground of all that is: not by relating to it as an objective fact to be known, but through direct participation in that larger reality. In addition, universal to many spiritual traditions, is the necessity for ethical conduct that helps to purify the practitioner's psychological life. Without this inner purification, many traditions teach that the highest stages of contemplation are impossible.

Later in this chapter, we will sketch a neuroscientific account of how some forms of contemplative practice may restructure the human brain and mind to bring it into a more attuned alignment with the outside world. These speculations will undoubtedly be far too simplistic, but they are cast in a framework that will be intelligible to the growing scientific understanding of the brain as a predictive organ while making contact with ancient wisdom teachings. Before we get there, however, we need first to become familiar with some of these specific teachings about the nature of mind and experience that are putatively disclosed upon dissolution of the basic influences endemic to second-order reality. For didactic purposes, we will rely heavily on cartographies of the mind drawn from the Buddhist Vajrayana tradition.

Maps of the Mind from the Tibetan Buddhist Tradition

According to one systematic set of teachings, known as the Kagyu Mahamudra lineage, the coarsest analysis of the mind reveals five distinct qualities of consciousness corresponding to the main types of sensory qualia: visual, auditory, olfactory, gustatory, and somatosensory.[10] Next, there is mental consciousness. Mental consciousness refers to the realm of thoughts, judgments, memory, and imagination. A very detailed visualization of an object that is not immediately present to the eye, for instance, occurs in mental consciousness, as does the arising of dreams. Taken together, the first six types of consciousness exhaust nearly all scientific investigations of consciousness, and they make up the vast bulk of our everyday phenomenology.

Beyond this, the Kagyu tradition recognizes several increasingly subtle dimensions of the mind to which most of us do not ordinarily have explicit conscious access. These more subtle dimensions of the mind are believed to actually be more stable and enduring, while the first six kinds of sensory and mental phenomena fluctuate in and out of awareness. Although we rarely experience the very subtle dimensions of the mind, these ancient teachings based on profound meditation practices treat them as the deeper strata that lie below the mind's surface: they are believed to remain relatively unperturbed through periods of waking and sleep, and even through those blank periods of dreamless sleep or under general anesthesia.

One of these deep layers is called the *klesha-mind*. The term refers to some of the most rudimentary formations of experience around the budding sense of a distinct *I*. The klesha-mind does not refer to narrative stories or thoughts that subsequently clothe this elementary sense of being a separate *I*. The klesha-mind encompasses instead an inchoate, preconceptual, and preverbal type of minimal/embodied selfhood, closer to pure witnessing. Contrast this with the narrative/autobiographical layers, which are manifestations of mental consciousness, often rich in conceptual and language networks. The klesha-mind is the beginning of that ethereal property of "point of view–ness," which can be fairly neutral or invested with varying degrees of feeling charge. The klesha-mind implicitly introduces a pervasive flavor of *me* and *mine,* which is like a very thin membrane that accompanies consciousness across many transformations of state.

Manifestations of the klesha-mind sprout, like organic shoots, from an even more subtle and foundational layer of the mind known as the substrate, an all-pervasive matrix. The substrate refers to the pure potential for the emergence of any and all specific experiences. Memory traces from the past are also believed to be stored within the substrate. According to this school of thought, archetypal symbols and images, which are not specific to just a single individual, but shared universally across cultures, also originate from the substrate. This is a claim that partially echoes hypotheses put forward by Carl Jung concerning the existence of a transpersonal, collective unconscious, as well as certain ideas championed by Géza Róheim, as discussed in Chapter 10.

Usually, the substrate is submerged far below awareness. The substrate would correspond most closely to the neuroscientifically-derived concept of an MPE, an internal "dark space" or "zero point" that becomes the backdrop for filling-in by discrete perceptions, feelings and thoughts.[11] Employing this terminology, our consciousness routinely dissolves into the substrate when we drift off to dreamless sleep. The experience of millennia of Tibetan meditation masters suggests that, although a deep mental dullness normally covers the substrate, it is possible to become conscious of this layer of the mind through prolonged meditative training.[12] Upon waking, the substrate is rapidly filled in by a blooming world of sensations described by the six types of consciousness.[13]

Lifting the Veil: Passing from Second-Order to First-Order Reality

Most contemporary scientific theories claim that the consciousness spectrum bottoms out definitively somewhere in the nebulous distance of dream worlds. By contrast, according to the Kagyu-Mahamudra understanding of the mind, the substrate is yet another layer that can be breached to reveal an even more foundational dimension of consciousness. The mind's truest nature, these teachings state, is realized only in penetrating *past* the substrate. We will now sketch these distant shores, a task that will simultaneously take us to the furthest edges of ontology (reality, as it may be, at the first-order level).

The substrate consciousness continues to be permeated by an extremely subtle, dualistic frame of reference. Clinging to the dark, empty space of the substrate, like an oily residue on film, is the sense that this space itself is "mine," even when it is shorn of all contents. According to this interpretation, many of the bare MPEs still contain this ineffable feeling of separateness, a very deeply rooted self-centeredness. According to the traditional Kagyu teachings, however, this sliver of separation spreads to further fracture experience into more explicit polarities, including self versus other, internal versus external, us versus them, good versus bad, and so on.

From this perspective, we could try to reinterpret some of the nonordinary states of consciousness reviewed in previous chapters. For example, most, if not all, cases of drug-induced ego dissolution would probably qualify as only a "partial" dissolution. That is, while the narrative self clearly becomes destructured, in most cases a minimal self continues to organize experience, with residual dualistic separation.[14] Or consider the process of falling asleep. Most frequently we "swoon" into a turgid haze, unaware that thoughts and perceptions have dissolved into the oblivion of the substrate. When the visionary phenomena of dreams reappear, most reliably during REM sleep, we are instantaneously engrossed in an assortment of dream events and characters that seem to be "out there" relative to our dream self. This propensity for projicience, a term we introduced in Chapter 4, could be understood as revealing the substrate's potential for structuring conscious experience on the basis of a dualistic blueprint. It is as if the substrate has built into it a predetermined "mold," so that when specific images and feelings fill in the empty potential of the substrate, the contents of consciousness adopt characteristic contours with separation between an "in-here me" and "out-there world." Of course, such an organization of the conscious-state space makes a good deal of evolutionary sense as it provides a useful interface for becoming an effective agent in the world.

It is that germinal preference or bias to organize experiences along the dualism of self and not-self that Tibetan meditation masters say conceals a more direct experience of the ground of consciousness. When the mind is shorn of that last vestige of duality, then a first-order reality that is intrinsically nondual (non-two) becomes consciously accessible. In this first-order reality, there are no remaining dualities between "in-here me" and "out-there world."

In Tibetan Buddhist traditions like Mahamudra and Dzogchen, but also in pre-Buddhist Bön traditions, various names have been used to refer to first-order reality, including the Base, Primordial Consciousness, or the Basic Space of Phenomena. There is no distinction made between the Base, as the ultimate nature of the mind or consciousness, and the Base as the ultimate nature of reality.[15] At this deepest level, phenomenology and ontology are simply inseparable—or once again, nondual.

Underneath the multiple obscuring veils of duality, the Base is nothing other than reality, as it is, when stripped of all subjective distortions. Based on one interpretation, the substrate consciousness is not different, in principle, from the Base. It is simply a distortion that forms on the Base and produces suffering for the individual who mistakenly clings to the belief that they are separate from the external world—a "trick" that nature evolved to enable a biologically adaptive interface for moving organisms.

This ultimate dimension of reality, according to Vajrayana teachings, is beyond all dichotomies. It is neither inside nor outside, not self or nonself, not material or mental. The Base is an incomprehensible mystery for the cognitive and conceptual mind, but when consciously realized it becomes that which is most directly obvious and real. We could call it the wider reality behind the original confused dream. In classical texts the Base is sometimes described as the infinite source of the mind's appearances, including the empty space into which all appearances fade. To begin to explore this further, you are invited to engage with In-Sight 13.1, using personal inquiry to get experiential glimpses of the Base by knowing-through-identity.

In-Sight 13.1
Picture a movie playing on an LCD screen. The characters react to and get swept up in the cinematic drama. This corresponds to what we have called second-order reality: all the actors have their own separate points of view through which they live out this second-order reality. But for you as the observer of the movie, although you can appreciate the dramatic qualities from that perspective, it is also possible to contemplate the first-order reality. From a first-order reality perspective, anything that can possibly happen in the movie, along with every character in the movie, is a specific, transient arrangement of pixel arrangements on a screen of liquid crystals. The luminosity of the screen is the medium or matrix from which nothing ever wavers.

Extending this analogy to your own world, the endless appearances—including all of the perceptions, thoughts, and feelings that will ever occur to you in life, across both waking and dreaming—are, from the perspective of first-order reality, changing ornamentations of the Base. The Base is the clear luminosity of the screen itself, the foundation of all consciousness, consciousness itself—that very capacity to experience images, including all of the transformations that the images undergo as they arise, transform, and dissolve across the consciousness spectrum. The ever-changing contents of consciousness are the plethora of pixel arrangements that constitute an ever-moving mosaic.

According to Dzogchen texts, true enlightenment (*nirvana*) is an unwavering experiential realization that each of us, in our deepest core, is the very luminosity and clarity of the underlying Base. A conscious realization of the Base is referred to as *rigpa*—while to remain ignorant of the Base is a mental dullness known as *ma rigpa*.[16] An accomplished Dzogchen master stays in *rigpa* no matter what segment of the consciousness spectrum they happen to be in, or what they are doing. Each individual is understood to be a unique expression of the Base, as it comes to recognize itself or as it becomes mired in dualisms. For practitioners of this path, such a realization might begin as an intellectual comprehension or a very brief glimpse that lasts for only a very short time. Eventually, the practitioner moves from isolated moments of epiphany and begins to have a more and more enduring and unshakeable conviction of *rigpa* as an existential truth, which permeates all levels of experience and informs one's daily actions.

With the Base as the bedrock of all reality, everything is said to have "one taste," meaning that all appearances of an external universe, other people, thoughts, joys, and sorrows are in their entirety experienced as nothing other than so many creative ornamentations of the Base. Nothing that ever arises or can arise in experience—and not even the dissolution of all mental appearances into the substrate—ever wavers from the Base as the root matrix or "basic space of phenomena."[17] The dualistic mode of perception (or *avidya*) is what happens when we, as biological beings, quite automatically begin mindlessly reacting to everything that happens as if it all had an independent existence, with a thick boundary interposed as a wall between inside and outside, between self and world.

When it is not being actively distorted by any illusory misinterpretations of duality, the Base as ultimate reality is *both* personal and universal. To quote one renowned Dzogchen master, the Base is the "fundamental ground of existence, both at the universal level and at the level of the individual, the two being essentially the same; to realize one is to realize the other. If you realize yourself, you realize the nature of the universe."[18] In other words, the deeply ingrained belief that there actually exists a qualitative difference between subjective and objective is at root a perceptual error. The Base, from this understanding, is the shared, innermost core of all living beings.

For practitioners, none of this is abstract, theoretical terminology. *Base, primordial consciousness,* and related terms are rather so many words crafted to allow for using human language to indirectly point to an ineffable. experiential realization that spontaneously arises. This occurs when all processes of reification and splitting of experience into dualistic poles, all constrictions of experience, have been utterly relaxed, including those that are so subtle that they completely escape one's awareness.[19]

Facets of First-Order Reality

This is a good point to stop and reiterate the fundamental ideas from Tibetan Vajrayana. The first is that reality, as it is ordinarily experienced, is actually a second-order reality. Second-order reality is a deeply embedded misinterpretation of self and world, characterized by a splitting of experience into a dualistic perspective that creates a convincing, virtual-reality-like simulation.

Next, this tradition makes a powerful claim that it is possible to transcend this way of organizing experience and to have direct, experiential access to first-order reality. Practices like lucid dream yogas, seeing through all phenomenal appearances as dreamlike projections, Tögal visions, and "effortless" styles of meditation act to dissolve the numerous perceptual errors, habits, and psychological defenses that normally block directly experiencing first-order reality. These practices all aim to retrieve the rawest form of experience before their shaping by cognitive and emotional biases. When an organism finally gives up all processes (conscious and unconscious) involved in "re-presenting" self and world, through a subjective ego tunnel, then first-order reality simply "presents" itself as it *actually* is. Of course, the direct experience of reality is still psycho-physical and no less mediated through the nervous system than before: there is no suggestion that some "super-natural" form of perceiving is suddenly opened. Even the most advanced Tibetan meditation master does not suddenly begin to perceive gravitational waves or the strong and weak nuclear forces described by physics which are clearly a part of first-order reality but not accessible to a human being's perceptual capacities.

The Vajrayana tradition is clear, however, in saying that when a person becomes consciously aware of the Base, they experience first-order reality as having certain intrinsic qualities.[20] These are classically treated as multiple facets of a jewel, since first-order reality is far too vast and mysterious to ever be captured by any definitive descriptors. We have already mentioned the property of nonduality, yet another experiential aspect is that the Base is without boundaries of any sort. The Base is also said to be characterized by self-luminosity / radiance. This property refers to the fact that the Base is consciousness itself. Luminosity / sentience is its nature, as much as the wetness of the water or the mirror's capacity to reflect whatsoever is put before it without the mirror being changed by the reflections.

Zoran Josipovic, a neuroscientist and expert on nondual traditions, refers to this intrinsic luminosity as *nonconceptual reflexivity*.[21] He defines it as a direct knowing capacity without the mediation of any higher-order concepts or cognitive representations. As understood by Dzogchen, the self-luminosity quality of the Base is more foundational and prior to what psychologists term meta-consciousness—the capacity to become aware of being aware and to use cognitive labels and language to express that fact. Meta-consciousness entails a higher-order representation of a raw fact, while the self-luminosity of the Base has no need for any kind of representation, but is direct and immediate. The Base simply *is* awareness; awareness and receptivity is the very nature of the Base, without requiring itself to be "re-presented." It is prior to concepts and propositions. This makes it different from ordinary pre-reflective consciousness, which is still structured and shaped by implicit conceptual representations.

Another quality of the Base is its intrinsic *loving kindness,* or compassion. The Base is experienced as interconnected with everything. From a nondual perspective, a person who inflicts harm on another living being is essentially inducing self-harm, since the "other" is a particular manifestation of the same Base that forms the innermost core of the attacker as well. Dzogchen teaches that absolute compassion for all living beings is not something that needs to be explicitly cultivated per se (although that is useful at certain stages of one's practice), but something that emerges effortlessly once an authentic experiential realization of the Base has been stabilized. Tibetan Buddhism teaches that emptiness and compassion form two indispensable wings of authentic enlightenment, as enlightened masters become living, conscious conduits for radiating this pulsating, benevolent loving kindness throughout the world.[22]

Sketching a Scientific Account

We can consider the phenomenological observations gathered from a wide variety of nondual contemplative traditions, while putting to the side issues of ontology and metaphysics. In what follows we will offer some initial frameworks for understanding nondual awareness as a distinct form of organizing conscious experience.

Neurological Bases of Interior and Exterior Fracturing

Since the brain largely "talks to itself," there needs to be some neurophysiological basis for disentangling its autonomous dynamics from the external world. In other words, there are neurobiological roots to the dualistic fracturing that is experienced as a split between the inner and outer, the self and not-self.

When brain activity is tracked over time, it alternates between periods when the brain is oriented more toward maintaining its own endogenous activity (talking to itself), and other moments when it is more responsive to influences from the outside ("listening").[23] The most obvious example is the twenty-four-hour waxing and waning in the consciousness spectrum, as we wake up from completely interior dream worlds and get enmeshed in a consensual reality.

The dynamic tension between inside and outside is evident at many levels of the nervous system, including at the scale of whole-brain networks. About a dozen resting-state networks are defined by patterns of correlated functional activity even at rest, including the default-mode network (DMN) recruited when your mind wanders randomly or revisits autobiographical memories or reflects on your narrative models of selfhood. The DMN helps to shut out the outside world and shift the neurophenomenological "center of gravity" into the interior.

The DMN is ordinarily configured into a competitive relation with other resting-state networks, like the central executive network (CEN) and the salience network (SN).[24] These latter networks are oriented to the external world and organizing behaviors that allow a person to engage with concrete tasks. For this reason, they are also called extrinsic or task-positive networks. The DMN, the prototypical intrinsic or task-negative network, and the CEN and SN often relate in a see-saw relationship: when the DMN is active, the CEN and SN are relatively quiet, but when external demands call for it, the CEN and SN spring into action, inhibiting the DMN.

The competition between these brain networks maintains a dualistic separation between self and world, but it also helps to manage neural resources so that a person can flexibly shift between the interior and exterior as the situation demands. The experiential counterpart to this antagonistic brain network dynamic is a split between experiencing the inner self and the external world. If it is true that contemplation can bring a person to a more fundamental consciousness of nonduality, then we should expect to find alterations in these competitive brain network interactions.

Brain Reorganization Underlying Nonduality

Zoran Josipovic conducted the first study to investigate competition between intrinsic and extrinsic brain networks during nondual meditation.[25] Josipovic recruited experienced meditators in the Tibetan Buddhist tradition, with four thousand hours of meditation experience on the low end and thirty-seven thousand hours on the higher end.

Josipovic set out to examine differences in brain connectivity across three different experimental conditions: a passive state of mind-wandering, a "dualistic" focused attention meditation, and nondual awareness meditation proper. During focused attention meditation, a split between self and world is maintained as the practitioner concentrates on an external object. Nondual meditation, by contrast, is characterized by effortless nongrasping, a letting-go of all manner of doing or mentally constructing anything at all. Josipovic refers to the effortlessness of nondual meditation as being like "ceasing the effort of searching for one's keys upon finding them in one's pocket." One simply recognizes and eases into the background space of nonconceptual, self-luminous awareness while minimizing priors to the maximum degree possible.

Consistent with his original hypothesis, relative to passive mind-wandering, nondual meditation greatly relieved antagonism between the intrinsic and extrinsic brain networks, helping to balance the neurological see-saw. In contrast, the focused attention meditation actually increased the difference between the two brain networks.

Unfortunately, very few studies have focused specifically on nondual meditation, but a recent meta-analysis supports some of the effects we just highlighted.[26] Aggregating many fMRI studies of various meditation types, including individuals of varying proficiency, it reports that on the whole, meditation is associated with decreased DMN activation. One brain area within the DMN, the posterior cingulate cortex (PCC), seems to be most reliably inhibited by meditation, while extrinsic networks like the CEN increased in activation as a function of meditation. Among the most advanced meditators, the DMN and CEN had shifted toward *cooperative* rather than antagonistic patterns of connectivity. This was true during meditation and in resting-state scans outside of meditation.

The authors hypothesized that extensive meditation practice gradually loosens the tendency to structure experiences in a dualistic way. They suggested that the expertise gained through formal meditation sessions (which are "states") led to persistent neural changes, creating "a new normal" that settled outside of meditation periods (which is a "trait" level change). These scientific findings appear consistent with what different nondual traditions have described, where an experience of nonduality occurs first as a brief glimpse. Over time, such infrequent and transient realizations are thoroughly assimilated into one's everyday experience, until they become persistent at the trait level, with nothing specific that the practitioner needs to practice or "do."[27]

Second-Order and First-Order Reality: A Neuroscience-Informed View

We will now use a modern neuroscientific lens to evaluate claims that nonduality is in fact closer to a more ultimate reality, drawing from recently advanced research models.[28]

Much of today's cognitive neuroscience postulates that the brain constructs predictive models of self and world across hierarchically organized and nested layers of neuronal networks.[29] In terms of brain anatomy, a neuronal gradient extends from structures strongly tethered to the momentary state of the world beyond the brain (like the primary sensory areas) to the vast association areas that traffic in abstract predictions, unmoored from the present and combining multiple sensory modalities and possible action sequences.[30] For example predictive models of selfhood are organized along hierarchical levels ranging from minimal/embodied forms, rooted in sensations from the gastrointestinal, cardiovascular, and other physiological systems, up to narratively organized reaches of the autobiographical self.[31]

One way to visualize gradients of predictive hierarchies is using as a diagram known as a Pythagoras tree. The base of the tree represents predictions that relate to concrete here-and-now phenomena. The upper branches are high on a property called counterfactual depth. Counterfactually deep predictions are increasingly abstract and adrift from the concrete present, rising into the heights of dreamlike fantasy.

The models in the upper reaches of the Pythagorean tree are smeared out in time and space, covering fairly broad and increasingly imaginative hypotheses concerning self and world. The upper reaches also receive a massive number of convergent inputs from the brain's limbic areas which impart a strong emotional coloring to these imaginative as-if simulations. These upper layers of the Pythagorean tree canopy envelop fresh experiencing of the world through a sort of haze of idiosyncratic and subjective distortions.

The DMN sits near the top of the brain's predictive hierarchy, being most distanced from the present moment and the most prone to constantly exploring a large repertoire of predictions.[32] These predictions, heavily influenced by past histories of emotional reinforcement, eventually get imposed from the top down onto lower down areas. Left on its own, the brain spends much of its time and energy coming up with these counterfactually rich, high level predictions and sending them down to lower-level brain areas.[33] A first-person experience of such top-down predictions consists of the spontaneous flights of thought and mind wandering that are so familiar to many of us.[34] When we sleep, the churning away of counterfactual, abstract predictive models is experienced in the form of dreams, as when the brain simulates anticipated life situations and emotional challenges. This seemingly ever-present neuro-mental noise is what many meditative traditions call the untrained "monkey mind."

One of the most reliable effects of meditation is that it quells DMN activity, silencing the constant issuing of top-down predictions. In

Pythagoras tree used as an analogy for hierarchical model building of reality
inside the brain. The brain elaborates self and world models from predictions
that are temporally precise, constrained to the here-and-now, and then increasingly spatiotemporally smeared and abstract at the top reaches (canopy) of the
tree. Ascending the tree translates into greater counterfactual depth and reliance
on fantastical conjectures. Meditative depth, by contrast, becomes more
profound as one descends the tree from the upper layers and into embodied
presence confined to the freshness of the present moment. Top: Guillaume
Jacquenot / Wikimedia Commons / GNU Free Documentation License. Middle
and Bottom: Herbee / Wikimedia Commons.

particular, the PCC is inhibited with meditation, which appears to be a neural signature of loosening the solidity of overly rigid predictions that easily consume attention.[35] With the softening of the DMN's hold, and the gradual quieting of rich nodes within that brain network like the PCC, which is inhibited with meditation, it becomes easier to dis-identify from troublesome thoughts and feelings.[36]

In their many-to-(n)one model, Laukkonen and Slagter see meditation practice as a continuing process of pruning the upper branches of the hierarchical tree.[37] The meditative practice of returning to the present moment, over and over again, slowly retrains the brain to shift from abstract models of reality to being in direct contact with the present moment as it is.

When it is pursued diligently, meditation radically disrupts rehearsed and rigidified habits of encountering reality through old priors rather than being attuned to the actual state of the world, in objective terms. Perceiving and acting through habitual prior structures is responsible for what the Buddhist tradition would characterize as *avidya,* and the Christian contemplatives would label as fallen perception. Contemplative practices hold the potential to reverse passive capture by maladaptive models from the past and begin to effect a long process of realignment with a more unfiltered and undistorted experience of reality.

Laukkonen and Slagter envision three broad levels of the predictive hierarchy. The top level, richest in abstraction and counterfactual depth, corresponds to thoughts and conceptual narratives. The most entrenched priors on this level correspond to core beliefs: the beliefs that are overweighted in terms of shaping experience. The most entrenched core beliefs are predictions that are so habitual (for example, "my self is separate from the world and others") that a person is unaware of their presence. That individual might, however, be aware of the consequences of such beliefs, in terms of their impact on one's overall view of the world (for example, "it is a dangerous place"), themselves (for example, "I am a victim of circumstances"), and their feelings (for example, of chronic angst).

Below the uppermost level, which is primarily mental and organized into narratives, lies sensing, which is closer to the present moment. Below the sensing level is a still lower level that imposes predictions about the self and world as being distinct entities. What the Laukkonen and Slagter model suggests is that, with increasing meditation depth, one drops down the gradient and becomes more and more open to encounter reality as it truly is, in a fresh meeting that has lowered the confidence normally invested in high-level priors from the past. The gradual silencing of predictions opens one up to a more spontaneous and liberated meeting with the true nature of the world, an ever closer approach to the underlying first-order reality. That first-order reality is met no longer through conceptual thought or representations but in a much more immediate way, in what resembles an ever increasing alignment (or communion) among brain, body, and world. In Laukkonen and Slagter words: "Obviously, having a mind that is constructed through past experiences is advantageous in many situations, as it is organized to align the individual's needs with the possibilities in its environment. Yet, a mind that is too restricted—that too often occupies habitual modes of thinking and feeling—and is not flexibly adapted to changing situations, can be maladaptive."[38]

In summary, a couple of things are important to appreciate about the claims of the many-to-(n)one model. First, contemplative practice is hypothesized to loosen and de-automate primarily the higher (or deeper, counterfactual) levels of predictive hierarchies rather than lower-level ones that are more proximal to the concreteness of the present moment. Presumably, this is accomplished by alleviating the precision of high-level, abstract priors. Relieving the precision of high-level priors endows the hierarchical levels that are below them to gain more flexibility and range, as ascending bottom-up signals enjoy a higher level of freedom and have a larger share in shaping conscious experience. From an isolated brain that talks mostly to itself, in the form of fantastical conjectures, contemplative practice gives a strong nudge toward a state of brain-body-world interconnection. Certain lower-level priors are probably unaffected by meditation—for example, even meditation masters probably experience a range of visual illusions that are explained by priors operating on low-level visual circuits.

Second, the relaxation of hierarchically deep, counterfactually rich priors probably causes temporary violations of the principle that the nervous system is always minimizing free-energy (surprise). In other words, at least for some amount of time, it appears that practices like meditation can suspend that tendency. These breaches of the free-energy principle permit the brain to resettle into a different level of stable organization that is potentially more attuned to first-order reality. During the time when priors are relaxed, uncertainty is actually increased rather than decreased.

A recent report by an Argentine research group demonstrates that long-term meditation reliably increases the degree of unpredictability (entropy) evident in brain signals.[39] A similar principle is believed to operate when individuals, in a supportive and adequate clinical context, take psychedelic substances. As Carhart-Harris and Friston have noted, what follows in the wake of these periodic breaches of free energy, is that "normal brain function resumes and systems reset, potentially (but by no means absolutely) in a healthier way (ideally), more in tune with each other and the sensorium."[40]

Questions of Technique

Although Laukkonen and Slagter propose that various meditative practices eventually lead to a similar destination, it remains to be seen whether the kinds of nondual meditation, as taught in the Dzogchen tradition, for example, offer a more direct way to minimize the weight of entrenched priors when compared to more gradual meditation paths. In Dzogchen, direct realization of nonduality is not arrived at through minimizing phenomenal experience, as in other forms of Buddhism. These forms teach preliminary stages of mental quietude and stilling, as in the deep stages of *shamatha* meditation. The path of Dzogchen is entirely different, however, being aimed at activating a "primordial wisdom" through which the self-luminous quality of the Base is made explicit during ordinary wakeful experience.

The benefit of formulating theories like the many-to-(n)one theory is that they create new possibilities for formulating and testing specific scientific claims. As an example, if the above hypothesis is accurate—that meditation dissolves reliance on previously existing predictions—then this should be reflected in objective measures that are sensitive to prediction errors within neuronal hierarchies. Several research studies have explicitly tested this with subjects with varying degrees of meditation proficiency and across different meditation styles. There are now several findings consistent with the diminution of predictive coding. For example, one study measured the startle response—a primitive defensive reflex to unexpected, loud sounds—and

found that proficient meditators were less likely than control participants to demonstrate a repetition-induced suppression of startle.[41] The
proficient meditators, it would seem, did not generate as many expectations regarding upcoming sounds.

Another study measured brain responses evoked by sounds and
found that nondual meditation, relative to focused-attention meditation, diminished reliance on prior predictions, while increasing the
monitoring of auditory inputs.[42] This pattern was observed solely in
expert Tibetan Buddhist meditators, not in novice meditators, high-
lighting the value of taking into account variability that is driven by
meditation type as well as the amount of expertise. A concerted effort
to understand nondual phenomenology and neural functioning can
create novel opportunities for a future science of consciousness and
inform long-standing questions in contemplative communities, such
as how to better measure the enduring changes and transformations
facilitated by practices in daily life.

The discussion so far has been heavily oriented around meditation,
but in the future, it will be important for a more inclusive contemplative science to also consider other contemplative practices, like inner
prayer. Since Buddhism has dominated much of the conversation
within this nascent field, it is hoped that the gradual invitation to
deeply engage with other traditions, like Christian and Sufi, will further enrich future theorizing.[43] The Christian prayer of the heart, for
instance, as discussed at several points throughout this book, is likely
to have at least some overlapping mechanisms of action with open
monitoring and nondual forms of meditation. Repeated practice of
contemplative prayer slowly dissolves fixation with recurrent and
attention-consuming thoughts (*logismoi*) which likely leads to loosening rigid predictive hierarchies and heightens neural plasticity.
When pursued over a long time, the person praying the prayer, at least
in terms of their narrative/autobiographical self, is more and more
silenced, finally allowing the prayer to happen autonomously *through*
the individual, with a shift toward a more embodied and directly
sensed/felt mode of existing.

Today, the sciences of consciousness can perhaps slowly begin
to appreciate the insight behind many of the world's great wisdom
traditions, that our ordinary experience of reality conceals a truer

encounter with the world. As the modern understanding of neuroscience and psychology advances, we may increasingly be able to understand these powerful intuitions as more than fanciful metaphors.

Implicit and Explicit Nondual Awareness

Zoran Josipovic has proposed adding an implicit-explicit gradient of nondual awareness to models of consciousness that traditionally contain the two dimensions of global arousal (for example, awake versus deep sleep) and conscious content (for example, high as in waking or absolute minimal level as in MPEs).[44] The various transformations of consciousness that were reviewed in previous chapters can be specified by the combination of vectors, each one representing one of the three properties.

An ordinary waking state is determined by high global levels of brain arousal, relatively high conscious content, and only implicit nondual awareness. During nonlucid, deep, dreamless sleep, global arousal dims to levels below a critical threshold for consciousness, with minimal phenomenal content and a completely implicit ground of nonduality.

Josipovic postulates a family of transitional states, like those of ego dissolution and MPEs, where nonduality has yet to exceed a threshold of becoming fully realized. Drug-induced ego dissolution entails a moderate relaxation of core beliefs about self and a slightly more explicit realization of nonduality, but this state still involves residual separation in experiencing, where mental contents appear to manifest "out there." A neural signature of psychedelic states appears to be the diminution of DMN activity and the relaxation of overweighted beliefs that make up the narrative/autobiographical self.

An MPE, such as might be experienced in profound sensory deprivation or during lucid, imageless dreams, still contains a subtle sense of "me in-here." In both the drug-induced ego dissolution and the MPE examples, one still relies on a predictive self-world model with some residual degree of separateness. It is only in the explicit zone of the gradient of nonduality that the deepest foundational level of consciousness springs into full awareness, accompanied by the large-scale reorganization of competitive intrinsic and extrinsic brain networks.

In their topographic reorganization model of meditation, Austin Clinton Cooper and his colleagues suggest that the switch from an implicit to an explicit mode of nondual organization is characterized by a large-scale realignment process that gradually synchronizes multiple layers of self and world representations.[45] They define the dualistic mode of neurophenomenological organization as an exaggerated reliance on abstract, counterfactual kinds of predictions, in determining one's experience of lived reality. Excessive DMN and PCC activity reinforces passive captivity to counterfactual and abstract mental models, heavily mired in verbal networks, that reinforce separation of self and world, and decouple the narrative / autobiographical self from the minimal / embodied self.

The minimal / embodied self, in contrast to the narrative one, is naturally connected and continuous with the rest of the world. Despite core beliefs about separation between self and world, self and world are intertwined to form a holistic (nondual) unity that is more like how the color of food dye cannot be separated from the water in which it spreads. The brain and the larger world exist in a relationship of nestedness and this brain-world relation seems critical to the emergence of interior conscious experience.[46]

As the psychologist Eugene Gendlin has emphasized, our bodies are engaged in constant and ongoing series of energetic exchanges with the world that most of the time remain only tacit, ever-shifting background feelings.[47] For this reason, the phenomenologist Claire Petitmengin suggests that reestablishing contact with the fresh flow of experience requires repeatedly coming back into a felt sense of the body. Reminiscent of the unique style of "non-action" emphasized by Dzogchen and Heidegger's concept of *Gelassenheit,* she writes: "This "letting go" requires an opening up of the attention span, often described as associated with the displacement of the attentional center of gravity from the head to the body."[48]

With ever deeper letting-go, which is facilitated by effortless nondual meditation, there is increasingly stronger synchronization between the minimal / embodied self and the larger world within which the person is nested. This likely accounts for the visceral sense of deep connectedness with all of nature, including with other people, which spontaneously facilitates an all-encompassing sense of compassion.

Reconsidering Dual Aspect Monism and the Limits of Knowledge

In Chapter 1, we introduced the dual-aspect monist position. We postulated a unified psycho-physical reality which, through the limitations of observation, bifurcates into an outside dimension (studied through third-person methods) and an inside dimension (studied through first-person methods). We suggested, in Chapter 3, that space and time form the connecting bridge between these two observational modes. The dual aspect monist position naturally accommodates the proposition that reality is fundamentally interconnected, relational and nondual.

What are the deepest levels of knowing this mysterious underlying reality? If the intuitions of the various ancient religious and mystical schools or the scientific speculations sketched above have any validity to them, then it would indeed seem that a knowing-through-identity is the way into the farthest reaches of Jung and Pauli's *unus mundus* (One World) mentioned at the book's outset.[49] Essentially, sociologist and priest Andrew Greeley formulated just this hypothesis in a little book with big ideas on the limits of knowledge. "The real is encountered directly and as it is," he wrote, adding that "this knowledge is direct and immediate and needs neither propositions nor images . . . it is possible for the Real to 'rush in,' without having to operate through any of the intermediary tools that other forms of knowledge must utilize. . . . The person may be distinct from the rest of reality, but he is not discontinuous with it; there may be boundaries between self and world, but they are permeable and fuzzy."[50]

Perhaps these speculations will strike the mainstream of mind scientists as too extreme, but lately there seems to be an increasing tolerance for postulating more daring possibilities. We note, for example, that one of the leading consciousness researchers, Christof Koch, recently wrote a book arguing for consciousness as being identical to the interior feeling of life itself. Koch believes that consciousness is not a computable function.[51] The British philosopher Philip Goff has challenged physicalist models of reality, while offering a robust intellectual defense of the proposition that matter is not separate from consciousness.[52]

A growing consensus within mainstream brain science depicts consciousness as akin to a controlled hallucination.[53] We have reviewed much of the evidence in favor of this perspective throughout this book and certainly it plays no small part in the arguments that we have made here. At last, however, we would like to qualify that growing perspective by pointing that it is not the whole story. Virtual selves are not necessarily condemned to inhabit private, closed-in models of external reality. The potential exists for a distinct mode of being in which large-scale reorganization, as a consequence of spiritually transformative disciplines, retunes a person to enter into a more direct and dynamic dance with the rest of life: a reflection of the reality that exists "beyond the dream" and which is characterized by reciprocity, mutuality, and a hierarchically nested organization. This wider ecological context can perhaps better accommodate emerging findings about the neural bases of human consciousness.[54]

These developments could signal the dawn of a worldview that is surprisingly consonant with many of the best intuitions of the world's ancient spiritual traditions.[55] If that is true, then there might be truly original and novel opportunities to apply the latest scientific tools and add to the insights into the human mind made by contemplatives and mystics across millennia. As we move to this book's Coda, it is with great optimism for this vitally important endeavor. We close with some parting thoughts on how this merger of ancient and modern perspectives could prove to be a watershed event.

Coda

Toward an Optimistic Neurophenomenology of the Future

This book has peered into the depths of human consciousness as viewed through two lenses, an outside-in and an inside-out one, which together combine the subjective and objective. By looking through both lenses, we want to benefit from "stereoscopic vision" that prevents detailed knowledge about the brain from neglecting phenomenology, and ensure that neither the quantitative nor the qualitative eclipses the other.

We made the case that the world's spiritual traditions have many important lessons to convey that a contemporary science of consciousness would do well to notice. But there is another reason why the cumulative wisdom of the world's mystical traditions can be complementary to modern science: while questions about value and meaning sit uncomfortably within materialist frameworks of reality, which are focused on establishing facts, these are central to the great religions, and they can help to redress a too-limited vision of the human mind.

A version of the narrative that "brains dream reality" into existence can lead one to a bleak vision of what it means to be human. Taken out of context, the profusion of metaphors comparing consciousness to a virtual-reality simulation can promote nihilism. If our lived reality is really *nothing more* than a convincing and immersive constrained hallucination, then where does this leave us concerning the meaning of it all? If your skull encases a sophisticated dream-producing organ, then is having a meaningful life simply a matter of finding maximum amounts of pleasurable jolts while minimizing losses? Another complicating factor needs to be mentioned. Biology is malleable and adapts over time, such that habituation ensures a progressive weakening of pleasurable jolts. Over time, an organism would have to resort to more and more desperate measures to obtain weaker reinforcements, engaging in a cruel game that obeys the predictable law of diminishing returns. And if the self is just another virtual entity, an ultimate illusion, then why bother engaging in this hedonic pursuit at all?

This is precisely where the inspiring vision of human beings and their true potential bequeathed to us by millennia of the world's spiritual traditions can most enrich scientific theories of the human mind. That ancient wisdom, in its concern with quality rather than quantity, picks up the thread where science leaves off. It strives to embody the distant shores of human potential, rather than settle for describing and explaining how the brain/mind *usually* organizes itself. Not satisfied with accepting the limitations of humans, the contemplative undertakes a concerted effort to transform one's own consciousness at the deepest levels, overcoming conditioned and automatic forms of thinking, feeling, and behaving. The authentic spiritual path calls people to a lifelong endeavor, the aim of which is to enter into an experience of life that transcends a private "ego tunnel"—to attune the human being more directly to a fundamental level of reality and forge a vivid and continual awareness of connection with all of life. For a lasting sense of liberation from mental suffering, one must go beyond a perception that the self-in-world as ordinarily experienced is a virtual-reality show. Liberation comes with having direct knowledge (from the inside out) that this self need not be experienced as an isolated, closed-in entity, separate from the world "out there," as it is often portrayed by mainstream brain sciences. Rather, human beings have a latent potential to open up, to become creatively and freely responsive to a world and cosmos in seamless continuity with a state of being that is intrinsically wholesome, spiritually enriching, and intrinsically valuable, containing fundamental worth and goodness. "Being one with all that is" and interconnected with other people and living creatures can be a way of life, an ever increasing conscious realization and participation in first-order reality, rather than a dry, intellectual fact. When it is lived, such a conscious realization naturally produces compassion as divisions between self and other, us and them, are transcended and give way to an encompassing, ongoing sense of wonder, of awe, of mystery—in short, of the sacred.

In the Tibetan Buddhist tradition, authentic enlightenment is depicted as a bird with two wings, one representing emptiness and the other compassion.[1] Emptiness is achieved by peering deeply into the constructed nature of phenomenal experience. To gain insight into how deeply constructed the experiences of self and world are one must understand how dreamlike and illusory that second-order or apparent reality actually is. Everything is transient and in flux, so the heaviness of depression no longer feels so heavy as it might have before. But it is impossible to fly with one wing. On its own, the experience of ego dissolution and the annihilation of one's subjective "ego tunnel" would be a terrifying prospect, offering little in the way of inspiration, if it were not accompanied by a deeply felt sense of connectedness to all that is. It is only with a second wing in place, representing unconditional compassion—the feeling of connectedness with all of life—that this metaphorical bird of enlightenment can truly soar. This realization of absolute unity is what some philosophers have called communitas.[2] Together these wings can reach heights that afford a full, panoramic view of reality as it is—the all-encompassing view that can rescue us from the brink of nihilism.

The voices of generations of spiritual masters from different traditions are united on this matter. Their collective wisdom recognizes a first-order reality that is intrinsically knowing and beneficent. In classic religious texts, this ultimate reality is sometimes metaphorically depicted as sustaining everything within its caring embrace. This is an incredibly optimistic image, with a message that can revitalize a neurophenomenology of the future.

The VR helmet can be lifted, so to speak, as one transforms the inner epistemic structures one uses to contact and construct an experience of reality. The words of one esoteric writer and psychonaut P. D. Ouspensky describe his experience of that transformation. After exerting enormous effort to dissolve his own priors, the moment came when he saw the world anew, with fresh eyes: "Everything is alive," he reports saying to himself, 'there is nothing dead, it is only we who are dead. If we become alive for a moment, we shall feel that everything is alive, that all things live, feel and can speak to us."[3] Many of us have experienced flashes of this insight, but a more thoroughgoing renewal of consciousness can allow us to sustain it when we need it most.

This is a tremendously more awesome "big picture" than anything that a one-sided focus on dreamed realities can offer. A new era of neurophenomenology, which honors the depths of what it is to be human, is beginning to barely glimpse the possibilities that are just starting to come into view.[4] A constructive dialogue between science and spiritual traditions can draw unanticipated benefits for both partners in that conversation, and for anyone who listens to it carefully.

Notes

Preface

1. Francisco Varela, Evan Thompson, and Eleanor Rosch, *The Embodied Mind: Cognitive Science and Human Experience* (Cambridge, MA: MIT Press, 1993).

2. David B. Yaden and Roland R. Griffiths, "The Subjective Effects of Psychedelics Are Necessary for Their Enduring Therapeutic Effects," *ACS Pharmacology & Translational Science* 4, no. 2 (December 2020): 568–572.

3. We will often use the term *brain/mind* to signal the fundamental unity of these two perspectives. *Body/mind* might have been more appropriate, since the nervous system extends throughout the entire body, and biologists now know that "cognition" or intelligence is evident also in non-neural cells. For ease of usage, however, and to avoid confusing readers, we will mostly stick with the *brain/mind* dyad.

4. Roger Walsh, "What Is Wisdom? Cross-Cultural and Cross-Disciplinary Syntheses," *Review of General Psychology* 19, no. 3 (September 2015): 278–293.

5. David T. Bradford, "Brain and Psyche in Early Christian Asceticism," *Psychological Reports* 109, no. 2 (October 2011): 461–520.

6. Stanislav Grof, *Realms of the Human Unconscious: Observations from LSD Research* (London: Souvenir Press, 1979).

7. Theresa M. Carbonaro et al., "Survey Study of Challenging Experiences after Ingesting Psilocybin Mushrooms: Acute and Enduring Positive and Negative Consequences," *Journal of Psychopharmacology* 30, no. 12 (December 2016): 1268–1278.

1. The Matter of the Mind

1. Charles S. Sherrington, *Man on His Nature* (New York: Cambridge University Press, 1942), 178.

2. Norman D. Cook, Gil B. Carvalho, and Antonio Damasio, "From Membrane Excitability to Metazoan Psychology," *Trends in Neurosciences* 37, no. 12 (December 2014): 698–705.

3. Brian Boyd, *Vladimir Nabokov: The Russian Years* (Princeton: Princeton University Press, 1990), 11.

4. For an introduction to many of these philosophical positions on the nature of mind, but with a heavy materialistic bias, see Paul Churchland, *Matter and Consciousness,*3rd ed. (Cambridge, MA: MIT Press, 2013).

5. Francis Crick, *The Astonishing Hypothesis: The Scientific Search for the Soul* (New York: Touchstone, 1994), 4.

6. It must be stated, however, that twentieth-century physics has created a deep conundrum about the very question of what "matter" actually is. We know now that whatever it might be, it is not nearly as substantial as outdated notions of dense materiality might have implied. As the physicist Carlo Ravelli succinctly put it in the title of one of his books, "reality is not what it seems." The discrete particles, the solid "things" that are sometimes imagined to underlie the materiality of the world seem instead to be specific local manifestations of probabilistic fields of energy. In the world of consciousness too, reality is not what it seems.

7. See, for example, John C. Eccles and Karl Popper, *The Self and Its Brain: An Argument for Interactionism* (New York: Taylor & Francis, 2013); and Wilder Penfield, *Mystery of the Mind: A Critical Study of Consciousness and the Human Brain* (Princeton: Princeton University Press, 2016).

8. William James, *Principles of Psychology,* 2 vols. (New York: Henry Holt, 1890).

9. John R. Searle, *The Rediscovery of the Mind* (Cambridge, MA: MIT Press, 1992), 95.

10. Searle, *Rediscovery of the Mind,* 227.

11. Juergen Fell, "Identifying Neural Correlates of Consciousness: The State Space Approach," *Consciousness and Cognition* 13, no. 4 (December 2004): 709–729.

12. William James, *The Varieties of Religious Experience* (New York: The Modern Library, 1902), 378–379.

13. Stephen P. LaBerge et al., "Lucid Dreaming Verified by Volitional Communication during REM Sleep," *Perceptual and Motor Skills* 52, no. 3 (June 1981): 727–732.

14. Charles T. Tart, *Altered States of Consciousness: A Book of Readings* (New York: Wiley, 1969).

15. J. Allan Hobson, *The Dream Drugstore: Chemically Altered States of Consciousness* (Cambridge, MA: MIT Press, 2001).

16. David Lewis-Williams, *The Mind in the Cave: Consciousness and the Origins of Art* (London: Thames & Hudson, 2002), 125.

17. Simon Hanslmayr et al., "The Role of Alpha Oscillations in Temporal Attention," *Brain Research Reviews* 67, no.1–2 (June 2011): 331–343.

18. B. Alexander Diaz et al., "The Amsterdam Resting-State Questionnaire Reveals Multiple Phenotypes of Resting-State Cognition," *Frontiers in Human Neuroscience* 7 (August 2013): article 446.

19. Andreas Mavromatis, *Hypnagogia: The Unique State of Consciousness between Wakefulness and Sleep* (London: Thyrsos Press, 2010).

20. David Gelernter, *Tides of Mind: Uncovering the Spectrum of Consciousness* (New York: W. W. Norton, 2016).

21. Arthur J. Deikman, "Bimodal Consciousness," *Archives of General Psychiatry* 25, no. 6 (December 1971): 481–489.

22. J. Allan Hobson, "REM Sleep and Dreaming: Towards a Theory of Protoconsciousness," *Nature Reviews Neuroscience* 10 (October 2009): 806.

23. Gerald G. May, *Dark Night of the Soul: A Psychiatrist Explores the Connection between Darkness and Spiritual Growth* (New York: HarperCollins, 2004), 153–154.

24. Harald Walach, "Inner Experience—Direct Access to Reality: A Complementarist Ontology and Dual Aspect Monism Support a Broader Epistemology," *Frontiers in Psychology* 11 (April 2020): 640.

25. Kalu Rinpoche, *Luminous Mind: The Way of the Buddha* (Somerville, MA: Wisdom Publications, 1997).

26. John Anthony McGuckin, *The Book of Mystical Chapters: Meditations on the Soul's Ascent, from the Desert Fathers and Other Early Christian Contemplatives* (Boston: Shambhala, 2002).

27. Evan Thompson, *Mind in Life: Biology, Phenomenology, and the Sciences of the Mind* (Cambridge, MA: Harvard University Press, 2010).

28. B. Alan Wallace, *Contemplative Science: Where Buddhism and Neuroscience Converge* (New York: Columbia University Press, 2007).

29. Francisco J. Varela, "Neurophenomenology: A Methodological Remedy for the Hard Problem," *Journal of Consciousness Studies* 3, no. 4 (April 1996): 330–349.

30. Franklin Merrell-Wolff, *Franklin Merrell-Wolff's Experience and Philosophy: A Personal Record of Transformation of Consciousness and a Discussion of Transcendental Consciousness* (Albany: State University of New York Press, 1994).

31. Philip Sherrard, *Christianity: Lineaments of a Sacred Tradition* (Brookline, MA: Holy Cross Orthodox Press, 1998).

32. See, for example, Julian Urrutia et al., "Psychedelic Science, Contemplative Practices, and Indigenous and Other Traditional Knowledge Systems: Towards Integrative Community-Based Approaches in Global Health," *Journal of Psychoactive Drugs* 55 no. 5 (September 2023): 523–538; and Roger Walsh, "What Is Wisdom? Cross-Cultural and Cross-Disciplinary Syntheses," *Review of General Psychology* 19, no. 3 (September 2015): 278–293.

33. Evan Thompson, *Waking, Dreaming, Being: Self and Consciousness in Neuroscience, Meditation, and Philosophy* (New York: Columbia University Press, 2014).

34. B. Alan Wallace, *The Taboo of Subjectivity: Toward a New Science of Consciousness* (New York: Oxford University Press, 2004).

35. The Dalai Lama, *Sleeping, Dreaming, and Dying: An Exploration of Consciousness* (Somerville, MA: Wisdom Publications, 1997).

36. Harald Atmanspacher and Dean Rickles, *Dual-Aspect Monism and the Deep Structure of Meaning* (New York: Routledge, 2022).

37. Harald Atmanspacher, "Dual-Aspect Monism a la Pauli and Jung," *Journal of Consciousness Studies* 19, no. 9–10 (January 2012): 96–120.

38. The connecting bridge or association, as we will argue in Chapter 3, is rooted in space and time.

39. Max Velmans, "Reflexive Monism: Psychophysical Relations among Mind, Matter, and Consciousness," *Journal of Consciousness Studies* 19, no. 9–10 (January 2012): 143–165.

2. A Brain That Talks to Itself

1. Marcus Raichle, "Two Views of Brain Function," *Trends in Cognitive Sciences* 14, no. 4 (April 2010): 180–190.

2. Maurice Merleau-Ponty, *Phenomenology of Perception* (Delhi: Motilal Banarsidass, 1996), 55.

3. Romain Brette, "Is Coding a Relevant Metaphor for the Brain?" *Behavioral and Brain Sciences* 42 (July 2018): 21–22, e215.

4. Georg Northoff, *The Spontaneous Brain: From the Mind-Body to the World-Brain Problem* (Cambridge, MA: MIT Press, 2018).

5. Cristopher M. Niell and Michael P. Stryker, "Modulation of Visual Responses by Behavioral State in Mouse Visual Cortex," *Neuron* 65, no. 4 (February 2010): 472–479.

6. Marcus Raichle, "Two Views of Brain Function," *Trends in Cognitive Sciences* 14, no. 4 (April 2010): 180–190.

7. There are certain similarities here between how the brain comes to know about the world and the principles evident in digestion. To assimilate food particles fed from the outer world, the organism first decomposes them into an elementary set of molecules for absorption and then construction of larger, macro-molecular components that bear the organism's own immunological stamp. "The explanation," as the neurobiologist Walter Freeman wrote, "for this manner of function of both the neural and the digestive systems is essentially the same: the world is infinitely complex, and the self can only know and incorporate what the brain makes within itself." Walter J. Freeman, "Neurodynamic Models of Brain in Psychiatry," *Neuropsychopharmacology* 28 (June 2003): S56.

8. Andy Clark, "Whatever Next? Predictive Brains, Situated Agents, and the Future of Cognitive Science," *Behavioral and Brain Sciences* 36, no. 3 (June 2013): 181–204.

9. Donald Hoffman, *The Case against Reality: Why Evolution Hid the Truth From Our Eyes* (New York: W. W. Norton, 2019).

10. Gustavo Deco, Viktor K. Jirsa, and Anthony R. McIntosh, "Emerging Concepts for the Dynamical Organization of Resting-State Activity in the Brain," *Nature Reviews Neuroscience* 12 (December 2010): 43–56.

11. As one example, a team at the University of Rochester conducted an experiment in which they grafted one specific type of glial cell (an astrocyte) from the human brain into the brains of mice. The researchers discovered that the mice with the "rebooted" glial cells ran through mazes faster than their peers with standard murine astrocytes, demonstrating that these cells play a critical role in learning and memory. See Xiaoning Han et al., "Forebrain Engraftment by Human Glial Progenitor Cells Enhances Synaptic Plasticity and Learning in Adult Mice," *Cell Stem Cell* 12, no. 3 (March 2013): 342–353.

12. R. Douglas Fields, *The Other Brain: From Dementia to Schizophrenia, How New Neuroscience Discoveries Are Revolutionizing Medicine and Science* (New York: Simon & Schuster, 2009).

13. Richard Rappoport, *Nerve Endings: The Discovery of the Synapse* (New York: W. W. Norton, 2005).

14. Gerald Edelman and Joseph A. Gally, "Reentry: A Key Mechanism for Integration of Brain Function," *Frontiers in Integrative Neuroscience* 7 (August 2013): article 63.

15. Ryan T. Canolty and Robert T. Knight, "The Functional Role of Cross-Frequency Coupling," *Trends in Cognitive Sciences* 14, no. 11 (November 2010): 506–515.

16. Susana Martinez-Conde, Jorge Otero-Millan, and Stephen L. Macknik, "The Impact of Microsaccades on Vision: Towards a Unified Theory of Saccadic Function," *Nature Reviews Neuroscience* 14 (January 2013): 83–96.

17. Jon H. Kaas, "Evolution of Somatosensory and Motor Cortex in Primates," *Anatomical Record* 281A, no. 1 (November 2004): 1148–1156.

18. Donald Pfaff, *How Brain Arousal Mechanisms Work: Paths Toward Consciousness* (New York: Cambridge University Press, 2018).

19. Jaak Panksepp, *Affective Neuroscience: The Foundations of Human and Animal Emotions* (New York: Oxford University Press, 1998).

20. Lawrence Ward, "The Thalamus: Gateway to the Mind," *WIREs Cognitive Science* 4, no.6 (November 2013): 609–622.

21. Although the term "limbic system" has fallen out of favor among contemporary neuroscientists, and has been criticized for its excessive vagueness, it can arguably continue to provide a useful anatomical short-hand, as long as we are careful not to fall into the trap of simplifying it as an evolutionary primitive and unsophisticated part of the brain.

22. Walter J. Freeman, *How Brains Make Up Their Minds* (New York: Columbia University Press, 2000), 35.

23. Mark Solms and Oliver Turnbull, *The Brain and the Inner World: An Introduction to the Neuroscience of Subjective Experience* (New York: Routledge, 2018).

24. Rodney J. Douglas and Kevin A.C. Martin, "Mapping the Matrix: The Ways of the Neocortex," *Neuron* 56, no. 2 (October 2007): 226–238; and Paul L. Nunez, *Brain, Mind, and The Structure of Reality* (New York: Oxford University Press, 2010)

25. Lars Muckli and Lucy S. Petro, "Network Interactions: Non-Geniculate Input to V1," *Current Opinion in Neurobiology* 23, no. 2 (April 2013): 195–201.

26. Olaf Sporns, *Networks of the Brain* (Cambridge, MA: MIT Press, 2011).

27. Adam Sillito and Helen E. Jones, "Corticothalamic Interactions in the Transfer of Visual Information," *Philosophical Transactions of the Royal Society of London. Series B, Biological Sciences* 357, no. 1428 (December 2002): 1739–1752; S. Murray Sherman and R. W. Guillery, "The Role of the Thalamus in the Flow of Information to the Cortex," *Philosophical Transactions of the Royal Society of London. Series B, Biological Sciences* 357, no. 1428 (December 2002): 1695–1708.

28. Randy L. Buckner and Fenna M. Krienen, "The Evolution of Distributed Association Networks in the Human Brain," *Trends in Cognitive Sciences* 17, no. 12 (December 2013): 648–665.

29. Rodolfo Llinás, "Intrinsic Electrical Properties of Mammalian Neurons and CNS Function: A Historical Perspective," *Frontiers in Cellular Neuroscience* 8 (November 2014): 320.

30. David J. Linden, *The Accidental Mind: How Brain Evolution Has Given Us Love, Memory, Dreams, and God* (Cambridge, MA: Harvard University Press, 2007).

31. Francisco Varela et al., "The Brainweb: Phase Synchronization and Large-Scale Integration," *Nature Reviews Neuroscience* 2 (April 2001): 229–239.

32. Rafael Yuste, "From the Neuron Doctrine to Neural Networks," *National Review of Neuroscience* 16, no. 8: 487–497.

33. Steven Bressler and Emmanuelle Tognoli, "Operational Principles of Neurocognitive Networks," *International Journal of Psychophysiology* 60, no. 2 (May 2006): 139–148.

34. Gyorgy Buzsaki, *Rhythms of the Brain* (New York: Oxford University Press, 2006).

35. Tobias H. Donner and Markus Siegel, "A Framework for Local Cortical Oscillation Patterns," *Trends in Cognitive Sciences* 15, no. 5 (May 2011): 191–199.

36. In turn, the timing and probability of a single neuron spiking can be dependent on the cycle of a high-frequency oscillation. An individual neuron may be much more likely to fire a spike during the peak than during the trough of a fast oscillation produced by the local neuronal area in which it's embedded.

37. Sergei Gephstein et al., "Spatially Distributed Computation in Cortical Circuits," *Science Advances* 8, no. 16 (April 2022): eabl5865.

38. Gyorgy Buzsaki, *The Brain from Inside Out* (New York: Oxford University Press, 2019).

39. Georg Northoff, *The Spontaneous Brain: From the Mind-Body to the World-Brain Problem* (Cambridge, MA: MIT Press, 2018).

40. Marcus E. Raichle, "The Brain's Dark Energy," *Science* 314, no. 5803 (November 2006): 1249–1250.

41. Kurt Goldstein, *The Organism: A Holistic Approach to Biology Derived from Pathological Data in Man* (New York: American Book, 1939; New York: Zone Books, 2000), 95–96.

42. Martijn P. van den Heuvel and Hilleke E. Hulshoff Pol, "Exploring the Brain Network: A Review on Resting-State fMRI Functional Connectivity," *European Neuropsychopharmacology* 20, no. 8 (August 2010): 519–534.

43. Alfredo Fontanini and Donald B. Katz, "Behavioral States, Network States, and Sensory Response Variability," *Journal of Neurophysiology* 100, no. 3 (September 2008): 1160–1168.

44. Georg Northoff, *The Spontaneous Brain: From the Mind-Body to the World-Brain Problem* (Cambridge, MA: MIT Press, 2018).

3. Looking Directly at the Mind

1. Georg Northoff et al., "Is Temporo-Spatial Dynamics the 'Common Currency' of Brain and Mind? In Quest of 'Spatiotemporal Neuroscience,'" *Physics of Life Reviews* 33 (July 2020): 34–54.

2. William James, *The Principles of Psychology*, vol. 1 (1890; New York: Dover, 1950).

3. B. Alan Wallace, *Fathoming the Mind: Inquiry and Insight in Dudjom Lingpa's Vajra Essence* (Somerville, MA: Wisdom Publications, 2018).

4. Selen Atasoy et al., "Harmonic Brain Modes: A Unifying Framework for Linking Space and Time in Brain Dynamics," *The Neuroscientist* 24, no. 3 (June 2018): 277–293.

5. Robin L. Carhart-Harris et al., "The Entropic Brain: A Theory of Conscious States Informed by Neuroimaging Research with Psychedelic Drugs," *Frontiers in Human Neuroscience* 8 (February 2014): 20; and Robin L. Carhart-Harris, "The Entropic Brain—Revisited," *Neuropharmacology* 142 (November 2018): 167–178.

6. Christof Koch et al., "Neural Correlates of Consciousness: Progress and Problems," *Nature Reviews Neuroscience* 17 (May 2016): 308.

7. Ngapka Chogyam, and Khandro Dechen, *Roaring Silence: Discovering the Mind of Dzogchen* (Boston: Shambhala, 2002).

8. Antti Revonsuo, *Inner Presence: Consciousness as a Biological Phenomenon* (Cambridge, MA: MIT Press, 2010).

9. Gerald Edelman, Naturalizing Consciousness: A Theoretical Framework," *Proceedings of the National Academy of Sciences USA* 100, no. 9 (April 2003): 5520–5524.

10. Evan Thompson and Francisco J. Varela, "Radical Embodiment: Neural Dynamics and Consciousness," *Trends in Cognitive Sciences* 10, no. 1 (October 2001): 418–425.

11. Francisco J. Varela, "Neurophenomenology: A Methodological Remedy for the Hard Problem," *Journal of Consciousness Studies* 3 (April 1996): 330–349, 340.

12. Varela, "Neurophenomenology," 347.

13. Aviva Berkovich-Ohana et al., "The Hitchhiker's Guide to Neurophenomenology—The Case of Studying Self Boundaries with Meditators," *Frontiers in Psychology* 100, no. 3 (July 2020): 1680.

14. Natalie Depraz, "Epoché in Light of Samatha-Vipassana Meditation: Chögyam Trungpa's Buddhist Teaching Facing Husserl's Phenomenology," *Journal of Consciousness Studies* 26 (January 2019): 49–69.

15. Natalie Depraz, Francisco J. Varela, and Pierre Vermersch, "The Gesture of Awareness: An Account of Its Structural Dynamics," in *Investigating*

Phenomenal Consciousness: New Methodologies and Maps, ed. Max Velmans (Amsterdam: John Benjamins, 2000), 121–136.

16. Donald D. Price and James J. Barrell, *Inner Experience and Neuroscience: Merging Both Perspectives* (Cambridge, MA: MIT Press, 2012).

17. Claire Petitmengin, "Describing One's Subjective Experience in the Second Person: An Interview Method for the Science of Consciousness," *Phenomenology and the Cognitive Sciences* 5, no. 3–4 (November 2006): 229–269.

18. Claire Petitmengin. "On the Veiling and Unveiling of Experience: A Comparison between the Micro-Phenomenological Method and the Practice of Meditation," *Journal of Phenomenological Psychology* 52, no. 1 (August 2021): 36–77, 41.

19. Eugene T. Gendlin, *Focusing* (New York: Random House, 1982).

20. Evan Thompson, *Mind in Life: Biology, Phenomenology, and the Sciences of the Mind* (Cambridge, MA: Harvard University Press, 2010).

21. The Dalai Lama, *The Universe in a Single Atom: The Convergence of Science and Spirituality* (New York: Morgan Road Books, 2005), 136.

22. Antoine Lutz, John D. Dunne, and Richard J. Davison, "Meditation and the Neuroscience of Consciousness: An Introduction," in *The Cambridge Handbook of Consciousness,* ed. Philip David Zelazo, Morris Moscovitch, and Evan Thompson (New York: Cambridge University Press), 499–551.

23. Petitmengin. "On the Veiling and Unveiling of Experience."

24. John Welwood, *Toward a Psychology of Awakening: Buddhism, Psychotherapy, and the Path of Personal and Spiritual Transformation* (Boulder: Shambhala, 2002), 106.

25. Khenpo Tsultrim Gyamtso, *Stars of Wisdom: Analytical Meditation, Songs of Yogic Joy, and Prayers of Aspiration* (Boston: Shambhala, 2010).

26. Lutz, Dunne, and Davison, "Meditation and the Neuroscience of Consciousness."

27. See Martin Laird, *Into the Silent Land: A Guide to the Christian Practice of Contemplation* (New York: Oxford University Press, 2006); and B. Alan Wallace, *Mind in the Balance: Meditation in Science, Buddhism, and Christianity* (New York: Columbia University Press, 2009).

28. Gregory Palamas, *Holy Hesychia: The Stillness That Knows God* (Southover: Pleroma Publishing, 2016).

29. Rinpoche Thrangu, *Essentials of Mahamudra: Looking Directly at the Mind* (Somerville, MA: Wisdom Publications, 2004).

30. Matthieu Ricard, *On the Path to Enlightenment: Heart Advice from the Great Tibetan Masters* (Boston: Shambhala, 2013).

31. B. Alan Wallace, *Dreaming Yourself Awake: Lucid Dreaming and Tibetan Dream Yoga for Insight and Transformation* (Boston: Shambhala, 2012).

32. B. Alan Wallace, *Stilling the Mind: Shamatha Teachings from Dudjom Lingpa's Vajra Essence* (Somerville, MA: Wisdom Publications, 2011).

33. Ngapka Chogyam, and Khandro Dechen, *Roaring Silence: Discovering the Mind of Dzogchen* (Boston: Shambhala, 2002).

34. Rinpoche Thrangu, *Essentials of Mahamudra,* 106.

35. Khenchen Thrangu, *Vivid Awareness: The Mind Instructions of Khenpo Gangshar* (Boston: Shambhala, 2011).

36. B. Alan Wallace, *Hidden Dimensions: The Unification of Physics and Consciousness* (New York: Columbia University Press, 2007).

37. B. Alan Wallace, *The Attention Revolution: Unlocking the Power of the Focused Mind* (Somerville, MA: Wisdom Publications, 2006).

38. Lutz, Dunne, and Davison, "Meditation and the Neuroscience of Consciousness."

39. Dakpo Tashi Namgyal, *Clarifying the Natural State: A Principal Guidance Manual for Mahamudra* (Hong Kong: Rangjung Yeshe, 2001).

4. The 500-Million-Year-Old Virtual-Reality Simulator

1. Antti Revonsuo, "Consciousness, Dreams, and Virtual Realities," *Philosophical Psychology* 8, no. 1 (June 2008): 35–58, 52.

2. Steven Lehar, *The World in Your Head: A Gestalt View of the Mechanism of Conscious Experience* (New York: Psychology Press, 2012), 33.

3. Steven Lehar, "Gestalt Isomorphism and the Primacy of Subjective Conscious Experience: A Gestalt Bubble Model," *Behavioral and Brain Sciences* 26 (August 2003): 375–408.

4. Antti Revonsuo, *Inner Presence: Consciousness as a Biological Phenomenon* (Cambridge, MA: MIT Press, 2010).

5. Thomas Metzinger, "Empirical Perspectives from the Self-Model Theory of Subjectivity: A Brief Summary with Examples," *Progress in Brain Research* 168 (December 2007): 215–245.

6. Thomas Metzinger, *The Ego Tunnel: The Science of the Mind and the Myth of the Self* (New York: Basic Books, 2009).

7. Klaus Hepp, "Space, Time, Categories, Mechanics, and Consciousness: On Kant and Neuroscience," *Journal of Statistical Physics* 180 (April 2020): 896–909.

8. Carlo Roveli, *Reality Is Not What It Seems: The Journey to Quantum Gravity* (New York: Riverhead Books, 2017).

9. This topic will get a much more thorough treatment in Chapter 13, since many of the world's great spiritual traditions make a slightly different

argument regarding the knowability of ultimate reality. For now, we will stick to the second-order reality, the one which gets "dreamed." This useful terminology originates from Reginald A. Ray, *Secret of the Vajra World: The Tantric Buddhism of Tibet* (Boston: Shambhala, 2001).

10. Andy Clark, "Whatever Next? Predictive Brains, Situated Agents, and the Future of Cognitive Science," *Behavioral and Brain Sciences* 36, no. 3 (June 2013): 181–204.

11. The benefits of this experience through model building is that it allows an organism to adapt to a very large variation in physical environments having a high degree of complexity and ambiguity.

12. Keisuke Suzuki et al., "A Deep-Dream Virtual Reality Platform for Studying Altered Perceptual Phenomenology," *Scientific Reports* 7 (November 2017): 15982.

13. Lehar, *The World in Your Head*, 91.

14. J. Allan Hobson and Karl Friston, "Waking and Dreaming Consciousness: Neurobiological and Functional Considerations," *Progress in Neurobiology* 98 (May 2012): 82–98, 83.

15. Thomas Metzinger, *Being No One: The Self-Model Theory of Subjectivity* (Cambridge, MA: MIT Press), 140.

16. Donald Hoffman, *The Case against Reality: Why Evolution Hid the Truth from Our Eyes* (New York: W. W. Norton, 2019).

17. Jerome Y. Lettvin et al., "What the Frog's Eye Tells the Frog's Brain," *Proceedings of the IRE* 47, no. 11 (November 1959): 1940–1951.

18. Stuart Anstis, "Visual Filling-In," *Current Biology* 20, no. 16 (August 2010): R664–666.

19. Rufin VanRullen, "Perceptual Cycles," *Trends in Cognitive Sciences* 20, no. 10 (October 2016): 723–735.

20. Julien Dubois and Rufin VanRullen, "Visual Trails: Do the Doors of Perception Open Periodically?" *PLoS Biology* 9, no. 5 (May 2011): e1001056.

21. Charles Sherrington, *The Integrative Action of the Nervous System* (Oxford: Oxford University Press, 1947).

22. Antti Revonsuo, *Inner Presence: Consciousness as a Biological Phenomenon* (Cambridge, MA: MIT Press, 2010).

23. Rodika Sokoliuk and Rufin VanRullen, "The Flickering Wheel Illusion: When α Rhythms Make a Static Wheel Flicker," *The Journal of Neuroscience* 33, no. 33 (August 2013): 13498–13504.

24. Steven Lehar, "Gestalt Isomorphism and the Primacy of Subjective Conscious Experience: A Gestalt Bubble Model," *Behavioral and Brain Sciences* 26 (August 2003): 375–408.

25. Max Velmans, "Is the World in the Brain, or the Brain in the World?" *Behavioral and Brain Sciences* 26, no. 4 (August 2003): 427–429, 428.

26. Donald Pfaff, *How Brain Arousal Mechanisms Work: Paths toward Consciousness* (New York: Cambridge University Press, 2018).

27. Charles Sherrington, *The Brain and Its Mechanisms. The Rede Lecture* (London: University Press, 1933).

28. Jacob von Uexküll, *A Foray into the Worlds of Animals and Humans, with A Theory of Meaning* (Minneapolis: University of Minnesota Press, 2013).

29. Dreamscapes, obviously, are entirely subjective, reconstructed from memory and imagination (see Chapter 7).

30. Radmila Moacanin, *The Essence of Jung's Psychology and Tibetan Buddhism: Western and Eastern Paths to the Heart* (Somerville, MA: Wisdom Publications, 2003).

31. Ngapka Chogyam and Khandro Dechen, *Spectrum of Ecstasy* (Ramsey, NJ: Aro Books, 2003), 107–108.

32. Revonsuo, *Inner Presence.*

33. Originally translated into English simply as *Wonderment* by the Buddhist scholar Herbert Guenther. It has also been rendered as *Maya Yoga* by Keith Dowman.

34. Longchenpa, *Finding Rest in Illusion: The Trilogy of Rest,* vol. 3 (Boulder, CO: Shambhala, 2018), 84.

35. Keith Dowman, *Maya Yoga: Longchenpa's Finding Comfort and Ease in Enchantment* (Kathmandu: Vajra Publications, 2010).

36. Rinpoche Thrangu, *Essentials of Mahamudra: Looking Directly at the Mind* (Somerville, MA: Wisdom Publications, 2004).

37. See, for example, Roberto Assagioli, *Transpersonal Development: Beyond Psychosynthesis* (Forres, Scotland: Inner Way Productions, 2007); and Abraham Maslow, *Religions, Values, and Peak-Experiences* (New York: Penguin Books, 1994).

5. Adding Feelings to the Mix

1. Jaak Panksepp and Lucy Biven, *The Archeology of Mind: Neuroevolutionary Origins of Human Emotions* (New York: W. W. Norton, 2012).

2. Thomas Keating, *Invitation to Love: The Way of Christian Contemplation* (London: Bloomsbury, 2011).

3. David J. Anderson and Ralph Adolphs, "A Framework for Studying Emotions across Species," *Cell* 157, no. 1 (March 2014): 187–200.

4. Aslihan Selimbeyoglu and Josef Parvizi, "Electrical Stimulation of the Human Brain: Perceptual and Behavioral Phenomena Reported in the Old and New Literature," *Frontiers in Human Neuroscience* 4 (May 2010): 46.

5. Derek A. Denton, *The Primordial Emotions: The Dawning of Consciousness* (New York: Oxford University Press, 2005).

6. Derek A. Denton et al., "The Role of Primordial Emotions in the Evolutionary Origin of Consciousness," *Consciousness and Cognition* 18, no. 2 (June 2009): 500–514.

7. Jaak Panksepp, *Affective Neuroscience: The Foundations of Human and Animal Emotions* (New York: Oxford University Press, 1998).

8. Antonio Damasio, *Descartes' Error: Emotion, Reason and the Human Brain* (New York: Penguin Books, 1994).

9. Satoshi Ikemoto and Jaak Panksepp, "The Role of Nucleus Accumbens Dopamine in Motivated Behavior: A Unifying Interpretation with Special Reference to Reward-Seeking," *Brain Research Reviews* 31, no. 1 (December 1999): 6–41.

10. R. Christopher Pierce and Vidhya Kumaresan, "The Mesolimbic Dopamine System: The Final Common Pathway for the Reinforcing Effect of Drugs of Abuse?" *Neuroscience and Biobehavioral Reviews* 30, no. 2 (August 2005): 215–238.

11. Simone C. Motta et al., "The Periaqueductal Gray and Primal Emotion Processing Critical to Influence Complex Defensive Responses, Fear Learning, and Reward Seeking," *Neuroscience and Biobehavioral Reviews* 76 (May 2017): 39–47.

12. Blaine S. Nashold Jr et al., "Sensations Evoked by Stimulation in the Midbrain of Man," *Journal of Neurosurgery* 30, no. 1 (January 1969): 14–24.

13. Jaak Panksepp and Lucy Biven, *The Archeology of Mind: Neuroevolutionary Origins of Human Emotions* (New York: W. W. Norton, 2012).

14. As quoted in Matthieu Ricard, *On the Path to Enlightenment: Heart Advice from the Great Tibetan Masters* (Boston: Shambhala, 2013), 167.

15. Panksepp, *Affective Neuroscience*.

16. Rebecca Todd et al., "Emotional Objectivity: Neural Representations of Emotions and Their Interaction with Cognition," *Annual Review of Psychology* 71 (January 2020): 25–48.

17. Keith Dowman, *Maya Yoga: Longchenpa's Finding Comfort and Ease in Enchantment* (Kathmandu: Vajra Books, 2010).

18. Chogyam Trungpa, *Transcending Madness: The Experience of the Six Bardos* (Boston: Shambhala, 1992).

19. Panksepp, *Affective Neuroscience*.

20. Jaak Panksepp and Georg Northoff, "The Trans-Species Core SELF: The Emergence of Active Cultural and Neuro-Ecological Agents through Self-Related Processing within Subcortical-Cortical Midline Networks," *Consciousness and Cognition* 18, no. 1 (March 2009): 193–215.

21. Todd E. Feinberg and Jon M. Mallatt, *The Ancient Evolutionary Origins of Consciousness: How the Brain Created Experience* (Cambridge, MA: MIT Press, 2016).

22. Vladimir Miskovic and Adam K. Anderson, "Modality General and Modality Specific Coding of Hedonic Valence," *Current Opinion in Behavioral Sciences* 19 (February 2018): 91–97.

23. Luiz Pessoa, *The Cognitive-Emotional Brain: From Interactions to Integration* (Cambridge, MA: MIT Press, 2013).

24. Rebecca M. Todd et al., "Psychophysical and Neural Evidence for Emotion-Enhanced Perceptual Vividness," *Journal of Neuroscience* 32, no. 33 (August 2012): 11201–11212.

25. Keith Dowman, *The Flight of the Garuda: The Dzogchen Tradition of Tibetan Buddhism* (Somerville, MA: Wisdom Publications, 2012), 13.

26. Mauricio Sierra et al., "Separating Depersonalisation and Derealisation: The Relevance of the 'Lesion Method,'" *Journal of Neurology, Neurosurgery & Psychiatry* 72, no. 4 (April 2002): 530–532; and Russell M. Bauer, "Visual Hypoemotionality as a Symptom of Visual-Limbic Disconnection in Man," *Archives of Neurology* 39, no. 11 (November 1982): 702–708.

27. This explanation of Capgras and Cotard syndromes is primarily drawn from Vilayanur Ramachandran and Sandra Blakeslee, *Phantoms in the Brain: Human Nature and the Architecture of the Mind* (London: Fourth Estate, 1998).

28. For more see Chogyam Trungpa, *Glimpses of Abhidharma: From a Seminar on Buddhist Psychology* (Boston: Shambhala, 1975); and Evan Thompson, *Waking, Dreaming, Being: Self and Consciousness in Neuroscience, Meditation and Philosophy* (New York: Columbia University Press, 2014).

29. Matthieu Ricard and Wolf Singer, *Beyond the Self: Conversations between Buddhism and Neuroscience* (Cambridge, MA: MIT Press, 2018).

30. Adam K. Anderson, "Toward an Objective Neural Measurement of Subjective Feeling States," *Psychology of Consciousness* 2, no. 1 (March 2015): 30–33.

31. Kent C. Berridge and Morten L. Kringelbach, "Neuroscience of Affect: Brain Mechanisms of Pleasure and Displeasure," *Current Opinion in Neurobiology* 23, no. 3 (June 2013): 294–303.

32. Stuart Kauffman, *Investigations* (New York: Oxford University Press), 111.

33. Khenchen Thrangu, *Vivid Awareness: The Mind Instructions of Khenpo Gangshar* (Boston: Shambhala, 2011).

34. Patrul Rinpoche, *The Heart Treasure of the Enlightened Ones: The Practice of View, Meditation, and Action* (Boston: Shambhala, 1993).

35. Rob Nairn, *Living, Dreaming, Dying: Wisdom for Everyday Life from the Tibetan Book of the Dead* (Boston: Shambhala, 2004).

36. Chogyam Trungpa, *The Myth of Freedom and the Way of Meditation* (Boston: Shambhala, 2002).

37. John Welwood, *Toward a Psychology of Awakening: Buddhism, Psychotherapy, and the Path of Personal and Spiritual Transformation* (Boulder: Shambhala, 2002).

38. Lisa Feldman Barrett, *How Emotions Are Made: The Secret Life of the Brain* (New York: Houghton Mifflin Harcourt, 2017).

39. Jaak Panksepp, "Neurologizing the Psychology of Affects: How Appraisal-Based Constructivism and Basic Emotion Theory Can Coexist," *Perspectives on Psychological Science* 2, no. 3 (September 2007): 281–296.

6. The Psychopathology of Everyday Life

1. Martin Laird, *A Sunlit Absence: Silence, Awareness, and Contemplation* (New York: Oxford University Press, 2011), 25–26.

2. Russell T. Hurlburt and Sarah A. Akhter, "Unsymbolized Thinking," *Consciousness and Cognition* 17, no. 4 (December 2008): 1364–1374.

3. Julie Tseng and Jordan Poppenk, "Brain Meta-State Transitions Demarcate Thoughts across Task Contexts Exposing the Mental Noise of Trait Neuroticism," *Nature Communications* 11, no. 1 (July 2020): 3480.

4. B. Alexander Diaz et al., "The Amsterdam Resting-State Questionnaire Reveals Multiple Phenotypes of Resting-State Cognition," *Frontiers in Human Neuroscience* 7 (August 2013): 446.

5. Matthew A. Killingsworth and Daniel T. Gilbert, "A Wandering Mind Is an Unhappy Mind," *Science* 330, no. 6006 (November 2010): 932.

6. Jonathan Smallwood and Jonathan W. Schooler, "The Science of Mind Wandering: Empirically Navigating the Stream of Consciousness," *Annual Review of Psychology* 66 (January 2015): 487–518.

7. Paul Seli, Evan F. Risko, and Daniel Smilek, "On the Necessity of Distinguishing between Unintentional and Intentional Mind Wandering," *Psychological Science* 27, no. 5 (January 2016): 685–691.

8. Adrian F. Ward and Daniel M. Wegner, "Mind-Blanking: When the Mind Goes Away," *Frontiers in Psychology* 27 (September 2013): 650.

9. Sepehr Mortaheb et al., "Mind Blanking Is a Distinct Mental State Linked to a Recurrent Brain Profile of Globally Positive Connectivity during Ongoing Mentation," *Proceedings of the National Academy of Sciences USA* 119, no. 41 (October 2022): e2200511119.

10. Marcus E. Raichle et al., "A Default Mode of Brain Function," *Proceedings of the National Academy of Sciences USA* 98, no. 2 (January 2001): 676–682.

11. Kieran C. R. Fox et al., "The Wandering Brain: Meta-Analysis of Functional Neuroimaging Studies of Mind-Wandering and Related Spontaneous Thought Processes," *NeuroImage* 111, no.1 (May 2015): 611–621; and Malia F. Mason et al., "Wandering Minds: The Default Network and Stimulus-Independent Thought," *Science* 315, no. 5810 (January 2007): 393–395.

12. Daniel S. Margulies et al., "Situating the Default-Mode Network along a Principal Gradient of Macroscale Cortical Organization," *Proceedings of the National Academy of Sciences USA* 113, no. 44 (October 2016): 12574–12579.

13. Daniel L. Schacter, Donna Rose Addis, and Randy L. Buckner, "Remembering the Past to Imagine the Future: The Prospective Brain," *Nature Reviews Neuroscience* 8 (September 2007): 657–661.

14. Andrea Scalabrini et al., "All Roads Lead to the Default-Mode Network-Global Source of DMN Abnormalities in Major Depressive Disorder," *Neuropsychopharmacology* 45 (August 2020): 2058–2069.

15. Douglas S. Mennin and David M. Fresco, "What, Me Worry and Ruminate about DSM-5 and RDoC? The Importance of Targeting Negative Self-Referential Processing," *Clinical Psychology* 20, no. 3 (September 2013): 258–267.

16. David A. Kessler, *Capture: Unraveling the Mystery of Mental Suffering* (New York: Harper Perennial, 2017).

17. Elisabeth Behr-Sigel, *The Place of the Heart* (Scarsdale, NY: St. Vladimir's Seminary Press, 2013).

18. David T. Bradford, "Brain and Psyche in Early Christian Asceticism," *Psychological Reports* 109, no. 2 (October 2011): 461–520.

19. John Anthony McGuckin, *The Book of Mystical Chapters: Meditations on the Soul's Ascent, from the Desert Fathers and Other Early Christian Contemplatives* (Boston: Shambhala, 2002).

20. At the simplest levels, saliency is determined by things like novelty, brightness, and motion. Quickly moving objects or bright and loud stimuli, for example, are naturally salient and draw attention. At slightly more complex levels, those things which we desire or fear are also highly salient.

21. Evagrius Ponticus, *The Praktikos & Chapters on Prayer* (Kalamazoo, MI: Cistercian Publications, 1981), 25.

22. Patrizio Paoletti and Tal Dotan Ben-Soussan, "Reflections on Inner and Outer Silence and Consciousness without Content According to the Sphere Model of Consciousness," *Frontiers in Psychology* 11 (August 2020): 1807.

23. Martin Laird, *Into the Silent Land: A Guide to the Christian Practice of Contemplation* (New York: Oxford University Press, 2006), 82.

24. Anthony Coniaris, *Confronting and Controlling Thoughts: According to the Fathers of the Philokalia* (Minneapolis: Light & Life Publishing, 2004).

25. Christopher C. H. Cook, *The Philokalia and the Inner Life: On Passions and Prayer* (Eugene, OR: Pickwick Publications, 2012).

26. Jill Bolte Taylor, *My Stroke of Insight: A Brain Scientist's Personal Journey* (London: Hodder & Stoughton, 2009).

27. Taylor, *My Stroke of Insight,* 151.

28. David T. Bradford, "Brain and Psyche in Early Christian Asceticism," *Psychological Reports* 109, no. 2 (October 2011): 461–520.

29. Mary Margaret Funk, *Thoughts Matter* (Collegeville, MN: Liturgical Press, 2013).

30. Ivan M. Kontzevitch, *The Acquisition of the Holy Spirit in Ancient Russia* (Platina, CA: St. Herman of Alaska Brotherhood, 1988), 40.

31. Tito Colliander, *Way of the Ascetics: The Ancient Tradition of Discipline and Inner Growth* (Scarsdale, NY: St. Vladimir's Seminary Press, 1985), 50.

32. Robin Amis, *A Different Christianity: Early Christian Esotericism and Modern Thought* (Albany: State University of New York Press, 1995).

33. Esther Thelen and Linda B. Smith, *A Dynamic Systems Approach to the Development of Cognition and Action* (Cambridge, MA: MIT Press, 1994).

34. Robin L. Carhart-Harris et al., "Canalization and Plasticity in Psychopathology," *Neuropharmacology* 226, no. 15 (March 2023): 109398.

35. Stanislav Grof, *Realms of the Human Unconscious: Observations from LSD Research* (New York: Viking, 1975).

36. Adele M. Hayes et al., "Network Destabilization and Transition in Depression: New Methods for Studying the Dynamics of Therapeutic Change," *Clinical Psychology Reviews* 41 (November 2015): 27–39.

37. Aaron T. Beck, "The Evolution of the Cognitive Model of Depression and Its Neurobiological Correlates," *American Journal of Psychiatry* 165, no. 8 (August 2008): 969–977.

38. Adele M. Hayes et al., "Network Destabilization and Transition in Depression: New Methods for Studying the Dynamics of Therapeutic Change," *Clinical Psychology Reviews* 41 (November 2015): 27–39.

39. Sonja Lyubomirsky and Susan Nolen-Hoeksema, "Self-Perpetuating Properties of Dysphoric Rumination," *Journal of Personality and Social Psychology* 65, no. 2 (August 1993): 339–349.

40. Seth G. Disney et al., "Neural Mechanisms of the Cognitive Model of Depression," *Nature Reviews Neuroscience* 12 (July 2011): 467–477.

41. Robin L. Carhart-Harris and Karl J. Friston, "REBUS and the Anarchic Brain: Toward a Unified Model of the Brain Action of Psychedelics," *Pharmacological Reviews* 71, no. 3 (July 2019): 316–344.

42. See Rupert King, "Light and Shadow in the Forest: A Phenomenological Exploration of Heidegger's Clearing (die Lichtung)," *Existential Analysis* 26, no. 1 (January 2015): 103–118.

43. Vasilis Dakos, Egbert H. van Nes, and Marten Scheffer, "Flickering as an Early Warning Signal," *Theoretical Ecology* 6 (April 2013): 309–317.

44. Martin Laird, *Into the Silent Land: A Guide to the Christian Practice of Contemplation* (New York: Oxford University Press, 2006).

45. Inbar Graiver, *Asceticism of the Mind: Forms of Attention and Self-Transformation in Late Antique Monasticism* (Toronto, ON: Pontifical Institute of Mediaeval Studies, 2018).

46. Saint John Climacus, *The Ladder of Divine Ascent* (Boston: Holy Transfiguration Monastery, 2019).

47. Kallistos Ware, *The Power of the Name: The Jesus Prayer in Orthodox Spirituality* (Oxford: SLG Press, 2019), 19.

48. Jean-Claude Larchet, *The Spiritual Unconscious* (Montreal: Alexander Press, 2020).

49. Steven C. Hayes et al., "Acceptance and Commitment Therapy: Model, Processes, and Outcomes," *Behavior Research and Therapy* 44, no. 1 (January 2006): 1–25.

50. Craig P. Polizzi and Steven J. Lynn, "Regulating Emotionality to Manage Adversity: A Systematic Review of the Relation between Emotion Regulation and Psychological Resilience," *Cognitive Therapy and Research* 45 (January 2021): 577–597.

51. Jean-Claude Larchet, *Therapy of Spiritual Illnesses: An Introduction to the Ascetic Tradition of the Orthodox Church,* vol. 3 (Montreal: Alexander Press, 2017).

52. James K.A. Smith, *You Are What You Love: The Spiritual Power of Habit* (Grand Rapids, MI: Brazos Press, 2016).

53. Kirk J. Schneider, *Awakening to Awe: Personal Stories of Profound Transformation* (Plymouth, UK: Jason Aronson, 2009).

54. David Foster Wallace, *This Is Water: Some Thoughts, Delivered on a Significant Occasion about Living a Compassionate Life* (New York: Little, Brown, 2009), 112.

55. Eric L. Garland et al., "Mindfulness Training Promotes Upward Spirals of Positive Affect and Cognition: Multilevel and Autoregressive Latent Trajectory Modeling Analyses," *Frontiers in Psychology* 6 (February 2015): 15.

7. The Stuff Dreams Are Made Of

1. Jiri Wackermann, "Experience at the Threshold of Wakefulness," *Consciousness and Cognition* 19, no. 4 (December 2010): 1093–1094.

2. Antti Revonsuo, "Consciousness, Dreams and Virtual Realities," *Philosophical Psychology* 8, no. 1 (June 2008): 35–58.

3. See Ritchie E. Brown et al., "Control of Sleep and Wakefulness," *Physiological Reviews* 92, no. 3 (July 2012): 1087–1187.

4. Rodolfo Llinás and Denis Paré, "Of Dreaming and Wakefulness," *Neuroscience* 44, no. 3 (April 1991): 521–535.

5. Isabelle Arnulf, "The 'Scanning Hypothesis' of Rapid Eye Movements during REM Sleep: A Review of the Evidence," *Archives Italiennes de Biologie* 149, no. 4 (December 2011): 367–382.

6. PierCarla Cicogna et al., "A Comparison of Mental Activity during Sleep Onset and Morning Awakening," *Sleep* 21, no. 5 (March 1998): 462–470; and Gerald Vogel, David Foulkes, and Harry Trosman, "Ego Functions and Dreaming during Sleep Onset," *Archives of General Psychiatry* 14, no. 3 (March 1966): 238–248.

7. See Andreas Mavromatis, *Hypnagogia: The Unique State of Consciousness between Wakefulness and Sleep* (London: Thyrsos Press, 2010).

8. Daniel L. Schacter, "The Hypnagogic State: A Critical Review of the Literature," *Psychological Bulletin* 83, no. 3 (May 1976): 452–481.

9. Donald O. Hebb, "Concerning Imagery," *Psychological Review* 75, no. 6 (November 1968): 466–477.

10. Oliver Sacks, *Hallucinations* (Toronto: Alfred Knopf, 2012).

11. Shelley R. Adler, *Sleep Paralysis: Night-Mares, Nocebos, and the Mind-Body Connection* (New Brunswick, NJ: Rutgers University Press, 2011).

12. J. Allan Cheyne, "Sleep Paralysis and the Structure of Waking-Nightmare Hallucinations," *Dreaming* 13, no. 3 (September 2003): 163–179.

13. Ronald K. Siegel, *Fire in the Brain: Clinical Tales of Hallucination* (New York: Plume, 1993).

14. Stephen LaBerge, "Lucid Dreaming: Psychophysiological Studies of Consciousness during REM Sleep," in *Sleep and Cognition,* ed. John F. Kihlstrom and Daniel L. Schacter (Washington, DC: APA Press), 124.

15. Bradley F. Boeve et al., "REM Sleep Behavior Disorder and Degenerative Dementia: An Association Likely Reflecting Lewy Body Disease," *Neurology* 51 (August 1998): 363–370.

16. Antti Revonsuo, *Inner Presence: Consciousness as a Biological Phenomenon* (Cambridge, MA: MIT Press, 2006).

17. Katja Valli et al., "Dreams Are More Negative Than Real Life: Implications for the Function of Dreaming," *Cognition and Emotion* 22, no. 5 (June 2008): 833–861.

18. Inge Strauch and Barbara Meier, *In Search of Dreams: Results of Experimental Dream Research* (Albany: State University of New York Press, 1996).

19. Revonsuo, *Inner Presence*, 187.

20. David Foulkes and Nancy H. Kerr, "Point of View in Nocturnal Dreaming," *Perceptual and Motor Skills* 78 (February 1994): 690.

21. For coverage of this topic, see PierCarla Cicogna et al., "Slow Wave and REM Sleep Mentation," *Sleep Research Online* 3, no. 2 (2000): 67–72; Carlo Cipolli et al., "Beyond the Neuropsychology of Dreaming: Insights into the Neural Basis of Dreaming with New Techniques of Sleep Recording and Analysis," *Sleep Medicine Reviews* 35 (October 2017): 8–20; and Patrick McNamara et al., "REM and NREM Sleep Mentation," *International Review of Neurobiology* 92 (September 2010): 69–86.

22. Amir Muzur et al., "The Prefrontal Cortex in Sleep," *Trends in Cognitive Sciences* 11, no. 1 (November 2002): 475–481.

23. PierCarla Cicogna and Marino Bosinelli, "Consciousness during Dreams," *Consciousness and Cognition* 10, no. 1 (March 2001): 26–41.

24. J. Allan Hobson, *The Dreaming Brain* (New York: Basic Books, 1988).

25. J. Allan Hobson, "REM Sleep and Dreaming: Towards a Theory of Protoconsciousness," *Nature Reviews Neuroscience* 10 (October 2009): 803–813.

26. See G. William Domhoff, *The Neurocognitive Theory of Dreaming: The Where, How, When, What, and Why of Dreams* (Cambridge, MA: MIT Press, 2022).

27. Harry T. Hunt, *The Multiplicity of Dreams: Memory, Imagination, and Consciousness* (New Haven: Yale University Press, 1989).

28. Audrius Beinorius, "Play of the Subconscious: On the Samskaras and Vasanas in Classical Yoga Psychology," *Acta Orientalia Vilnensia* 5 (2004): 168–184.

29. Chogyal Namkhai Norbu, *Dream Yoga and the Practice of Natural Light* (Boston: Shambhala, 2002).

30. G. William Domhoff, *The Emergence of Dreaming: Mind-Wandering, Embodied Simulation, and the Default Network* (New York: Oxford University Press, 2017).

31. Dalai Lama, *Sleeping, Dreaming, and Dying: An Exploration of Consciousness* (Somerville, MA: Wisdom Publications, 1997), 42.

32. G. William Domhoff and Kieran C. R. Fox, "Dreaming and the Default Network: A Review, Synthesis, and Counterintuitive Research Proposal," *Consciousness and Cognition* 33 (May 2015): 342–353.

33. Daniel L. Schacter, Donna Rose Addis, and Randy L. Buckner, "Remembering the Past to Imagine the Future: The Prospective Brain," *Nature Reviews Neuroscience* 8 (September 2007): 657–661.

34. G. William Domhoff, *The Emergence of Dreaming: Mind-Wandering, Embodied Simulation, and the Default Network* (New York: Oxford University Press, 2017), 175.

35. Georg Northoff, Andrea Scalabrini, and Stuart Fogel, "Topographic-Dynamic Reorganisation Model of Dreams (TRoD)—A Spatiotemporal Approach," *Neuroscience and Biobehavioral Reviews* 148 (May 2023): 105117.

36. Erin J. Wamsley and Robert Stickgold, "Dreaming and Offline Memory Processing," *Current Biology* 23, no. 7 (December 2010): R1010–1013.

37. Frederick Snyder et al., "Phenomenology of REMS Dreaming," *Psychophysiology* 4, no. 3 (1968): 375.

38. Audrius Beinorius, "Play of the Subconscious: On the Samskaras and Vasanas in Classical Yoga Psychology," *Acta Orientalia Vilnensia* 5 (2004): 168–184.

39. Rob Nairn, *Living, Dreaming, Dying: Wisdom for Everyday Life from the Tibetan Book of the Dead* (Boston: Shambhala, 2004).

40. B. Alan Wallace, *Dreaming Yourself Awake: Lucid Dreaming and Tibetan Dream Yoga for Insight and Transformation* (Boston: Shambhala, 2012).

41. Andrew Holecek, *Dreams of Light: The Profound Daytime Practice of Lucid Dreaming* (Boulder, CO: Sounds True, 2020).

42. John Suler, "Images of the Self in Zen Meditation," *Journal of Mental Imagery* 14 (January 1990): 197–204.

43. B. Alan Wallace, *Stilling the Mind: Shamatha Teachings from Dudjom Lingpa's Vajra Essence* (Somerville, MA: Wisdom Publications, 2011).

44. Stephen Parker et al., "Defining Yoga-Nidra: Traditional Accounts, Physiological Research, and Future Directions," *International Journal of Yoga Therapy* 23, no. 1 (September 2013): 11–16.

45. Prakash Chandra Kavi, "Conscious Entry into Sleep: Yoga Nidra and Accessing Subtler States of Consciousness," *Progress in Brain Research* 280 (April 2023): 43–60.

46. Richard Miller, *Yoga Nidra: A Meditative Practice for Deep Relaxation and Healing* (Boulder, CO: Sounds True, 2010).

47. Wallace, *Dreaming Yourself Awake*.

48. Dzogchen Ponlop, *Mind beyond Death* (Ithaca, NY: Snow Lion Publications, 2006).

49. Evan Thompson, *Waking, Dreaming, Being: Self and Consciousness in Neuroscience, Meditation, and Philosophy* (New York: Columbia University Press, 2014).

50. Tenzin Wangyal Rinpoche, *Wonders of the Natural Mind: The Essence of Dzogchen in the Native Bon Tradition of Tibet* (Ithaca, NY: Snow Lion Publications, 2000).

51. Dalai Lama, *Dzogchen: Heart Essence of the Great Perfection* (Ithaca, NY: Snow Lion Publications, 2004).

52. Holecek, *Dreams of Light.*

53. Dzogchen Ponlop, *Mind beyond Death.*

54. In laboratory studies of mirror gazing, participants report changes in perceptions of their reflected image including alterations and distortions in facial features, feelings of dissociation while viewing the image, changes in their body and sense of personal identity, and visions of non-human entities. See Giovanni B. Caputo et al., "Mirror- and Eye-Gazing: An Integrative Review of Induced Altered and Anomalous Experiences," *Imagination, Cognition, and Personality* 40, no. 4 (June 2021): 418–457.

55. As cited in Wallace, *Dreaming Yourself Awake,* 106.

56. David T. Saunders et al., "Lucid Dreaming Incidence: A Quality Effects Meta-Analysis of 50 Years of Research," *Consciousness and Cognition* 43 (July 2016): 197–215.

57. Stephen LaBerge and Howard Rheingold, *Exploring the World of Lucid Dreaming* (New York: Ballantine Books, 1990).

58. Karen K. Konkoly et al., "Real-Time Dialogue between Experimenters and Dreamers furing REM Sleep," *Current Biology* 31, no. 7 (April 2021): 1417–1426.

59. Ursula Voss et al., "Lucid Dreaming: A State of Consciousness with Features of Both Waking and Non-Lucid Dreaming," *Sleep* 32, no. 9 (September 2009): 1191–1200.

60. Ursula Voss et al., "Induction of Self Awareness in Dreams through Frontal Low Current Stimulation of Gamma Activity," *Nature Neuroscience* 17 (May 2014): 810–812.

61. Stephen LaBerge et al., "Pre-Sleep Treatment with Galantamine Stimulates Lucid Dreaming: A Double-Blind, Placebo-Controlled, Crossover Study," *PLoS One* 13, no. 8 (August 2018): e0201246.

8. Psychedelics, Sensory Isolation Tanks, and Dreamy Mental States

1. See also Jay T. Shurley, "Profound Experimental Sensory Isolation," *American Journal of Psychiatry* 117 (1960): 539–545.

2. Reprinted as John C. Lilly and Jay T. Shurley, "Experiments in Solitude, in Maximum Achievable Physical Isolation with Water Suspension of Intact, Healthy Persons," in *Psychophysiological Aspects of Space*

Flight, ed. Bernard E. Flaherty (New York: Columbia University Press, 1961), 246.

3. Jose Luis Diaz, "Sacred Plants and Visionary Consciousness," *Phenomenology and the Cognitive Sciences* 9 (April 2010): 159–170.

4. Rainer Kraehenmann, "Dreams and Psychedelics: Neurophenomenological Comparison and Therapeutic Implications," *Current Neuropharmacology* 15, no. 7 (October 2017): 1032–1042.

5. Lester Grinspoon and James B. Bakalar, *Psychedelic Drugs Reconsidered* (New York: Basic Books, 1979), 253.

6. John R. Smythies, "The Mescaline Phenomena," *British Journal for the Philosophy of Science* 3, no. 12 (February 1953): 339–347.

7. Albert Hofmann, *LSD, My Problem Child: Reflections on Sacred Drugs, Mysticism and Science* (San Jose, CA: Multidisciplinary Association for Psychedelic Studies, 2017), 47.

8. Camila Sanz et al., "The Experienced Elicited by Hallucinogens Presents the Highest Similarity to Dreaming within a Large Database of Psychoactive Substance Reports," *Frontiers in Neuroscience* 12 (January 2018): 7.

9. Benny Shanon, *The Antipodes of the Mind: Charting the Phenomenology of the Ayahuasca Experience* (Oxford: Oxford University Press, 2002).

10. Nicholas J. Wade and Josef Brozek, *Purkinje's Vision: The Dawning of Neuroscience* (Mahwah, NJ: Lawrence Erlbaum Associates, 2001).

11. See Bastiaan C. Ter Meulen et al., "From Stroboscope to Dream Machine: A History of Flicker-Induced Hallucinations," *European Neurology* 62, no. 5 (September 2009): 58–64.

12. Alan Richardson and Fiona McAndrew, "The Effects of Photic Stimulation and Private Self-Consciousness on the Complexity of Visual Imagination Imagery," *British Journal of Psychology* 81, no. 3 (August 1990): 390.

13. John R. Smythies, "The Stroboscopic Patterns. II. The Phenomenology of the Bright Phase and After-Images," *British Journal of Psychology* 50, no. 4 (May 1959): 305–324.

14. John R. Smythies, "The Stroboscopic Patterns. I. The Dark Phase," *British Journal of Psychology* 50, no. 2 (May 1959): 106–116.

15. John R. Smythies, "The Stroboscopic Patterns. III. Further Experiments and Discussion," *British Journal of Psychology* 51, no. 3 (August 1960): 247–255.

16. John Geiger, *Chapel of Extreme Experience: A Short History of Stroboscopic Light and the Dream Machine* (Brooklyn, NY: Soft Skull Press, 2003), 34.

17. Carsten Allefeld et al., "Flicker-Light Induced Visual Phenomena: Frequency Dependence and Specificity of Whole Percepts and Percept Features," *Consciousness and Cognition* 20, no. 4 (December 2011): 1344–1362.

18. Igor A. Shevelev et al., "Visual Illusions and Traveling Alpha Waves Produced by Flicker Frequency," *International Journal of Psychophysiology* 39, no. 1 (December 2000): 9–20.

19. Aldous Huxley, *The Doors of Perception/Heaven and Hell* (London: Granada, 1977).

20. See Geiger, *Chapel of Extreme Experience.*

21. Mical Raz, "Alone Again: John Zubek and the Troubled History of Sensory Deprivation Research," *Journal of the History of Behavioral Sciences* 49, no. 4 (August 2013): 379–395.

22. For the best summary of this research see Woodburn Heron, "The Pathology of Boredom," *Scientific American* 196, no. 1 (January 1957): 52–57.

23. Woodburn Heron, "The Pathology of Boredom," *Scientific American* 196, no. 1 (January 1957): 54.

24. Peter Suefeld and Jane S. P. Mocellin, "The 'Sensed Presence' in Unusual Environments," *Environment and Behavior* 19, no. 1 (January 1987): 33–52.

25. John Geiger, *The Third Man Factor: Surviving the Impossible* (New York: Weinstein Books, 2009).

26. John C. Lilly, *The Deep Self: Consciousness Exploration in the Isolation Tank* (Nevada City: Gateway Books and Tapes, 2007).

27. Woodburn Heron, "Cognitive and Physiological Effects of Perceptual Isolation," in *Sensory Deprivation: A Symposium Held at Harvard Medical School,* ed. Philip Solomon et al. (Cambridge, MA: Harvard University Press, 1965), 6–33.

28. Lotfi B. Merabet et al., "Visual Hallucinations during Prolonged Blindfolding in Sighted Subjects," *Journal of Neuro-Ophthalmology* 24, no. 2 (June 2004): 109–113.

29. John C. Lilly, "Mental Effects of Reduction of Ordinary Levels of Physical Stimuli on Intact, Healthy Persons," *Psychiatric Research Reports* 5 (June 1956): 1–9, 7.

30. Jay T. Shurley, "Mental Imagery in Profound Sensory Isolation," in *Hallucinations,* ed. Louis J. West (New York: Grune & Stratton, 1962), 153–157.

31. These are published as Lilly, *The Deep Self,* ch. 13.

32. John C. Lilly, *Programming and Metaprogamming in the Human Biocomputer* (New York: Julian Press, 1972).

33. Reginald A. Ray, *Secret of the Vajra World: The Tantric Buddhism of Tibet* (Boston: Shambhala, 2001).

34. Christopher Hatchell, *Naked Seeing: The Great Perfection, the Wheel of Time, and Visionary Buddhism in Renaissance Tibet* (New York: Oxford University Press, 2014).

35. Lopon Tenzin Namdak, *Bonpo Dzogchen Teachings* (Kathmandu: Vajra Publications, 2007), 198.

36. Robert Olds and Rachel Olds, *Luminous Heart of Inner Radiance: Drawings of the Tögal Visions* (Heart Seed Press, 2010). Another valuable resource for learning more about Tögal visionary phenomena is Martin Lowethal, *Dawning of Clear Light: A Western Approach to Tibetan Dark Retreat Meditation* (Charlottesville, VA: Hampton Roads Publishing, 2003).

37. Ian A. Baker, "Embodying Enlightenment: Physical Culture in Dzogchen Revealed in Tibet's Lukhang Murals," *Asian Medicine* 7, no. 1 (January 2012): 225–264.

38. Christopher Hatchell, *Naked Seeing: The Great Perfection, the Wheel of Time, and Visionary Buddhism in Renaissance Tibet* (New York: Oxford University Press, 2014).

39. Lloyd L. Avant, "Vision in the Ganzfeld," *Psychological Bulletin* 64, no. 4 (October 1965): 246–258.

40. James J. Gibson, *The Ecological Approach to Visual Perception* (New York: Psychology Press, 2015), 143.

41. Keiichiro Tsuji et al., "Detailed Analyses of Ganzfeld Phenomena as Perceptual Events in Stimulus-Reductive Situations," *Swiss Journal of Psychology* 63 (September 2004): 217–223.

42. Simon Niedenthal, "Learning from the Cornell Box," *Leonardo* 35, no. 3 (June 2002): 249–254.

43. Ciaran Benson, "Points of View and the Visual Arts: James Turrell, Antonio Damasio and the 'No Point of View Phenomenon,'" in *Theoretical Issues in Psychology,* ed. John R. Morss, Niamh Stephenson, and Hans Rappard (New York: Springer, 2001), 119–129.

44. Tal Dotan Ben-Soussan et al., "Fully Immersed: State Absorption and Electrophysiological Effects of the OVO Whole-Body Perceptual Deprivation Chamber," *Progress in Brain Research* 244 (February 2019): 165–184.

45. Michele Pellegrino et al., "The Cloud of Unknowing: Cognitive De-differentiation in Whole-Body Perceptual Deprivation," *Progress in Brain Research* 277 (February 2023): 109–140.

46. Vladimir Miskovic et al., "Electrophysiological and Phenomenological Effects of Short-Term Immersion in an Altered Sensory Environment," *Consciousness and Cognition* 70 (April 2019): 39–49.

47. Jiri Wackermann et al., "Ganzfeld-Induced Hallucinatory Experiences, Its Phenomenology and Cerebral Electrophysiology," *Cortex* 44 (June 2008): 1368.

48. Dieter Vaitl et al., "Psychobiology of Altered States of Consciousness," *Psychological Bulletin* 131 (January 2005): 98–127.

49. Jiri Wackermann et al., "Brain Electrical Activity and Subjective Experience during Altered States of Consciousness: Ganzfeld and Hypnagogic States," *International Journal of Psychophysiology* 46, no. 2 (November 2002): 123–146.

50. As quoted in Berand J. Baars, "A Scientific Approach to Silent Consciousness," *Frontiers in Psychology* 4 (October 2013): 678.

51. Longchenpa, *Finding Rest in the Nature of the Mind,* vol. 1 (Boulder, CO: Shambhala, 2017), 294.

52. This In-Sight was formulated jointly with Lama Surya Das.

53. Published as Wilder Penfield, "Some Mechanisms of Consciousness Discovered during Electrical Stimulation of the Brain," *Proceedings of the National Academy of Sciences USA* 44, no. 2 (February 1958): 51–66.

54. Wilder G. Penfield, "The Interpretive Cortex: The Stream of Consciousness in the Human Brain Can Be Electrically Re-Activated," *Science* 129 (June 1959): 1719–1725.

55. Wilder G. Penfield and Phanor Perrot, "The Brain's Record of Auditory and Visual Experience," *Brain* 86 (December 1963): 664.

56. Guillaume Herbet et al., "Disrupting Posterior Cingulate Connectivity Disconnects Consciousness from the External Environment," *Neuropsychologia* 56 (April 2014): 239–244.

57. See Robin Carhart-Harris, "Waves of the Unconscious: The Neurophysiology of Dreamlike Phenomena and Its Implications for the Psychodynamic Model of the Mind," *Neuro-Psychoanalysis* 9, no. 2 (2007): 183–211.

58. David Foulkes and Stephan Fleisher, "Mental Activity in Relaxed Wakefulness," *Journal of Abnormal Psychology* 84, no. 1 (February 1975): 66–75.

59. Harry T. Hunt and Cara M. Chefurka, "A Test of the Psychedelic Model of Altered States of Consciousness," *Archives of General Psychiatry* 33 (July 1976): 867–876.

60. Oliver Sacks, *Hallucinations* (Toronto: Alfred Knopf, 2012).

9. Dreaming Reality

1. Jurij Moskvitin, *An Essay on the Origin of Thought* (Athens: Ohio University Press, 1974).

2. Moskvitin, *An Essay on the Origin of Thought,* 52.

3. Moskvitin, *An Essay on the Origin of Thought,* 26.

4. Moskvitin, *An Essay on the Origin of Thought,* 30.

5. J. Allan Hobson, "REM Sleep and Dreaming: Towards a Theory of Protoconsciousness," *Nature Reviews Neuroscience* 10 (October 2009): 803–813.

6. Vincent A. Billock and Brian H. Tsou, "Seeing Forbidden," *Scientific American* 302, no.2 (February 2010): 72–77.

7. Nicholas J. Wade and Josef Brozek, *Purkinje's Vision: The Dawning of Neuroscience* (Mahwah, NJ: Lawrence Erlbaum Associates, 2001).

8. Heinrich Klüver, *Mescal and Mechanisms of Hallucination* (Chicago: University of Chicago Press, 1966).

9. Ian A. Baker, *Tibetan Yoga: Principles and Practices* (Rochester, VT: Inner Traditions, 2019).

10. H. Ümit Sayin, "Does the Nervous System Have an Intrinsic Archaic Language? Entoptic Images and Phosphenes," *NeuroQuantology* 12, no. 3 (September 2014): 427–445.

11. Vladimir Miskovic et al., "Perceptual Phenomena in Destructured Sensory Fields: Probing the Brain's Intrinsic Functional Architectures," *Neuroscience and Biobehavioral Reviews* 98 (March 2019): 265–286.

12. Vincent A. Billock and Brian H. Tsou, "Elementary Visual Hallucinations and Their Relationship to Neural Pattern-Forming Mechanisms," *Psychological Bulletin* 138, no. 4 (July 2012): 744–774.

13. Roland K. Siegel, "Hallucinations," *Scientific American* 237, no. 4 (October 1977): 132–141.

14. Dominic H. ffytche, "The Hodology of Hallucinations," *Cortex* 44, no. 8 (September 2008): 1067–1083.

15. Lynne Hume, *Portals: Opening Doorways to Other Realities through the Senses* (New York: Routledge, 2020).

16. Siegel, "Hallucinations."

17. Paul C. Bresloff et al., "What Geometric Visual Hallucinations Tell Us about the Visual Cortex," *Neural Computation* 14, no. 3 (March 2002): 473–491.

18. Siegel, "Hallucinations," 134.

19. Robert E. L. Masters and Jean Houston, *The Varieties of Psychedelic Experience: The Classic Guide to the Effects of LSD on the Human Psyche* (Rochester, VT: Inner Traditions, 2000), 158–159.

20. Marcus E. Raichle, "The Brain's Dark Energy," *Science* 314, no. 5803 (November 2006): 1249–1250.

21. Olaf Sporns, *Networks of the Brain* (Cambridge, MA: The MIT Press, 2011).

22. Amos Arieli et al., "Coherent Spatiotemporal Patterns of Ongoing Activity Revealed by Real-Time Optical Imaging Coupled with Single-Unit Recording in the Cat Visual Cortex," *Journal of Neurophysiology* 73, no. 5 (May 1995): 2072–2093.

23. Amos Arieli et al., "Dynamics of Ongoing Activity: Explanation of the Large Variability in Evoked Cortical Responses," *Science* 273 (September 1996): 1868–1871.

24. Tal Kenet et al., "Spontaneously Emerging Cortical Representations of Visual Attributes," *Nature* 425 (October 2003): 954–956.

25. Dario L. Ringach, "Spontaneous and Driven Cortical Activity: Implications for Computation," *Current Opinion in Neurobiology* 19 (August 2009): 440.

26. See Jozsef Fiser et al., "Small Modulation of Ongoing Cortical Dynamics by Sensory Input during Natural Vision," *Nature* 431 (September 2004): 573–578; and Jae-eun Kang Miller et al., "Visual Stimuli Recruit Intrinsically Generated Cortical Ensembles," *Proceedings of the National Academy of Sciences USA* 111, no. 38 (September 2014): E4053–4061.

27. Andreas K. Engel, "Intrinsic Coupling Modes: Multiscale Interactions in Ongoing Brain Activity," *Neuron* 80, no. 4 (November 2013): 867–886.

28. Rodolfo Llinás and Denis Paré, "Of Dreaming and Wakefulness," *Neuroscience* 44, no. 3 (April 1991): 521–535.

29. Rodolfo Llinás and Urs Ribary, "Perception as an Oneiric-Like State Modulated by the Senses," in *Large-Scale Neuronal Theories of the Brain*, ed. Christof Koch and Joel L. Davis (Cambridge, MA: MIT Press, 1994).

30. Andy Clark, "Whatever Next? Predictive Brains, Situated Agents, and the Future of Cognitive Science," *Behavioral and Brain Sciences* 36, no. 3 (June 2013): 181–204.

31. Jakob Hohwy, "The Self-Evidencing Brain," *Nôus* 50, no. 2 (June 2016): 259–285.

32. Francesco Donnarumma et al., "Action Perception as Hypothesis Testing," *Cortex* 89 (April 2017): 45–60.

33. Hermann von Helmholtz, *Science and Culture: Popular and Philosophical Lectures* (Chicago: University of Chicago Press, 1995).

34. Anil Seth, *Being You: A New Science of Consciousness* (New York: Dutton, 2021).

35. Karl J. Friston, "The Free-Energy Principle: A Unified Brain Theory?" *Nature Reviews Neuroscience* 11 (January 2010): 127–138.

36. Karl Friston, "A Theory of Cortical Responses," *Philosophical Transactions of the Royal Society of London. Series B, Biological Sciences* 360, no. 1456 (April 2005): 815–836.

37. Jakob Hohwy et al., "Predictive Coding Explains Binocular Rivalry: An Epistemological Review," *Cognition* 108, no. 3 (September 2008): 687–701.

38. Sepideh Sadaghiani and Andreas Kleinschmidt, "Functional Interactions between Intrinsic Brain Activity and Behavior," *Neuroimage* 80 (October 2013): 379–386.

39. Giovanni Pezzulo, Marco Zorzi, and Maurizio Corbetta, "The Secret Life of Predictive Brains: What's Spontaneous Activity For?" *Trends in Cognitive Sciences* 25 no. 9 (September 2021): 730–743.

40. Vladimir Miskovic et al., "Perceptual Phenomena in Destructured Sensory Fields: Probing the Brain's Intrinsic Functional Architectures," *Neuroscience and Biobehavioral Reviews* 98 (March 2019): 265–286.

41. J. Allan Hobson and Karl Friston, "Consciousness, Dreams, and Inference: The Cartesian Theater Revisited," *Journal of Consciousness Studies* 21 (January 2014): 6–32.

42. Andy Clark, *Surfing Uncertainty: Prediction, Action, and the Embodied Mind* (Oxford: Oxford University Press, 2016).

43. Stephen LaBerge, "Signal-Verified Lucid Dreaming Proves That REM Sleep Can Support Reflective Consciousness," *International Journal of Dream Research* 3, no. 1 (July 2010): 26–27.

44. Harriet Feldman and Karl J. Friston, "Attention, Uncertainty, and Free-Energy," *Frontiers in Human Neuroscience* 4 (December 2010): 215.

45. Louis J. West, "A General Theory of Hallucinations and Dreams," in *Hallucinations*, ed. Louis J. West (New York: Grune & Stratton, 1962), 275–290.

46. Daniel L. Schacter, Donna Rose Addis, and Randy L. Buckner, "Remembering the Past to Imagine the Future: The Prospective Brain," *Nature Reviews Neuroscience* 8 (September 2007): 657–661.

47. Gustavo Deco et al., "Emerging Concepts for the Dynamical Organization of Resting-State Activity in the Brain," *Nature Reviews Neuroscience* 12 (December 2010): 43–56.

48. Robin L. Carhart-Harris and Karl J. Friston, "REBUS and the Anarchic Brain: Toward a Unified Model of the Brain Action of Psychedelics," *Pharmacological Reviews* 71, no. 3 (July 2019): 316–344.

49. Katrin H. Preller et al., "Effective Connectivity Changes in LSD-Induced Altered States of Consciousness in Humans," *Proceedings of the National Academy of Sciences USA* 116, no. 7 (February 2019): 2743–2748.

50. Aldous Huxley, *The Doors of Perception/Heaven and Hell* (London: Granada, 1977).

51. Chris Letheby and Philip Gerrans, "Self Unbound: Ego Dissolution in Psychedelic Experience," *Neuroscience of Consciousness* 2017 (June 2017): nix016.

52. Robin L. Carhart-Harris, "How Do Psychedelics Work?" *Current Opinion in Psychiatry* 32, no.1 (January 2019): 16–21.

53. Raphaël Millière, "Looking for the Self: Phenomenology, Neurophysiology and Philosophical Significance of Drug-Induced Ego Dissolution," *Frontiers in Human Neuroscience* 11 (May 2017): 245.

54. Carhart-Harris and Friston, "REBUS and the Anarchic Brain, 319.

55. Jim Grigsby and David Stevens, "Memory, Neurodynamics, and Human Relationships," *Psychiatry* 65, no. 1 (Spring 2002): 13–34.

56. Draulio B de Araujo et al., "Seeing with the Eyes Shut: Neural Basis of Enhanced Imagery Following Ayahuasca Ingestion," *Human Brain Mapping* 33, no. 11 (November 2012): 2550–2560.

57. Olivia L. Carter et al., "Psilocybin Links Binocular Rivalry Switch Rate to Attention and Subjective Arousal Levels in Humans," *Psychopharmacology* 195 (September 2007): 415–424.

58. ffytche, "The Hodology of Hallucinations."

59. Wilder Penfield and Theodore Rasmussen, *The Cerebral Cortex of Man* (New York: Macmillan, 1950).

60. Gerwin Schalk et al., "Facephenes and Rainbows: Causal Evidence for Functional and Anatomical Specificity of Face and Color Processing in the Human Brain," *Proceedings of the National Academy of Sciences USA* 114, no. 46 (October 2017): 12285–12290.

61. Carsten Allefeld et al., "Flicker-Light Induced Visual Phenomena: Frequency Dependence and Specificity of Whole Percepts and Percept Features," *Consciousness and Cognition* 20, no. 4 (December 2011): 1344–1362.

62. Paul C. Bressloff et al., "Geometric Visual Hallucinations, Euclidean Symmetry and the Functional Architecture of Striate Cortex," *Philosophical Transactions of the Royal Society of London. Series B, Biological Sciences* 356 (March 2001): 299–330.

63. Daniel J. Schwartzmann et al., "Increased Spontaneous EEG Signal Diversity during Stroboscopically-Induced Altered States of Consciousness," *bioRxiv* 511766 (2019).

64. Ian A. Baker, "Embodying Enlightenment: Physical Culture in Dzogchen Revealed in Tibet's Lukhang Murals," *Asian Medicine* 7, no. 1 (January 2012): 225–264.

65. Shardza Tashi Gyaltsen, *Heart Drops of Dharmakaya: Dzogchen Practice of the Bon Tradition* (Ithaca, NY: Snow Lion Publications, 2002).

66. Christopher Hatchell, *Naked Seeing: The Great Perfection, The Wheel of Time, and Visionary Buddhism in Renaissance Tibet* (New York: Oxford University Press, 2014).

67. Lopon Tenzin Namdak, *Bonpo Dzogchen Teachings* (Kathmandu: Vajra Publications, 2007), 198.

68. Tenzin Wangyal Rinpoche, *Wonders of the Natural Mind: The Essence of Dzogchen in the Native Bon Tradition of Tibet* (Ithaca, NY: Snow Lion Publications, 2000).

69. Chogyal Namkhai Norbu, *The Crystal and the Way of Light: Sutra, Tantra, and Dzogchen* (Ithaca, NY: Snow Lion Publications, 1999), 89.

70. Hatchell, *Naked Seeing*, 56.

71. Fabio Giommi et al., "The (In)flexible Self: Psychopathology, Mindfulness, and Neuroscience," *International Journal of Clinical and Health Psychology* 23 (October–December 2023): 100381.

10. The Imaginative Brain and the Origins of Myth

1. See also Erik Goodwyn, *Understanding Dreams and Other Spontaneous Images: The Invisible Storyteller* (New York: Routledge, 2018).

2. See David Lewis-Williams, *The Mind in the Cave: Consciousness and the Origins of Art* (London: Thames & Hudson, 2004).

3. Genevieve von Petzinger, *The First Signs: Unlocking the Mysteries of World's Oldest Symbols* (New York: Atria Paperback, 2017).

4. Christopher S. Henshilwood et al., "An Abstract Drawing from the 73,000-Year-Old Levels at Blombos Cave, South Africa," *Nature* 562 (September 2018): 115–118.

5. Ronald K. Siegel, *Fire in the Brain: Clinical Tales of Hallucination* (New York: Plume, 1993).

6. Jurij Moskvitin, *An Essay on the Origin of Thought* (Athens: Ohio University Press, 1974), 50.

7. Moskvitin, *An Essay on the Origin of Thought,* 133.

8. James David Lewis-Williams et al., "The Signs of All Times: Entoptic Phenomena in Upper Paleolithic Art [and Comments and Reply]," *Current Anthropology* 29, no. 2 (April 1988): 201–245. Lewis-Williams' theory of course elicited both endorsements and critiques—the large number of commentaries on the target article provide an extensive overview.

9. Another striking example of this is the art discovered on the tombs at Newgrange, dating to the Neolithic period, approximately 3,000 years ago. Even a cursory comparison of the swirls imprinted on the walls of the Newgrange tomb passage and the functional architecture of the human visual cortex suggests numerous structural similarities. Like a spider spinning its web, the secretion of cultural artifacts might unintentionally reveal the inner structure of the nervous system that was responsible for producing the said artifacts.

10. Oliver Sacks, *Hallucinations* (New York: Vintage Books, 2013), 131.

11. See Tom Froese et al., "Turing Instabilities in Biology, Culture, and Consciousness? On the Enactive Origins of Symbolic Material Culture," *Adaptive Behavior* 21, no. 3 (June 2013): 139–214.

12. Peter Suedfeld, *Restricted Environmental Stimulation: Research and Clinical Applications* (New York: John Wiley & Sons, 1980).

13. Lynne Hume, *Portals: Opening Doorways to Other Realities through the Senses* (New York: Routledge, 2020).

14. Yulia Ustinova, "Cave Experiences and Ancient Greek Oracles," *Time and Mind* 2, no. 3 (2009): 265–286.

15. Shahar Arzy et al., "Why Revelations Have Occurred on Mountains? Linking Mystical Experiences and Cognitive Neuroscience," *Medical Hypotheses* 65, no. 5 (2005): 841–845.

16. See Yulia Ustinova, *Caves and the Ancient Greek Mind: Descending Underground in Search for Ultimate Truth* (Oxford: Oxford University Press, 2009).

17. See Henry Spiller et al., "The Delphic Oracle: A Multidisciplinary Defense of the Gaseous Vent Theory," *Journal of Toxicology* 40, no. 2 (2002): 189–196; and Henry Spiller et al., "Gaseous Emissions at the Site of Delphic Oracle: Assessing the Ancient Evidence," *Clinical Toxicology* 46, no. 5 (June 2008): 487–488.

18. John R. Hale, "Questioning the Delphic Oracle," *Scientific American* 289, no. 2 (August 2003): 66–73.

19. How we interpret these facts can be quite consequential in how much meaning is ascribed to the cognitive content of visionary phenomena. A modern, reductionist, and purely materialist interpretation might seek to denigrate the oracle's activities as *nothing but* a particular biochemical state of her brain caused by the action of ethylene gas. The ancient Greek understanding was much more nuanced by comparison. Plutarch did not shy away from suggesting the role played by the presence of the fragrant vapor or pneuma filling the innermost temple. The ancient Greeks believed that their gods used natural elements to communicate with, and through, humans, but this did nothing to denigrate the value of the messages. The most important part, Plutarch believed, was not the immediate trigger of the Pythia's prophesying but rather, the rigorous ascetic practices she pursued beforehand so as to be an appropriately responsive instrument for responding to the *pneuma*'s prompts.

20. Yulia Ustinova, "Consciousness Alteration Practices in the West from Prehistory to Late Antiquity," in *Altering Consciousness: Multidisciplinary Perspectives*, vol. 1, ed. Etzel Cardeña and Michael J. Winkelman (Santa Barbara, CA: Praeger, 2011), 45–72.

21. Raymond de Becker, *The Understanding of Dreams and Their Influence on the History of Man* (New York: Bell, 1968).

22. Peter Kingsley, *In the Dark Places of Wisdom* (Point Reyes Station, CA: Golden Sufi Center, 2010).

23. Gil Renberg, *Where Dreams May Come*, 2 vols. (Leiden, NL: Brill, 2017).

24. Kingsley, *In the Dark Places of Wisdom*, 180.

25. Mary Watkins, *Waking Dreams* (New York: Gordon and Breach, 1976).

26. Erik G. Goodwyn, *The Neurobiology of the Gods: How Brain Physiology Shapes the Recurrent Imagery of Myth and Dreams* (New York: Routledge, 2012).

27. Louise Goupil and Tristan A. Bekinschtein, "Cognitive Processing during the Transition to Sleep," *Archives Italiennes de Biologie* 160 (June 2012): 140–154.

28. Géza Róheim, *The Gates of the Dream* (Madison, CT: International Universities Press, 1953).

29. We might add here an additional point: the fact that many scientific anthropologists appraised indigenous animistic worldviews as uneducated, subjective, and primitive speaks more to the modern person's own state of being chronically cut off from their own depths than it does to the supposedly low level of intellectual sophistication of traditional (pre-scientific) cultures.

30. Aurelia Spadafora and Harry T. Hunt, "The Multiplicity of Dreams: Cognitive-Affective Correlates of Lucid, Archetypal, and Nightmare Dreaming," *Perceptual and Motor Skills* 71 (October 1990): 627–644.

31. Lynne Hume, "The Dreaming in Contemporary Aboriginal Australia," in *Indigenous Religions: A Companion*, ed. Graham Harvey (New York: Cassell, 2000), 125–138.

32. Gerardo Reichel-Dolmatoff, *Amazonian Cosmos: The Sexual and Religious Symbolism of the Tukano Indians* (Chicago: University of Chicago Press, 1971).

33. Jonathan Weinel, *Inner Sound: Altered States of Consciousness in Electronic Music and Audio-Visual Media* (New York: Oxford University Press, 2018).

34. Robert E. L. Masters and Jean Houston, *The Varieties of Psychedelic Experience: The Classic Guide to the Effects of LSD on the Human Psyche* (Rochester, VT: Inner Traditions, 2000).

35. Harry T. Hunt, *The Multiplicity of Dreams: Memory, Imagination, and Consciousness* (New Haven: Yale University Press, 1989).

36. Rudolf Otto, *The Idea of the Holy* (Eugene, OR: Wipf & Stock, 2021).

37. Christopher MacKenna, "From the Numinous to the Sacred," *Journal of Analytical Psychology* 54, no. 2 (April 2009): 167–182.

38. Lionet Corbett, *The Religious Function of the Psyche* (New York: Routledge, 2002).

39. The numinous, however, is always ambivalent. In addition to its potential to aid psychological integrative tendencies and to exert a healing function, it can also, when it is actively defended against, precipitate overwhelming feelings of terror and dissociation. See J. Allan Cheyne, "The Ominous Numinous: Sensed Presence and 'Other' Hallucinations," *Journal of Consciousness Studies* 8 (May 2001): 133–150.

40. The famous twentieth-century poet W. B. Yeats had a long-standing interest in visionary phenomena, waking dreams, and other non-ordinary

forms of consciousness. Yeats himself had visions of scintillating geometric forms which he strongly believed to express cosmic truths and secrets. He and his wife, Georgie Hyde-Lees, wrote a mysterious book further expounding on these themes and proposing a syncretic framework of different philosophies and esoteric teachings. In it, special significance was accorded to the spiraling form of a gyre, which seemed to them to underlie certain universal processes. W. B. Yeats, *A Vision: The Original 1925 Version,* eds. Catherine E. Paul and Margaret Mills Harper, *Collected Works of W. B. Yeats, Volume XIII.* (New York: Scribner, 2008).

41. Hunt, *The Multiplicity of Dreams,* 129.

42. Arthur J. Deikman, "Bimodal Consciousness," *Archives of General Psychiatry* 25, no. 6 (December 1971): 481–489.

43. Saul Bellow, "Nobel Prize Lecture," December 1976, https://www.nobelprize.org/prizes/literature/1976/bellow/lecture/.

44. Ronald Siegel, *Intoxication: The Universal Drive for Mind-Altering Substances* (Rochester, VT: Inner Traditions, 2005).

45. Tom Froese et al., "Turing Instabilities in Biology, Culture, and Consciousness? On the Enactive Origins of Symbolic Material Culture," *Adaptive Behavior* 21, no. 3 (June 2013): 139–214.

46. Michael Pollan, *How to Change Your Mind: What the New Science of Psychedelics Teaches Us About Consciousness, Dying, Addiction, Depression, and Transcendence* (New York: Penguin Press, 2019).

47. Oshin Vartanian et al., "The Creative Brain under Stress: Considerations for Performance in Extreme Environments," *Frontiers in Psychology* 11 (October 2020): 585969.

48. Kim P.C. Kuypers, "Out of the Box: A Psychedelic Model to Study the Creative Mind," *Medical Hypotheses* 115 (June 2018): 13–16.

49. Peter Suedfeld et al., "Enhancement of Scientific Creativity by Flotation Rest (Restricted Environmental Stimulation Technique)," *Journal of Environmental Psychology* 7 (September 1987): 219–231.

50. Robin L. Carhart-Harris et al., "The Entropic Brain: A Theory of Conscious States Informed by Neuroimaging Research with Psychedelic Drugs," *Frontiers in Human Neuroscience* 8 (February 2014): 20; and Robin L. Carhart-Harris, "The Entropic Brain—Revisited," *Neuropharmacology* 142 (November 2018): 167–178.

51. Selen Atasoy et al., "Connectome-Harmonic Decomposition of Human Brain Activity Reveals Dynamical Repertoire Re-Organization Under LSD," *Scientific Reports* 7 (December 2017): 17661.

52. Daniel J. Schwartzmann et al., "Increased Spontaneous EEG Signal Diversity during Stroboscopically-Induced Altered States of Consciousness," *bioRxiv* 511766 (2019).

53. Stanislav Grof and Joan Halifax, *The Human Encounter with Death* (New York: E. P. Dutton, 1978).

54. J. Allan Hobson, "REM Sleep and Dreaming: Towards a Theory of Protoconsciousness," *Nature Reviews Neuroscience* 10, no. 11 (October 2009): 803–813, 806.

55. Mary Watkins, *Invisible Guests: The Development of Imaginal Dialogues* (Fairfax, CA: Human Development Books, 2015).

11. Who Is Watching the Show?

1. Bernadette Roberts, *The Experience of No-Self: A Contemplative Journey,* rev. ed. (Albany: State University of New York Press, 1993), 22.

2. Roberts, *The Experience of No-Self,* 23.

3. Thomas Metzinger, *The Ego Tunnel: The Science of the Mind and the Myth of the Self* (New York: Basic Books, 2009).

4. Michael Kirchoff et al., "The Markov Blankets of Life: Autonomy, Active Inference and the Free Energy Principle," *Journal of the Royal Society, Interface* 15, no. 138 (January 2008): 20170792.

5. Patrizio Paoletti et al., "Tackling the Electro-Topography of the Selves through the Sphere Model of Consciousness," *Frontiers in Psychology* 13 (May 2022): 836290.

6. Raphaël Millière, "Psychedelics, Meditation, and Self-Consciousness," *Frontiers in Psychology* 9 (September 2018): 1475.

7. Patrizio Paoletti and Tal Dotan Ben-Soussan, "Reflections on Inner and Outer Silence and Consciousness without Contents According to the Sphere Model of Consciousness," *Frontiers in Psychology* 11 (August 2020): 1807.

8. Georg Northoff and Jaak Panksepp, "The Trans-Species Concept of Self and the Subcortical-Cortical Midline System," *Trends in Cognitive Sciences* 12, no. 7 (July 2008): 259–264.

9. Andrea Scalabrini et al., "The Self and Its Internal Thought: In Search of a Psychological Baseline," *Consciousness and Cognition* 97 (January 2022): 103244.

10. Paoletti et al., "Tackling the Electro-Topography of the Selves."

11. Jill Bolte Taylor, *My Stroke of Insight: A Brain Scientist's Personal Journey* (London: Hodder & Stoughton, 2009), 41.

12. Taylor, *My Stroke of Insight,* 42.

13. Sarah-J. Blakemore et al., "Central Cancellation of Self-Produced Tickle Sensation," *Nature Neuroscience* 1 (November 1998): 635–640.

14. Kalina Christoff et al., "Specifying the Self for Cognitive Neuroscience," *Trends in Cognitive Sciences* 15, no. 3 (March 2011): 104–112.

15. Robin L. Carhart-Harris and Karl J. Friston, "The Default-Mode, Ego-Functions and Free-Energy: A Neurobiological Account of Freudian Ideas," *Brain* 133 (April 2010): 1265–1283.

16. Mark Solms, *The Hidden Spring: A Journey to the Source of Consciousness* (New York: W. W. Norton, 2022).

17. Marcello Constantini and Patrick Haggard, "The Rubber Hand Illusion: Sensitivity and Reference Frame for Body Ownership," *Consciousness and Cognition* 16, no. 2 (June 2007): 229–240.

18. Bigna Lenggenhager et al., "Video Ergo Sum: Manipulating Bodily Self-Consciousness," *Science* 317, no. 5841 (August 2007): 1096–1099.

19. Metzinger, *The Ego Tunnel*, 77.

20. Margaret S. Mahler, Fred Pine, and Anni Bergman, *The Psychological Birth of the Human Infant: Symbiosis and Individuation* (New York: Routledge, 2018).

21. Anil Seth, "Interoceptive Inference, Emotion, and the Embodied Self," *Trends in Cognitive Sciences* 17, no. 11 (November 2013): 565–573.

22. A. D. (Bud) Craig, *How Do You Feel? An Interoceptive Moment with Your Neurobiological Self* (Princeton: Princeton University Press, 2015).

23. Karlfried Graf Dürckheim, *Daily Life as Spiritual Exercise: The Way of Transformation* (London: Allen & Unwin, 1971).

24. Walter J. Freeman, *How Brains Make Up Their Minds* (New York: Columbia University Press, 2000).

25. Jaak Panksepp and Lucy Biven, *The Archeology of Mind: Neuroevolutionary Origins of Human Emotions* (New York: W. W. Norton, 2012).

26. Rodolfo Llinás, *I of the Vortex: From Neurons to Self* (Cambridge, MA: MIT Press, 2001).

27. Jaak Panksepp, "The Periconscious Substrates of Consciousness: Affective States and the Evolutionary Origins of the Self," *Journal of Consciousness Studies* 5 (May 1998): 566–582.

28. Marie Vandekerckhove and Jaak Panksepp, "A Neurocognitive Theory of Higher Mental Emergence: From Anoetic Affective Experiences to Noetic Knowledge and Autonoetic Awareness," *Neuroscience and Biobehavioral Reviews* 35, no. 9 (October 2011): 2017–2025.

29. Raphaël Millière, "The Varieties of Selflessness," *Philosophy and the Mind Sciences* 1 (March 2020): 1–41.

30. Michael Pollan, *How to Change Your Mind: What the New Science of Psychedelics Teaches Us About Consciousness, Dying, Addiction, Depression, and Transcendence* (New York: Penguin Press, 2019), 277.

31. Raphaël Millière, "Looking for the Self: Phenomenology, Neurophysiology and Philosophical Significance of Drug-Induced Ego Dissolution," *Frontiers in Human Neuroscience* 11 (May 2017): 245.

32. Jasmine T. Ho, Katrin H. Preller, and Bigna Lenggenhager, "Neuro-pharmacological Modulation of the Aberrant Bodily Self through Psyche-delics," *Neuroscience and Biobehavioral Reviews* 108 (January 2020): 526–541.

33. James J. Gattuso et al., "Default Mode Network Modulation by Psy-chedelics: A Systematic Review," *Neuropharmacology* 26, no. 3 (October 2022): 155–188.

34. Jennifer M. Windt, "The Immersive Spatiotemporal Hallucination Model of Dreaming," *Phenomenology and the Cognitive Sciences* 9, no. 2 (May 2010): 295–316, 301.

35. Daphne Simeon and Jeffrey Abugel, *Feeling Unreal: Depersonaliza-tion and the Loss of the Self* (New York: Oxford University Press, 2013), 4, 7.

36. Michael Levin, "The Computational Boundary of a 'Self': Develop-mental Bioelectricity Drives Multicellularity and Scale-Free Cognition," *Frontiers in Psychology* 10 (December 2019): 2688.

37. Yochai Ataria et al., "How Does it Feel to Lack a Sense of Bound-aries? A Case Study of a Long-Term Mindfulness Meditator," *Consciousness and Cognition* 37 (December 2015): 133–147.

38. In one study, for example, a minority of individuals report persistent adverse effects associated with an 8-week long course of mindfulness-based cognitive therapy. The negative effects that persisted were associated with feel-ings of dissociation and dysregulated arousal. See Willoughby B. Britton et al., "Defining and Measuring Meditation-Related Adverse Effects in Mindfulness-Based Programs," *Clinical Psychological Science* 9, no. 6 (May 2021): 1185–1204.

39. David B. Yaden et al., "The Varieties of Self-Transcendent Experience," *Review of General Psychology* 21, no. 2 (June 2017):143–160.

40. Daniel J. Siegel, *The Mindful Brain: Reflection and Attunement in the Cultivation of Well-Being* (New York, NY: W.W. Norton & Company, Inc, 2007).

41. Ann Taves, "Mystical and Other Alterations in Sense of Self: An Ex-panded Framework for Studying Nonordinary Experiences," *Perspectives on Psychological Science* 15, no. 3 (May 2020): 669–690.

42. Matthias Forstmann et al., "Transformative Experience and Social Connectedness Mediate the Mood-Enhancing Effects of Psychedelic Use in Naturalistic Settings," *Proceedings of the National Academy of Sciences USA* 117, no. 5 (January 2020): 2338–2346.

43. Aki Nikolaidis et al., "Subtypes of the Psychedelic Experience Have Reproducible and Predictable Effects on Depression and Anxiety Symptoms," *Journal of Affective Disorders* 324, no. 1 (March 2023): 239–249.

44. Brian L. Lancaster, "The Mythology of Anatta: Bridging the East-West Divide," in *The Authority of Experience: Readings on Buddhism and Psychology,* ed. John Pickering (New York: Routledge, 1997), 170–204.

45. Ayya Khema, *Who Is My Self? A Guide to Buddhist Meditation* (Somerville, MA: Wisdom Publications, 1997).

46. Joaquin Perez-Remon, *Self and Non-Self in Early Buddhism* (The Hague: Mouton, 1980), 304.

47. Christos Yannaras, *Person and Eros* (Brookline, MA: Holy Cross Orthodox Press, 2007).

48. Thomas Keating, *The Human Condition: Contemplation and Transformation* (Mahwah, NJ: Paulist Press, 1999); and Thomas Merton, *New Seeds of Contemplation* (New York: New Directions Books, 2007).

49. Jean-Claude Larchet, *Therapy of Spiritual Illnesses: An Introduction to the Ascetic Tradition of the Orthodox Church,* vol. 1 (Montreal: Alexander Press, 2017).

50. Gerald E. H. Palmer, Philip Sherrard, and Kallistos Ware, *The Philokalia: The Complete Text,* vol. 1 (London: Faber and Faber, 1983), 170.

51. See Larchet, *Therapy of Spiritual Illnesses,* ch. 3.

52. One might compare the poetic quotations provided here from the Holy Fathers of the Christian East to Longchenpa's poetic metaphors of ordinary life as a waking dream in Chapter 4, or indeed, to the following famous lines from the Buddha's *Diamond Sutra:* "Thus shall you think of all this world: a star at dawn, a bubble in a stream, a flash of lightning in a summer cloud, a flickering lamp, a phantom, and a dream."

53. David T. Bradford, *The Spiritual Tradition in Eastern Christianity: Ascetic Psychology, Mystical Experience and Physical Practices* (Leuven, Belgium: Peeters, 2016).

54. Rowan Williams, *Looking East in Winter: Contemporary Thought and the Eastern Christian Tradition* (Dublin: Bloomsbury, 2021), 67.

55. Williams, *Looking East in Winter,* 99.

56. Fabio Giommi et al., "The (In)flexible Self: Psychopathology, Mindfulness, and Neuroscience," *International Journal of Clinical and Health Psychology* 23 (October–December 2023): 100381.

57. Saga Briggs, *How to Change Your Body: The Science of Interoception and Healing through Connection to Yourself & Others* (Santa Fe, NM: Synergetic Press, 2023).

58. Stanley Keleman, *Living Your Dying* (New York: Random House, 1975).

59. Yochai Ataria, "Traumatic Memories as Black Holes: A Qualitative-Phenomenological Approach," *Qualitative Psychology* 1, no. 2 (August 2014): 123–140.

60. Georg Northoff, "Neuroscience and Whitehead I: Neuro-Ecological Model of Brain," *Axiomathes* 26 (April 2016): 219–252.

61. Jean Gebser, *The Ever-Present Origin* (Athens: Ohio University Press, 1986).

12. The Emptiness of Being

1. Fariba Bogzaran, "Lucid Art and Hyperspace Lucidity," *Dreaming* 13, no. 1 (March 2003): 29–42, 33.

2. Linda L. Magallon, "Awake in the Dark: Imageless Lucid Dreaming," *Lucidity Letter* 6, no. 1 (June 1987): 86–90.

3. Jamie Sleigh, Catherine Warnaby, and Irene Tracey, "General Anaesthesia as Fragmentation of Selfhood: Insights from Electroencephalography and Neuroimaging," *British Journal of Anaesthesia* 121, no. 1 (July 2018): 235–236.

4. Wilson Van Dusen, *The Natural Depth in Man* (New York: Harper & Row, 1972), 74.

5. Thomas Metzinger, "Minimal Phenomenal Experience: Meditation, Tonic Alertness, and the Phenomenology of 'Pure' Consciousness," *Philosophy and the Mind Sciences* 1, no. 1 (March 2020): 1–44.

6. Robert K. C. Forman, "What Does Mysticism Have to Teach Us about Consciousness?" *Journal of Consciousness Studies* 5, no. 2 (January 1998): 185–201.

7. Talis Bachmann and Anthony G. Hudetz, "It Is Time to Combine the Two Main Traditions in the Research on the Neural Correlates of Consciousness: $C = L \times D$," *Frontiers in Psychology* 5 (August 2014): 940.

8. Robert K. C. Forman, *The Innate Capacity: Mysticism, Psychology, and Philosophy* (New York: Oxford University Press, 1998).

9. Chogyal Namkhai Norbu, *Dream Yoga and the Practice of Natural Light* (Ithaca, NY: Snow Lion Publications, 1992).

10. Ramakrishna Rao, "Applied Yoga Psychology Studies of Neurophysiology of Meditation," *Journal of Consciousness Studies* 18 (January 2011): 161–198.

11. Martin Laird, *An Ocean of Light: Contemplation, Transformation, and Liberation* (New York: Oxford University Press, 2019).

12. Walter T. Stace, *Mysticism and Philosophy* (New York: J. B. Lippincott, 1960).

13. Toby J. Woods et al., "The Path to Contentless Experience in Meditation: An Evidence Synthesis Based on Expert Texts," *Phenomenology and the Cognitive Sciences* (June 2022): 1–38.

14. Bertram D. Lewin, "Inferences from the Dream Screen," *International Journal of Psycho-Analysis* 29 (January 1948): 224–231, 224.

15. David B. Cohen, "Failure to Recall Dream Content: Contentless vs Dreamless Reports," *Perceptual and Motor Skills* 34 no. 3 (June 1972): 1000–1002.

16. Francesca Siclari et al., "Assessing Sleep Consciousness within Subjects using a Serial Awakening Paradigm," *Frontiers in Psychology* 4 (August 2013): 542.

17. Valdas Noreika et al., "Early-Night Serial Awakenings as a New Paradigm for Studies on NREM Dreaming," *International Journal of Psychophysiology* 74, no. 1 (October 2009): 14–18.

18. Francesca Siclari et al., "The Neural Correlates of Dreaming," *Nature Neuroscience* 20 (April 2017): 872–878.

19. Peter Fazekas, Georgina Nemeth, and Morten Overgaard, "White Dreams Are Made of Colours: What Studying Contentless Dreams Can Teach Us about the Neural Basis of Dreaming and Conscious Experiences," *Sleep Medicine Reviews* 43 (February 2019): 84–91.

20. Annalotta Scheinin et al., "Foundations of Human Consciousness: Imaging the Twilight Zone," *Journal of Neuroscience* 41, no. 8 (February 2021): 1769–1778.

21. Sleigh, Warnaby, and Tracey, "General Anaesthesia as Fragmentation of Selfhood."

22. Jennifer M. Windt, "How Deep Is the Rift between Conscious States in Sleep and Wakefulness? Spontaneous Experience over the Sleep-Wake Cycle," *Philosophical Transactions of the Royal Society of London. Series B, Biological Sciences* 376, no. 1817 (February 2021): 20190696.

23. Allison G. Harvey and Nicole Tang, "(Mis)Perception of Sleep in Insomnia: A Puzzle and a Resolution," *Psychological Bulletin* 138, no. 1 (January 2012): 77–101.

24. To these examples of consciousness-without-content, we should add the recent discovery of "mind blanking." As was noted in Chapter 7, these episodes occur during waking hours, when brain activity transiently resembles the neural organization evident during deep sleep.

25. Jennifer M. Windt, "Consciousness in Sleep: How Findings From Sleep and Dream Research Challenge Our Understanding of Sleep, Waking, and Consciousness," *Philosophy Compass* 15, no.4 (March 2020): e12661.

26. Stephen Laberge and Donald J DeGracia, "Varieties of Lucid Dreaming Experience," in *Individual Differences in Conscious Experience,* eds. Robert G. Kunzendorf and Benjamin Wallace (Amsterdam: John Benjamins Publishing, 2000), 269–308.

27. See https://dreamstudies.org/2010/05/13/exploring-the-void-in-lucid -dreaming/.

28. Robert Waggoner, *Lucid Dreaming: Gateway to the Inner Self* (Needham, MA: Moment Point Press, Inc., 2009). See also George Gillespie, "Dreams and Dreamless Sleep," *Dreaming* 12 (December 2002): 199–207.

29. Fariba Bogzaran, "Lucid Art and Hyperspace Lucidity," *Dreaming* 13 (March 2003): 29–42.

30. Cyril Costines, Tilmann Lhündrup Borghardt, and Marc Wittmann, "The Phenomenology of 'Pure' Conasciousness As Reported by an Experienced Meditator of the Tibetan Buddhist Karma Kagyu Tradition," *Philosophies* 6, no.2 (June 2021): 50.

31. Toby Woods et al., "Subjective Experiences of Committed Meditators Across Practices Aiming for Contentless States," *Mindfulness* 14 (June 2023): 1457–1478.

32. Jennifer M. Windt, Tore Nielsen, and Evan Thompson, "Does Consciousness Disappear in Dreamless Sleep?" *Trends in Cognitive Sciences* 20, no. 12 (December 2016): 871–882.

33. Karen K. Konkoly et al., "Real-Time Dialogue between Experimenters and Dreamers during REM Sleep," *Current Biology* 31, no. 7 (April 2021): 1417–1426.

34. Adrian Owen, *Into the Gray Zone: A Neuroscientist Explores the Border between Life and Death* (New York: Scribner, 2017).

35. Tim Bayne et al., "Are There Levels of Consciousness?" *Trends in Cognitive Sciences* 20, no. 6 (June 2016): 405–413.

36. Windt, Nielsen, and Thompson, "Does Consciousness Disappear in Dreamless Sleep?"

37. Stanislas Dehaene and Jean-Pierre Changeux, "Experimental and Theoretical Approaches to Conscious Processing," *Neuron* 70, no. 2 (April 2011): 200–227.

38. William James, "Does 'Consciousness' Exist?" *Journal of Philosophy, Psychology, and Scientific Methods* 1, no. 18 (September 1904): 477–491, 485.

39. Metzinger, "Minimal Phenomenal Experience."

40. See John C. Lilly, *The Deep Self: Consciousness Exploration in the Isolation Tank* (Nevada City: Gateway Books and Tapes, 2007), ch. 13.

41. Torsten Norlander et al., "The Experience of Flotation-Rest as a Function of Setting and Previous Experience of Altered States of Consciousness," *Imagination, Cognition, and Personality* 20, no. 2 (October 2000): 161–178.

42. John C. Lilly, *Center of the Cyclone: Looking into Inner Space* (Oakland: Ronin Publishing, 1972).

43. Alexander Shulgin and Ann Shulgin, *Tihkal: The Continuation* (Berkeley: Transform Press, 1997).

44. See Anna O. Ermakova et al., "A Narrative Synthesis of Research with 5-MeO-DMT," *Journal of Psychopharmacology* 36, no. 3 (March 2022): 273–294; and Joseph Barsuglia et al., "Intensity of Mystical Experiences Occasioned by 5-MeO-DMT and Comparison with a Prior Psilocybin Study," *Frontiers in Psychology* 9 (December 2018): 2459.

45. As quoted in William James, *Varieties of Religious Experience* (New York: Penguin, 1987), 347.

46. Robert K. C. Forman, *Mysticism, Mind, Consciousness* (Albany: State University of New York Press, 1999).

47. Enzo Tagliazucchi, "Language as a Window into Altered States of Consciousness Elicited by Psychedelic Drugs," *Frontiers in Pharmacology* 13 (March 2022): 812227.

48. Zoran Josipovic and Vladimir Miskovic, "Non-Dual Awareness and Minimal Phenomenal Experience," *Frontiers in Psychology* 11 (August 2020): 2087.

49. Evan Thompson, *Waking, Dreaming, Being: Self and Consciousness in Neuroscience, Meditation, and Philosophy* (New York: Columbia University Press, 2014).

50. Arvind Sharma, *Sleep as a State of Consciousness in Advaita Vedanta* (Albany: State University of New York Press, 2004).

51. Ruben E. Laukkonen, "Cessations of Consciousness in Meditation: Advancing a Scientific Understanding of Nirodha Samapathi," *Progress in Brain Research* 280 (April 2023): 61–87.

52. See Christoper Chapple, "The Unseen Seer and the Field: Consciousness in Samkhaya and Yoga," in *The Problem of Pure Consciousness,* ed. Robert C. Forman (New York: Oxford University Press, 1997), 53–70; and Ramakrishna Rao, "Applied Yoga Psychology Studies of Neurophysiology of Meditation," *Journal of Consciousness Studies* 18 (January 2011): 161–198.

53. Bina Gupta, *The Disinterested Witness: A Fragment of Advaita Vedanta Phenomenology* (Evanston, IL: Northwestern University Press, 1998).

54. Robert Shankman, *The Experience of Samadhi: An In-Depth Exploration of Buddhist Meditation* (Boston: Shambhala, 2008).

55. Jeremy Yamashiro, "Brain Basis of Samadhi: The Neuroscience of Meditative Absorption," *New School Psychology Bulletin* 13, no. 1 (December 2015): 1–10.

56. B. Alan Wallace, *Contemplative Science: Where Buddhism and Neuroscience Converge* (New York: Columbia University Press, 2007).

57. Chogyal Namkhai Norbu, *Dream Yoga and the Practice of Natural Light* (Boston: Shambhala, 2002).

58. Tenzin Wangyal Rinpoche, *The Tibetan Yogas of Dream and Sleep: Practices for Awakening* (Boulder, CO: Shambhala, 2022), 163.

59. Walter T. Stace, *Mysticism and Philosophy* (New York: J. B. Lippincott, 1960).

60. Saint John of the Cross, *The Dark Night of the Soul* (New York: Frederick Ungar, 1957), 51–54.

61. For an introduction see Jon Gregerson, *The Transfigured Cosmos: Four Essays in Eastern Orthodox Christianity* (Brooklyn: Angelico Press, 2020).

62. Saint Nicodemus the Hagiorite, *Nicodemos of the Holy Mountain: A Handbook of Spiritual Counsels* (New York: Paulist Press, 1989).

63. Thomas Johnson-Medland, *Turning Within: An Introductory Manual of Meditation for Christians Living in the World* (Devon, UK: Praxis Institute Press, 1998).

64. David T. Bradford, "Comparable Process Psychologies in Eastern Christianity and Early Buddhism," *Chromatikon* 7 (2011): 87–102.

65. George Maloney, *Prayer of the Heart: The Contemplative Tradition of the Christian East* (Notre Dame, IN: Ave Maria Press, 2008); and Stephen Muse, *Treasure in Earthen Vessels: Prayer & The Embodied Life* (Waymart, PA: St. Tikhon's Monastery Press, 2017).

66. Evgeniia Kadloubovsky and Gerald E. H. Palmer, *Writings from the Philokalia on Prayer of the Heart* (London: Faber and Faber, 1992), 159.

67. Thomas Matus, *Yoga and the Jesus Prayer Tradition: An Experiment in Faith* (New York: Paulist Press, 1984).

68. Andrew Louth, "Light, Vision, and Religious Experience in Byzantium," in *The Presence of Light: Divine Radiance and Religious Experience*, ed. Matthew T. Kapstein (Chicago: University of Chicago Press, 2004), 85–104.

69. As quoted in Andrew D. Mayes, *Diving for Pearls: Exploring the Depths of Prayer with Isaac the Syrian* (Collegeville, MN: Cistercian Publications / Liturgical Press, 2021), 77.

70. Georg Nothoff, *The Spontaneous Brain: From the Mind-Body to the World-Brain Problem* (Cambridge, MA: MIT Press, 2018).

71. Metzinger, "Minimal Phenomenal Experience."

72. To see a simulation video of this system in action, see Ivan Dimkovic, "Artificial Brain Simulation: Ascending Reticular Activating System and Thalamocortical Networks," DigiCortex, https://www.youtube.com/watch?v=acdL-8wZhLc.

73. Ravinder Jerath et al., "A Unified 3D Default Space Consciousness Model Combining Neurological and Physiological Processes That Underlie Conscious Experience," *Frontiers in Psychology* 6 (August 2015): 1204.

74. Antti Revonsuo, *Inner Presence: Consciousness as a Biological Phenomenon* (Cambridge, MA: MIT Press, 2006), 171.

75. Ravinder Jerath et al., "Conceptual Advances in the Default Space Model of Consciousness," *World Journal of Neuroscience* 8, no. 2 (May 2018): 254–269.

76. Ulf Winter et al., "Content-Free Awareness: EEG-fcMRI Correlates of Consciousness as Such in an Expert Meditator," *Frontiers in Psychology* 10 (February 2020): 3064.

77. Avijit Chowdhury et al., "Investigation of Advanced Mindfulness Meditation 'Cessation' Experiences using EEG Spectral Analysis in an Intensively Sampled Case Study," *Neuropsychologia* 190 (November 2023): 108694.

78. Ravinder Jerath et al., "Widespread Depolarization during Expiration: A Source of Respiratory Drive?" *Medical Hypotheses* 84, no. 1 (January 2015): 31–37.

13. Nonduality and Reality Beyond the Dream

1. Christof Koch et al., "Neural Correlates of Consciousness: Progress and Problems," *Nature Reviews Neuroscience* 17 (May 2016): 307–321, 308.

2. See also Tal D Ben-Soussan and Patrizio Paoletti, "Life in Light of the Sphere Model of Consciousness: a Bio-Electrophysiological Perspective on (Well-)Being and the Embodied Self," *Current Opinion in Behavioral Sciences* (February 2024): 101344.

3. Ian Whicher, "Nirodha, Yoga Praxis and the Transformation of the Mind," *Journal of Indian Philosophy* 25, no. 1 (February 1997): 1–67.

4. Reginald A. Ray, *Secret of the Vajra World: The Tantric Buddhism of Tibet* (Boston: Shambhala, 2001).

5. David T. Bradford, "Comparable Process Psychologies in Eastern Christianity and Early Buddhism," *Chromatikon* 7 (2011): 87–102.

6. Jean-Claude Larchet, *Therapy of Spiritual Illnesses: An Introduction to the Ascetic Tradition of the Orthodox Church,* vol. 1 (Montreal: Alexender Press, 2017).

7. As quoted in Panayiotis Nellas, *Deification in Christ: The Nature of the Human Person* (New York: St. Vladimir's Seminary, 1987), 87.

8. See Claire Petitmengin, "On the Veiling and Unveiling of Experience: A Comparison between the Micro-Phenomenological Method and the Practice of Meditation," *Journal of Phenomenological Psychology* 52, no. 1 (August 2021): 36–77.

9. Klaus Hepp, "Space, Time, Categories, Mechanics, and Consciousness: On Kant and Neuroscience," *Journal of Statistical Physics* 180 (April 2020): 896–909.

10. Khenchen Thrangu Rinpoche, *Everyday Consciousness and Primordial Awareness* (Ithaca, NY: Snow Lion Publications, 2011).

11. Ravinder Jerath et al., "A Unified 3D Default Space Consciousness Model Combining Neurological and Physiological Processes That Underlie Conscious Experience," *Frontiers in Psychology* 6 (August 2015): 1204.

12. B. Alan Wallace, *Stilling the Mind: Shamatha Teachings from Dudjom Lingpa's Vajra Essence* (Somerville, MA: Wisdom Publications, 2011).

13. B. Alan Wallace, *Fathoming the Mind: Inquiry and Insight in Dudjom Lingpa's Vajra Essence* (Somerville, MA: Wisdom Publications, 2018).

14. Raphaël Millière, "Psychedelics, Meditation, and Self-Consciousness," *Frontiers in Psychology* 9 (September 2018): 1475.

15. B. Alan Wallace, *Hidden Dimensions: The Unification of Physics and Consciousness* (New York: Columbia University Press, 2007).

16. Sogyal Rinpoche, *The Tibetan Book of Living and Dying* (New York: HarperCollins, 2002).

17. Longchen Rabjam, *The Precious Treasury of the Basic Space of Phenomena* (Junction City, CA: Padma Publishing, 2001).

18. Chogyal Namkhai Norbu, *The Crystal and the Way of Light: Sutra, Tantra, and Dzogchen* (Ithaca, NY: Snow Lion Publications, 1999), 89.

19. Keith Dowman, *Old Man Basking in the Sun: Longchenpa's Treasury of Natural Perfection* (Kathmandu: Vajra Publications, 2006).

20. Zoran Josipovic, "Implicit-Explicit Gradient of Nondual Awareness or Consciousness As Such," *Neuroscience of Consciousness* 2021, no. 2 (October 2021): niab031.

21. Zoran Josipovic, "Nondual Awareness: Consciousness-As-Such as Non-Representational Reflexivity," *Progress in Brain Research* 244 (February 2019): 273–298.

22. Matthieu Ricard, *On the Path to Enlightenment: Heart Advice from the Great Tibetan Masters* (Boston: Shambhala, 2013).

23. Simon Hanslmayr et al., "The Role of Alpha Oscillations in Temporal Attention," *Brain Research Reviews* 67 (June 2011): 331–343.

24. Olaf Sporns, *Networks of the Brain* (Cambridge, MA: MIT Press, 2016).

25. Zoran Josipovic, "Neural Correlates of Nondual Awareness in Meditation," *Annals of the New York Academy of Sciences* 1307 (January 2014): 9–18.

26. Austin Clinton Cooper et al., "Beyond the Veil of Duality: Topographic Reorganization Model of Meditation," *Neuroscience of Consciousness* 2022, no. 1 (October 2022): niac013.

27. John Myrdhin Reynolds, *The Golden Letters: The Three Statements of Garab Dorje, the First Dzogchen Master* (Ithaca, NY: Snow Lion Publications, 1996).

28. Much of this discussion will draw from Reuben E. Laukkonen and Heleen A. Slagter, "From Many to (N)one: Meditation and the Plasticity of the Predictive Mind," *Neuroscience and Biobehavioral Reviews* 128 (September 2021): 199–217.

29. Andy Clark, "Whatever Next? Predictive Brains, Situated Agents, and the Future of Cognitive Science," *Behavioral and Brain Sciences* 36, no. 3 (June 2013): 181–204.

30. Michael Gilead, Yaacov Trope, and Nira Liberman, "Above and Beyond the Concrete: The Diverse Representational Substrates of the Predictive Brain," *Behavioral and Brain Sciences* 43 (July 2019): 1–74.

31. Chris Letheby and Philip Gerrans, "Self Unbound: Ego Dissolution in Psychedelic Experience," *Neuroscience of Consciousness* 2017, no. 1 (June 2017): nix016.

32. Daniel S. Margulies et al., "Situating the Default-Mode Network along a Principal Gradient of Macroscale Cortical Organization," *Proceedings of the National Academy of Sciences USA* 113, no. 44 (October 2016): 12574–12579.

33. Giovanni Pezzulo, Marco Zorzi, and Maurizio Corbetta, "The Secret Life of Predictive Brains: What's Spontaneous Activity For?" *Trends in Cognitive Sciences* 25 no. 9 (September 2021): 730–743.

34. Andrea Scalabrini et al., "The Self and Its Internal Thought: In Search of a Psychological Baseline," *Consciousness and Cognition* 97 (January 2022): 103244.

35. Cooper et al., "Beyond the Veil of Duality."

36. On a neuronal level this disidentification or stepping back from capture by existing models is possibly reflected in the disassembly of large-scale neuronal networks that otherwise entrain local populations of neurons. See Jim Grigsby and David Stevens, *Neurodynamics of Personality* (New York, NY: The Guilford Press, 2000).

37. Laukkonen and Slagter, "From Many to (N)one."

38. Laukkonen and Slagter, "From Many to (N)one," 200.

39. Rocío-Martínez Vivot et al., "Meditation Increases the Entropy of Brain Oscillatory Activity," *Neuroscience* 431 (April 2020): 40–51.

40. Robin L. Carhart-Harris and Karl J. Friston, "REBUS and the Anarchic Brain: Toward a Unified Model of the Brain Action of Psychedelics," *Pharmacological Reviews* 71, no. 3 (July 2019): 338.

41. Elena Antanova et al., "More Meditation, Less Habituation? The Effect of Mindfulness Practice on the Acoustic Startle Reflex," *PLoS One* 10, no. 5 (May 2015): e0123512.

42. Enrico Fucci et al., "Differential Effects of Non-Dual and Focused Attention Meditations on the Formation of Automatic Perceptual Habits in Expert Meditators," *Neuropsychologia* 119 (October 2018): 92–100.

43. Llewellyn Vaughan-Lee, *Prayer of the Heart in Christian & Sufi Mysticism* (Point Reyes, CA: Golden Sufi Center, 2017).

44. Josipovic, "Implicit-Explicit Gradient."

45. Austin Clinton Cooper et al., "Beyond the Veil of Duality: Topographic Reorganization Model of Meditation," *Neuroscience of Consciousness* 2022, no. 1 (October 2022): niac013.

46. Georg Northoff and David Smith, "The Subjectivity of Self and Its Ontology: From the World-Brain Relation to Point of View in the World," *Theory & Psychology* 33, no. 4 (March 2022): 485–514.

47. Eugene T. Gendlin, *Focusing-Oriented Psychotherapy: A Manual of the Experiential Method* (New York: Guilford Press, 1996).

48. Petitmengin, "On the Veiling and Unveiling of Experience," 45.

49. Harald Atmanspacher, "Dual-Aspect Monism a la Pauli and Jung," *Journal of Consciousness Studies* 19, no. 9–10 (January 2012): 96–120.

50. Andrew M. Greeley, *Ecstasy: A Way of Knowing* (Englewood Cliffs, NJ: Prentice-Hall, 1974), 61.

51. Christof Koch, *The Feeling of Life Itself: Why Consciousness Is Widespread but Can't Be Computed* (Cambridge, MA: MIT Press, 2020).

52. Philip Goff, *Galileo's Error: Foundations for a New Science of Consciousness* (New York: Pantheon Books, 2019).

53. Anil Seth, *Being You: A New Science of Consciousness* (New York: Dutton, 2021).

54. Georg Northoff, *The Spontaneous Brain: From the Mind-Body to the World-Brain Problem* (Cambridge, MA: MIT Press, 2018).

55. Daniel J. Siegel, *Mind: A Journey to the Heart of Being Human* (New York: W. W. Norton, 2017).

Coda

1. See Matthieu Ricard, *On the Path to Enlightenment: Heart Advice from the Great Tibetan Masters* (Boston: Shambhala, 2013). In the writings of the fourth-century Syriac Christian sage Saint Ephrem, we find the poetic description of the two wings being truth and love. See Sebastian Brock, *The*

Luminous Eye: The Spiritual World Vision of Saint Ephrem the Syrian (Kalamazoo, MI: Cistercian Publications, 1992).

2. Edith Turner, *Communitas: The Anthropology of Collective Joy* (New York: Palgrave Macmillan, 2012).

3. As cited in Maurice Nicoll, *Psychological Commentaries on the Teachings of Gurdjieff and Ouspensky,* vol. 5 (Boulder, CO: Shambhala, 1984), 1548.

4. See, for example, Jeremy Lent, *The Web of Meaning: Integrating Science and Traditional Wisdom to Find Our Place in the Universe* (Gabriola Island, BC: New Society Publishers, 2021); and Iain McGilchrist, *The Matter with Things: Our Brains, Our Delusions and the Unmaking of the World* (London: Perspectiva Press, 2021).

Acknowledgments

Sadly, this final piece of text saluting the many individuals who contributed to the creation of this book must be written by one of its authors alone (VM), and must begin with deep gratitude to the other. Many of the seeds of its ideas were sown in conversations with my dear friend, colleague, and coauthor, Steven Jay Lynn.

Steve was a gifted conversationalist with an insatiable curiosity and a wonderful sense of humor. We had a number of mutual interests, so our dialogues were more in the nature of creative play and we stretched these sounding sessions out, over the years, in various locales from Northern California to a rustic, woodsy cabin in upstate New York. He was steadfast in his support and encouragement, without which I likely would not have persevered in completing the manuscript. It was especially poignant that Steve passed away unexpectedly the very week that we received the typeset page proofs and missed seeing the final fruits. I shall greatly miss his companionship and treasure this book as a testimonial to our friendship and its enduring influence, trusting that the ideas expressed herein will be for a wider good.

Together we had already planned, in this space, to extend our sincere thanks to the individuals who provided insightful comments and help throughout the various stages of our manuscript preparation. In particular, Susan Lee Cohen championed the early cause of our book and recognized its potential. Lama Surya Das enlightened us in many illuminating discussions in the early stages of the book's inception.

Acknowledgments

Zoran Josipovic gave us the benefit of his erudite and skillful critique, and more specifically, his excellent knowledge of the finer nuances of the Dzogchen tradition and practices. Andrew Kinney at Harvard University Press for believing in the project and being an early advocate, and Sharmila Sen for her bold advocacy and shepherding of the book to its finish line. Also, Samantha Mateo for all of her efforts at keeping us organized and on track. Julia Kirby for her superb and incisive editorial help.

Finally, I (VM) would like to thank the monastic brothers and sisters of New Skete Monasteries for all of their support, friendship, patience and encouragement.

Index

The letter *f* following a page number denotes a figure.

Index

Index